Philosophy of the Social Science

This volume is a unique contribution to the philosophy of the social sciences, presenting the results of cutting-edge philosophers' research alongside critical discussions by practicing social scientists. The book is motivated by the view that the philosophy of the social sciences cannot ignore the specific scientific practices according to which social scientific work is being conducted, and that it will be valuable only if it evolves in constant interaction with theoretical developments in the social sciences. With its unique format guaranteeing a genuine discussion between philosophers and social scientists, this thought-provoking volume extends the frontiers of the field. It will appeal to all scholars and students interested in the interplay between philosophy and the social sciences.

PROFESSOR MANTZAVINOS holds the Chair of Economics and Philosophy at Witten/Herdecke University, Germany. He is the author of *Wettbewerbstheorie* (Berlin, 1994), *Individuals, Institutions, and Markets* (Cambridge, 2001) and *Naturalistic Hermeneutics* (Cambridge, 2005).

Philosophy of the Social Sciences

Philosophical Theory and Scientific Practice

C. Mantzavinos

CAMBRIDGE UNIVERSITY PRESS

CAMBRIDGE UNIVERSITY PRESS
Cambridge, New York, Melbourne, Madrid, Cape Town, Singapore,
São Paulo, Delhi

Cambridge University Press
The Edinburgh Building, Cambridge CB2 8RU, UK

Published in the United States of America
by Cambridge University Press, New York

www.cambridge.org
Information on this title: www.cambridge.org/9780521739061

First published 2009

Printed in the United Kingdom at the University Press, Cambridge

A catalogue record for this publication is available from the British Library

Library of Congress Cataloguing in Publication data
Philosophy of the social sciences : philosophical theory and scientific
    practice / [edited by] C. Mantzavinos.
      p.   cm.
    ISBN 978-0-521-51774-4 (hardback) – ISBN 978-0-521-73906-1 (pbk.)
    1. Social sciences–Philosophy. I. Mantzavinos, Chrysostomos. II. Title.
    H61.15.P485 2009
    300.1–dc22
    2009020693

ISBN 978-0-521-51774-4 hardback
ISBN 978-0-521-73906-1 paperback

Contents

Contributors

JAMES ALT
Harvard University
Department of Government

MICHAEL E. BRATMAN
Stanford University
Department of Philosophy

NANCY CARTWRIGHT
London School of Economics
Department of Philosophy, Logic
and Scientific Method

PIERRE DEMEULENAERE
Université de Paris Sorbonne
Department of Philosophy and
Sociology

GERD GIGERENZER
Director, Max Planck Institute for
Human Development, Berlin

WERNER GÜTH
Director, Max Planck Institute for
Economic Research, Jena

HARTMUT KLIEMT
Frankfurt School of Finance and
Management

JACK KNIGHT
Duke University
Department of Political Science
and School of Law

DANIEL LITTLE
University of Michigan-Dearborn
Department of Philosophy

STEVEN LUKES
New York University
Department of Sociology

C. MANTZAVINOS
Witten/Herdecke University
Chair of Economics and
Philosophy

SANDRA MITCHELL
University of Pittsburgh
Department of History and
Philosophy of Science

DAVID PAPINEAU
King's College London
Department of Philosophy

PHILIP PETTIT
Princeton University
University Center for Human
Values

DIEGO RIOS
Witten/Herdecke University
Economics and Philosophy

DAVID-HILLEL RUBEN
Birkbeck College, University
of London and NYU in London

JOHN R. SEARLE
University of California, Berkeley
Department of Philosophy

IAN SHAPIRO
Yale University
Department of Political Science

ROBERT G. SHULMAN
Yale University
Department of Molecular
Biophysics and Biochemistry

ERNEST SOSA
Rutgers University
Department of Philosophy

MARK TURNER
Case Western Reserve University
Department of Cognitive Science

JAMES WOODWARD
California Institute of Technology
Division of the Humanities and
Social Sciences

Acknowledgements

It is a pleasant duty to thank those who have helped me in the process of editing this book. I would first like to thank all the authors, some of them friends, for being so collaborative. Special thanks go to Diego Rios – with whom I have discussed all the strategic steps that led to the book – for his advice and his goodwill. For helpful discussions on chapters of the book, I would also like to thank my graduate students, Pablo Abitbol and Catherine Herfeld. I am also grateful to Darrell Arnold who was responsible for linguistic corrections of my own texts, my secretary, Julia Pusch who provided her unfailing help for the project, and my student, Julia Köhn, to whom I am particularly grateful for her excellent help with the preparation of the manuscript during the final stages.

For their active interest in the project and their excellent collaboration, I would like to express my gratitude to my editors at Cambridge University Press, Chris Harrison and especially John Haslam, who was responsible for the book. Thanks are also due to Carrie Cheek and Karen Matthews for their help during the editing process. I would also like to thank my production editors Christopher Hills and Jodie Barnes of Cambridge University Press and Paula Devine for copyediting the book.

A major step towards preparing the volume was the organisation of a conference that I initiated at Herdecke in June 2007. I would particularly like to thank the Witten/Herdecke University and the Fritz Thyssen Foundation for their generous financial support.

Lastly, I want to thank my wife and my family for their love and care during the time I have been working on this project.

Introduction

C. Mantzavinos

Philosophy of science examines "scientific knowledge." It tries to illuminate the specific characteristics of science, the way it is produced, the historical dimensions of science, and the normative criteria at play in appraising science. The discussions mostly take place with reference to the natural sciences, which are still at the core of the philosophy of science as a discipline. The examples used are often taken from one of the natural sciences (usually physics); and it is characteristic that the training of most contemporary philosophers of science has been – at least partly – in one of the natural sciences. The philosophy of the social sciences, on the other hand, traditionally deals with such problems as the role of understanding (*Verstehen*) in apprehending social phenomena, the status of rational choice theory, the role of experiments in the social sciences, the logical status of game theory, as well as whether there are genuine laws of social phenomena or rather social mechanisms to be discovered, the historicity of the social processes, etc.

The aim of this volume is to push the frontiers of the philosophy of the social sciences as a sub-discipline of the philosophy of science by presenting the results of cutting-edge research in the main fields, along with their critical discussion by practicing social scientists. The enterprise is motivated by the view that the philosophy of the social sciences cannot ignore the specific scientific practices according to which scientific work is being conducted in the social sciences and will only be valuable if it evolves in constant interaction with the theoretical developments in the social sciences. Since a great number of basic concepts of the philosophy of the social sciences have become increasingly sophisticated and technical, and even philosophically minded social scientists do not follow the philosophical discussion on a number of issues – like intentionality, reductionism, shared agency etc. – there is a real need for interaction between the two communities. This volume is designed to close this gap and to foster an exchange between philosophers and philosophically minded social scientists on philosophical concepts and the practices of apprehending social phenomena.

1

With this in mind, the format of the volume is the following: It includes ten chapters by philosophers, who draw from their broader research agenda, but focus on one or more specific issues. Social scientists who are philosophically minded, but who nevertheless employ the standard scientific practices of their respective disciplines provide comments on the chapters. This format guarantees a genuine discussion of the issues, engaging both philosophers and social scientists in productive dialogue that provides insights into the three main areas of the philosophy of the social sciences. The book is designed so that its three parts correspond to those three areas.

The first area concerns *Basic Problems of Sociality* (Part I). The social sciences deal with the interactions of individuals and the products of those interactions – obviously from very different angles. In a nutshell, the problems that a social scientist deals with are problems of sociality, and the philosophy of the social sciences attempts to shed some light on those problems. Social ontology, broadly defined to include issues such as collective intentionality, shared agency, the reality of group agents, etc., delineates the field of philosophical work that deals with what exists in the social world.

The second area concerns the *Laws and Explanation in the Social Sciences* (Part II). When problems of social interaction are studied by social scientists, a series of problems emerge concerning the appropriate method of study and the epistemological status of the obtained knowledge. A few of the notorious problems concern whether there are any laws in the social sciences and whether there are genuine social scientific explanations or rather *Verstehen* (understanding). Methodology of the social sciences, broadly defined to include issues such as how social scientific knowledge relates to knowledge that is produced by the natural and life sciences, the degree of complexity of social phenomena, issues related to how to proceed to policy advice based on the empirical findings of the social sciences, etc., is the field of philosophical work that deals with the method of the study of what exists in the social world.

The third area concerns *How Philosophy and the Social Sciences Can Enrich Each Other* (Part III). The relationship between philosophy and the sciences is a difficult problem which remains unresolved. However, it seems that philosophy does not have a more epistemologically privileged position than the sciences and that there is rather a continuum between philosophy and the sciences. Besides, the application of scientific theses, research, and results must be both acceptable and imperative for philosophy. The scaffolding of philosophy erected on the social sciences is far from perfect – its exact shape and function is the third main area of research into the philosophy of the social sciences.

A detailed plan of the chapters and comments is provided at the beginning of each Part, so that the reader has a map of what awaits him and what he can look for in every Part of the book. The *Epilogue* contains a short reflection on the problem areas of the discipline and how they have been addressed in this volume.

Part I

Basic Problems of Sociality

Part I of this book starts with *John Searle*'s chapter on "Language and Social Ontology." This aims to work out the role of language in the creation, constitution, and maintenance of social reality and to answer the question, "What are the ontological implications of the very capacity to categorize linguistically?" The upshot of Searle's discussion is that all social reality and all social institutions presuppose language. Searle defends the view that the logical form of the creation of the institutional fact is always a Declaration and elaborates on how the theory of speech acts can be applied to institutional analysis. He extends the account he presented in *The Construction of Social Reality*, including a "power creation operator" in the conceptual analysis to do justice to the phenomenon of power, which is inimical to all institutions. In his comment, *Mark Turner* focuses on the implications of taking language for granted, as Searle indicates a number of authors do. Taking language for granted implies taking political ontology for granted, taking intentionality for granted, taking personal identity for granted, and taking counterfactuality and a number of other things for granted. Turner shows that theorizing about this impressive list of entities, mechanisms, and processes that are taken for granted when approaching the social world can be productively done with the help of the tools of modern cognitive science. Arguing that social reality and performance are conditioned by the nature of our basic mental operations and our most characteristic capacities, such as language, Turner suggests a view complementary to Searle's.

In Chapter 2, *Michael Bratman* tackles the issue of shared agency – an issue of obvious importance for social ontology. Beginning with an underlying model of individual planning agency called the planning theory, he develops a framework that aims to support theorizing about forms of modestly sized sociality. He then seeks a conceptual and metaphysical bridge from such individual planning agency to modest forms of sociality. Bratman explores the idea that this involves shared intention. He suggests a *constructivist* view of shared intention: Individual participants are guided by norms of individual planning agency; and, given the special contents of their intentions and their interrelations, these apply and conform to corresponding social norms on shared intention. According to Bratman, constructivism posits a kind of normative emergence – when individuals become aware of this normative emergence, they may go on to explicitly internalize these social norms and directly appeal to them in their practical reasoning. *Pierre Demeulenaere* asks in his comment: "Where is the social?" He points out Bratman's commitment to individualism since his aim is to proceed from individual intentions to shared intentions without introducing any irreducible social level independent of individual intention. Demeulenaere does endorse methodological individualism. Specified *negatively*, this is the rejection of any kind of "social" agency i.e., the rejection of the suggestion that "collective entities" or institutions are, as such, actors. Specified *positively*, it is the assertion that only individuals are actors, and any institution or group depends on individuals to be "active." He nevertheless criticizes the notion that the social level somehow "emerges" from the individual level. Even if one grants that shared intentions rest on individual intentions, it is inaccurate to view the social as *emerging* from the individual, the social being everywhere, pervading individual activities.

Philip Pettit addresses another issue of obvious importance for the social sciences: Are groups to be apprehended as agents? Pettit first sets out the requirements that systems of any kind must fulfil if they are to count as agents, arguing that they should display a purposive–representational pattern of behavior. He then looks at the way in which individuals might seek, on the basis of shared intention, to form a group agent, and he focuses on the capacity of a straw-vote assembly to display a broadly agential pattern of behavior. Pettit claims that the straw-vote assembly, which has been identified as a candidate for group agency, is hard to dismiss as an institutional possibility; it does satisfy the requirements of group agency. His conclusion is that groups can be real agents. *Diego Rios* situates the argument of Pettit's chapter in the broader context of the philosophical discussions on individualism. He recollects the powerful arguments of Hayek and Popper in defence of the thesis that groups are not true agents; and though he agrees with the main thrust of Pettit's argument, he raises the question of the applicability of his analysis in the hard case of fully opaque groups. These are characterized by the fact that *none* of the individual members is aware of the purposive global outcomes of the group. Rios suggests that further arguments are needed to make a convincing case that even these groups can be ascribed agential status.

These three interactions deal with problems of social ontology, highlighting the issue of sociality from different angles. The first interaction is centred around the role of language for the construction of social reality, both from an *a priori* and an empirical point of view. The second interaction deals with a traditional problem of the social sciences – how individuality and sociality are related. It focuses more specifically on the way that shared intentions can be conceptualized as resting on individual intentions; it is an attempt, in a way, to find the locus of the social. The third interaction tackles the issue of whether groups can be agents, a core issue of the individualism–collectivism debate, elaborating on the conditions that could make the ascription of agential status to groups plausible and acceptable. All three interactions deal thus with *Basic Problems of Sociality*.

1 Language and Social Ontology

John R. Searle

This chapter is concerned with the ontology of a certain class of social entities and the role of language in the creation and maintenance of such entities. The social entities I have in mind are such objects as the $20 bill in my hand, The University of California, and the President of the United States. I also include such facts as that Barack Obama is President of the United States; that the piece of paper I hold in my hand is a $20 bill; and that I am a citizen of the United States. I call such facts "institutional facts," and it will emerge that the facts are logically prior to the objects (because the object is only institutional if it is created by a certain linguistic operation that creates an institutional fact). Under the concept of social entity, I also mean to include such institutions as money, property, government, and marriage. I believe that where the social sciences are concerned, social ontology is prior to methodology and theory. It is prior in the sense that unless you have a clear conception of the nature of the phenomena you are investigating, you are unlikely to develop the right methodology and the right theoretical apparatus for conducting the investigation.

I have also a polemical aim for wishing to discuss social ontology and that is that I believe we have a long tradition, going back to the ancient Greeks, of misconstruing the role of language in the creation and con-stitution of social and political reality. It is characteristic of, I believe, all the authors known to me in our tradition, from the Greeks to the present, that they do not accurately see the role of language in the crea-tion, constitution, and maintenance of social reality. Social and politi-cal theorists assume that we are already language-speaking animals and then go on to discuss society, without understanding the significance of language for the very existence of human social reality. Every author I have read, from the ancient Greeks right through to such contempo-rary authors as Foucault, Bourdieu, and Habermas, takes language for granted. What I mean when I say they take language for granted is that in their discussions of social reality they are discussing people who already have a language. It may seem puzzling that I charge Foucault,

Bourdieu, and Habermas with taking language for granted, since they do discuss language and its relation to society. But it seems to me that in each case, they fail to tell us what language is. For example, Bourdieu remarks, correctly, that the ability to control the way that political issues are linguistically categorized is an important element in political power. But he does not tell us what is involved in being able to use language to categorize at all. What are the ontological implications of the very capacity to categorize linguistically? The worst offenders in this regard are the Social Contract theorists who simply assume that we are language-speaking animals and that we all get together in the state of nature and form a social contract. The point I make in this chapter is that once you have a language you already have a social contract. The social contract is built into the very essence of language. So by way of beginning this discussion, I am going to briefly examine the nature of language.

1   What Is Language?

Human language is an extension of prelinguistic forms of intentionality and I need to identify some of the relevant features of intentional states that form the basis for the evolution of language. We do not know how language evolved from prelinguistic forms of mental life, and because of the absence of fossil evidence, maybe we will never know how it evolved; but even if we do not know the details of the evolution of language we can still identify the conceptual distinctions between prelinguistic intentionality and linguistic forms of intentionality. At one time beasts more or less like ourselves, hominids, walked on the Earth in Africa and did not have language. Now we have language. What is it that we have that they did not have? More specifically: what conceptual resources are already available in prelinguistic intentionality, and what do we have to add to prelinguistic intentionality to get language?

To begin to answer this question, I have to say some things about intentionality in general. Intentional states and events are those mental states and events that are *directed at* or *about* objects and states of affairs in the world. They include not only intending, in the sense which I intend to go to the movies, but also perception and intentional action, belief, desire, the emotions and indeed any state that has a directed content. Intentional states have some remarkable features that already foreshadow corresponding features in language, and on the basis of which we could develop a language. Specifically intentional states typically have a propositional content in a certain psychological mode. Thus, for example, I can believe that it is raining, fear that it is raining, or hope that it is raining. In each case I have the same propositional content – that it is raining – but I have these in different psychological modes. And this corresponds in language to the distinction between the propositional content of the speech act and the corresponding speech act mode, the type of speech act that it is. Thus, I can order you to leave the room; I can ask whether you will leave the room; and I can predict that you will leave the room. In each case, we have the same propositional content, that you will leave the room, presented in different types of speech acts.

Because intentional states typically have a propositional content, they can represent how things are in the world, or how we would like them to be, or how we intend to make them be. A belief is supposed to represent how things are in the world, a desire represents how we would like them to be, an intention, how we intend to make them be. Let us introduce the notion of "conditions of satisfaction" to describe what is common to all these cases and then, leaving out all sorts of details, we can say that the essence of intentionality is representation of conditions of satisfaction. In each case, the intentional state represents its conditions of satisfaction: truth conditions in the case of belief, carrying out conditions in the case of intentions, and fulfillment conditions in the case of desires.

Another crucial feature of intentional states that carries over to language is that intentional states have different ways of fitting reality. The aim of a belief is to be true, the aim of an intention is to be carried out, the aim of the desire is to be fulfilled. We may think of beliefs therefore as supposed to represent how things are. They fit the world with what we can call the "mind-to-world" direction of fit (the state in the mind is supposed to represent how things are in the world) but desires and intentions are not supposed to represent how things are, but rather how we would like them to be, or how we intend to make them be and we may say therefore that they have the "world-to-mind" direction of fit (the state of the world is supposed to come to match how things are represented in the mind). The best test for the presence of the mind-to-world direction of fit is to ask, Can the state in question be literally true or false? Beliefs can be true or false, and thus they have the mind-to-world direction of fit. Desires and intentions cannot literally be true or false, and thus they do not have the mind-to-world direction of fit. All of this is going to carry over to language (with some absolutely crucial variations). I think in simple metaphors, and so I like to think of the mind (or word)-to-world direction of fit as going downward ↓ from the representation to reality, and the world-to-mind (or word) of direction of fit is going upward ↑, from the reality to the representation. And I will use upward and downward arrows to represent the two directions of fit. Statements, like beliefs, have the downward direction of fit ↓; orders and promises like desires and intentions, have the upward direction of fit ↑.

Our hominids also have conscious perceptions and actions and this will give them a set of perceptual and action categories for coping with their experiences. They will perceive objects, properties, and relations and they will act in a way that will manifest their own agency, and their capacity for experiencing causation. If they can recognize the same object on different occasions and distinguish one object from another, they are manifesting the categories of identity of an individuation. They thus can operate with a rather hefty set of Aristotelian and Kantian categories, even though of course they have no concepts corresponding to these categories.

We have been talking as if intentionality were a property only of individual minds, but of course in understanding society we have to introduce the notion of collective intentionality. When you and I are engaged in some sort of cooperative behavior such as preparing a meal together or having a conversation we have collective intentionality. A meeting like this where we are all gathered together to discuss common issues is a paradigm case of collective intentionality. All intentionality is in individual human and animal brains, but some of it is in the form of the first-person plural. It is not just that I am doing this and you are doing this, but we are doing this together; and this fact is represented in each of our heads in the form of collective intentionality.

So far we are imagining that our prelinguistic hominids have an inventory of prelinguistic intentional states, that have the remarkable features that they can represent states of affairs in the world and they can do so with different directions of fit. We also imagine that they have experiences like ours that manifest such categories as object, identity, property, relation etc. And they are capable of cooperating, thus they have the capacity for collective intentionality. What do we have to add to all of that to get language?

2   Meaning, Conventions and Syntax

There are lots of differences between the linguistic forms of intentionality and the prelinguistic forms, but for the purpose of our present discussion, which is about social ontology, the three crucial features of language which prelinguistic intentionality does not have are *meaning*, *convention*, and *syntactic structure*. I will now discuss each of these. Lots of prelinguistic animals have the capacity to communicate with other animals by way of signaling. The bees are the most famous case, but the bee language has some puzzling features, so let us take an even simpler case, the vervet monkey. These monkeys have different signals for different types of danger. They have one type of signal if the danger is from a leopard, another type if the danger is from a snake, and a third if the danger is from a hawk. We do not know what goes on in the mind of the monkey when it is making the signals but it is not hard to imagine what might have gone on in the mind of early humans who correspondingly had signals for indicating the presence of danger (or of food or fire). We need first to distinguish between making the signal as a physical act and giving the signal as a vehicle of communication of meaning. What is the difference between just making the signal and intending the signal to mean that a certain sort of danger is present? The signals correspond to one-word sentences and if we imagine hominids doing this we will imagine the utterances of the hominids, even though they do not yet have conventional words and sentences, could be translated into modern English as, for example, "Danger!" or "Food!" or "Fire!" But in each case we need to distinguish between just making the sound and making it as a way of conveying some meaning. The making of the sound by itself is the condition of satisfaction of the intention to make that very sound. But if we make the sound meaningfully then we add something crucial to the sound itself. What we add is that the sound itself must now have further conditions of satisfaction – in this case truth conditions. We can generalize this point: Meaningful utterances are those where the speaker intentionally imposes conditions of satisfaction on conditions of satisfaction. This, I will argue, is the essence of speaker meaning. The condition of satisfaction of the non-meaningful intentional utterance is simply that an utterance should be produced. But if the utterance is to be meaningful it must itself have further conditions of satisfaction, such as truth conditions or fulfillment conditions. It is the intentional imposition of these further semantic conditions of satisfaction onto the conditions of satisfaction already present in the intentional utterance that constitutes speaker meaning.

You can see this difference in the case of developed human languages if you think of the difference between saying something and meaning it, and saying the very same thing without meaning it. Suppose, for example, that I am practicing French pronunciation, and I say over and over "il pleut" but I do not actually mean that it is raining. I am just trying to pronounce the words correctly. In this case, the condition of satisfaction of my intention is simply that I should produce the sounds correctly. But suppose I go outdoors with a French-speaking friend; I notice the rain and I say "il pleut." In this case, I not only intend to make the sound, but I mean something by it. My meaning something consists in the intention that the sound should now have further conditions of satisfaction – in this case truth conditions – and indeed I invoke the conventions of French to determine the truth conditions in question. To our account of prelinguistic intentionality we have now added a crucial element – speaker meaning.

In order to understand fully the significance of this point we have to be clear about the distinction between *representation* and *expression*. Many animal signals simply serve to express, in the etymologically literal sense of the word, press out, some inner intentional state. Thus when a dog barks angrily, his barking is an expression of anger. But linguistically significant forms of communication involve more than just expressing intentional states, they involve representing states of affairs in the world. Thus in the existing human languages there is a crucial distinction between, for example saying, "Ouch!" and saying "I am in pain." The first cannot be literally true or false, though it can be sincere or insincere. The second can be literally true or false because it represents a state of affairs in the world. It does not just express an intentional state, though it does, by the way, also express an intentional state. Every statement is an expression of a belief. But the point of expressing the belief is not to tell you about the speaker's inner mental states of belief but to tell you about how things are in the world. If I say "It's raining" I do indeed express the belief that it's raining but the point of the utterance is not to tell you something autobiographical about myself, but to tell you about how things are in the real world. In analyzing social ontology the key element in language is the representing, not the expressing, function.

So the point of the present discussion is that in seeing how language relates to prelinguistic forms of intentionality we have to see that one of the essential features of language for our purposes is its capacity to be used to represent states of affairs in the world with the different directions of fit, and we do this by imposing conditions of satisfaction on conditions of satisfaction.

The next step is to imagine the introduction of conventions. Let us suppose that our early humans develop conventional ways of communicating these various contents. They might evolve different conventions for different types of contents to be communicated. These communications would correspond to one-word sentences in our language which might be such utterances as "Food!," "Fire!," "Danger!" The introduction of a standard or conventional way to communicate meaning is an enormous step forward, because it enables both speaker and hearer to have a reasonable expectation both that the speaker means something identifiable by the utterance and that the hearer can reasonably be assumed to understand the utterance. I intend no heavy metaphysical theory about convention. All I mean is that the creatures in question have evolved procedures with a *normative* character that give them justifiable expectations. The speaker has the expectation that if he utters such and such a sound it will be understood in a certain way. And the hearer has the expectation that if he hears a certain sound produced intentionally, he can be justified in assuming that some specific meaning was intended. And these expectations are grounded in the fact that the conventions in question are essentially *normative*. There is a right way and a wrong way to use the words.

So far we have only one-word sentences. The next step is to break them up into component parts. We already have in the perceptual experiences of the hominids in question the experiences of objects with their properties and relations. So the hominids already are equipped with certain traditional Kantian and Aristotelian categories. They do not yet have concepts corresponding to these categories but they are able to operate with such perceptual and volitional categories as object, property, relation, cause etc. and this is manifest in their actual conscious experiences.

If they already have speaker meaning and conventions, together with the perception of objects with the properties, relations, and causal connections, etc., it is not such a big step to imagine that they evolve expressions which correspond to objects, which can be used as referring expressions and they also evolve expressions for describing or characterizing various features of objects, and these can be used as predicating expressions. I am going through this rapidly but I want to convey that many of the resources essential for full-blown languages are already present in prelinguistic intentionality and if we add meaning, conventions, and the internal structure we are both building on a pre-existing apparatus and developing it in a remarkable direction which I now want to make explicit.

3   Commitments and Other Forms of Deontology

Once they have evolved conventions and once they can make explicit the content that they are communicating by breaking up the signal into the syntactical elements that compose complete sentences, they have an additional feature not present in prelinguistic intentionality. Their utterances will now involve special kinds of commitments. "Commitment" is the most general term for a class of phenomena by which humans are bound together with special kinds of reasons for action: these include rights, responsibilities, authorizations, obligations, permissions, and entitlements among others. Just to have a general term I call all of these "deontologies" from the Greek word for duty. Recognized deontologies are what makes human society possible. The development of linguistic commitments is the opening wedge to social deontology in general.

A belief or an intention or even a desire is already a sort of commitment. If for example your belief turns out to be false, you have to (that is, you are committed to) abandon the belief. But the commitment involved in making a statement is vastly greater and in a different category altogether from the commitment involved in just having a belief. If I believe it is raining and I discover that it is not raining, I have to give up my belief. But if I am using conventional sentences of a language to explicitly and intentionally make a statement to someone else that it is raining, then I have a much greater degree of commitment, indeed a different order of commitment altogether, from the commitment involved in just having a belief. I am, at the very least, committed to the truth of what I say, I am committed to sincerity, that is, to not lying, and I am committed to being able to provide reasons for my claim. Different sorts of speech acts have different kinds of commitments, but in every speech act there is some kind of commitment. This is most obvious in the case of statements and promises but it is also present even in orders and apologies. There is a deontology involved in every serious literal speech act performed according to the conventions of a language. Now this deontology has a crucial logical property which is going to be essential for the creation of social and institutional reality. The creation of a deontology of commitments, as well as rights, duties, obligations, etc. creates desire independent reasons for action. If I make a statement, for example, I have a desire independent reason for telling the truth because my utterance commits me to telling the truth.

Later on I will argue that the glue that holds society together consists of deontologies, that these are created by collective intentionality imposing certain types of functions, that I call "status functions," on people and objects, and that this operation is essentially linguistic in character. So far I have just tried to show that any use of language carries a simple form of this deontology, that every speech act intentionally, deliberately and consciously performed in accordance with the conventions of a language, involves the speaker in commitments, and that these commitments exemplify the deontology that I am talking about.

In order to make this clear I have to say a bit more about commitment.[1] There are two connected elements in our notion of commitment. The first is commitment as an undertaking which it is hard or impossible to reverse. The second element is the notion of obligation. In this sense a commitment to do something involves you in an obligation to do that thing, where the obligation is precisely a desire independent reason

---

[1] For a good discussion of commitment see Seamus Miller in Savas L. Tsohatzidis (2007).

of the sort that I have been talking about. In the first sense of commitment I can undertake a course of action which it is hard to reverse, but where I am under no obligation. For example, if I start driving to Los Angeles on Highway 5 I am committed to going by 5 because it is very hard to change highways to go by 101. But there is no obligation in this case. A speech act of making a promise combines both elements of commitments. I make an undertaking which it is difficult and often impossible to reverse, and the undertaking was one of placing myself under an obligation to do something.

Now a remarkable thing about language and the reason that it is much more than simply a signaling system is that it is a source of our characteristically human forms of deontology. The use of language essentially involves the creation of commitments of various kinds and these are deontic reasons, desire independent reasons for action.

This is exhibited by the distinctly human capacity for lying. Once they are in possession of a language humans have the capacity to deliberately tell lies. Lying is only possible because language gives them the capacity of committing themselves to telling the truth. This, by the way, is why the actor onstage, or the author of a novel, is not lying: their utterances are not commitments to telling the truth.

4   The Extension of Deontology

Given a language, even of the rudimentary sort that I have been describing, it is inevitable that the users of the language will create other sorts of deontologies in the way that they use language to treat objects and people in their environment. Once we have the sorts of commitment I have been describing it is not a very big step to imagine that the hominids are able to say such things as "This is my hut"; "This is my woman"; "That person is the leader." Such utterances are more than just representations of pre-existing states of affairs, *they lay claims*. If the people who make the claims can get others to accept the claims, then they have created a kind of deontology that goes beyond the deontology of the speech act. If other people grant that this is my hut, for example, then they grant that I have certain rights to this hut. The deontology of private property, marriage, and authority is a natural extension of prelinguistic forms of social life, once you have a language rich enough to create a deontology. But these forms have a very peculiar direction of fit.

For example, if I say "this is my property" and can get other people to recognize that it is my property then I have created a deontology of property rights by representing that deontology as existing. That is, I made it the case that I have property rights, and thus I achieved the upward or world-to-word direction of fit, but I did it by representing that state of affairs as already existing, by the word-to-world direction of fit. This is the decisive logical move in the creation of human civilization. We create money, government, private property, and marriage for example, by representations that have the double direction of fit.

Our paradigm examples of creating a reality by declaring it to exist are the performative utterances, such as "I now pronounce you husband and wife," "War is hereby declared," and "I promise to come to see you on Wednesday." In these cases there is an explicit performative verb, which enables the speaker to perform the act named by the verb. What I am now suggesting is that there are all sorts of cases involving the creation of a social reality which do not require explicit performative verbs, but where we can create a social reality by linguistically representing that social reality as existing. In speech act theory these speech acts with the double direction of fit are called Declarations (Searle 1979). They make something the case by declaring it to be the case. This, by the way, is not to be confused with two independent directions of fit, but one fitting that goes both ways. Performatives are a special type of Declaration.

So far, in this part of the argument I have made three claims:

First, once you have a language with speaker meaning conventions and internal syntactical structures you already have a system of commitments of various kinds: a system of deontologies.

Second, the deontologies already present in language are easily and inevitably extended to create such institutional facts as those involving private property, marriage, and power relations. It is not logically necessary that they should be so extended. One can imagine a society with no institutional facts beyond language, but I know of no such society (not even the Piraja). The advantages of creating such institutional facts make some such creation practically inevitable.

Third, the logical operation by which we create these deontic systems is formally the same as that of the performative utterances. They are Declarations. We make something the case by representing it as being the case. Such utterances have the double direction of fit.

5   The Creation of Institutional Facts: "X Counts as Y in C"

We have now reached the point where we can connect the present discussion to my earlier work on institutional reality. In *The Construction of Social Reality* (Searle 1995) I said that the creation of money, property, government, and marriage requires the iterated application of constitutive rules of the form "X counts as Y in C." For example this piece of paper counts as a $20 bill, Barack Obama counts as President of the

United States, I count as a professor at the University of California. This operation creates new functions by assigning a new status to a person or an object. In addition to such functions as the function of a knife or the function of a comb, functions which can be performed by virtue of the physical structure of the object in question, there are functions which can only be performed by virtue of the collective acceptance by the community that the person or the object has a certain status, and with that status a function which can only be performed by virtue of a collective acceptance of that status. I call these "status functions." They are the glue that holds human society together, because they carry a special type of deontology that makes society possible.

The whole point of doing this is to create new power relations. The deontology that goes with the creation of institutional facts carries with it a collective acceptance of power. Thus corresponding to the "X counts as Y in C" the procedure for creating institutional facts is a power creation operator "we accept (S has power [S does A])," and the whole system of rights, duties, and obligations etc. can be expressed using this power creation operator and the Boolean operations performed on it. For example if I have an obligation I have a negative power. If I am obligated to pay my traffic ticket then the form of my obligation is "We do *not* accept (S has power [S does *not* pay the fine])."

I argued in that work that language is essential to this operation because there is no fact of the matter about the X term having the Y status function, other than the fact that we do so represent it. The man is only President in so far as we represent him as President; the piece of paper is only a dollar bill in so far as we represent it as a dollar bill, but all such representations require language. We make something the case by representing it as being the case in language. And the logical form of that representation is that of a Declaration.

In order to do this we require collective acceptance of a deontology. The creation of the status function, which is the same thing as the creation of the institutional fact, only works to the extent that it is collectively accepted. And the whole point of doing this is to create deontic powers – rights, duties, obligations, authorities, permissions, etc. The upshot is that language creates institutional facts in a way that is both top-down and bottom-up. The top-down way is the way that I described in the *Construction*. Once you have a language you can create new institutional facts more or less at will provided that you can get people to accept the facts that you are creating. We could, for example, create a new organization right now just by performing certain sorts of speech acts. But there is also a bottom-up relation between language and ontology of society: once you have a language, the creation of institutional reality according to the "X counts as Y" formula is pretty much inevitable, because you have the mechanism for using the double direction of fit. The operation of the "counts as" mechanism is already Declarational. If we all accept that X counts as Y, where Y is a status function, then we make it the case that X is Y, by representing it as being the case.

6   Free-standing Y Terms and the Expansion of the Theory

I have benefited from the many criticisms and discussions of the account that I gave in *The Construction of Social Reality*. I want now to incorporate one of those criticisms into the present discussion. The criticism is that there can be cases where we create status functions without imposing status function on a pre-existing person or object. So we can for example create corporations, just like that, out of thin air, so to speak. And indeed, even money, my favorite example of a status function, does not require actual currency for its existence. The magnetic traces on the computer disks in banks *represent* the amount of money the account holders have, but the computer traces are not themselves money, nor need there be any currency which is represented by these traces. Barry Smith, one of the authors of this objection, calls these cases "free-standing Y terms."

There is a shorter answer to the objection, which I will state initially and then develop it into a longer answer, which will give us some further insight into the special status of language. The short answer to this objection is that if you combine the formula that states the rule "X counts as Y in C" with the formula that states the power creation operator "we accept (S has power [S does A])," then it turns out that the free-standing Y-terms always bottom out in actual human beings who have the powers in question, because they are represented as having them. So it is true you don't need a physical realization to have money, or a corporation, or chess pieces in blindfold chess, but you do have to have *owners* of money and *officers* and *shareholders* of corporations, and *players* in a game of chess, and the power creation operator operates over them. Institutional facts still bottom out in brute facts, but the brute facts in these cases are actual human beings and the sounds and marks that constitute the linguistic representations. To put the point succinctly: to create a minimal institutional reality you need exactly two things: human beings (or some sort of being with the relevantly similar cognitive capacities) and a language capable of representing the double direction of fit. The cases of the free-standing Y terms are always cases where these are the only two phenomena necessary for the specific form of institutional reality. For others such as private real property, licensed drivers, and married couples, you need material objects or human beings that have specific physical properties to which the status function can be assigned.

The longer answer to the objection is to continue the account of the special role of language that I have been developing in this chapter. Here is how it goes. The theory says that all institutional reality is created by representation. Somebody is President, or something is private property only if it is represented as such. But if that is the case then the purest cases of institutional facts will be where the representation is all that is needed. There need be no *Object* X which is represented as Y, rather we just represent the existence of the Y status function and so bring it into existence. Of course, the act of representing is a speech act and we can treat the speech act as the X term that creates the Y status function. For example, "This utterance as X counts as the creation of the ABC Corporation." But the Y status function, the corporation, continues to exist after the performance of the speech act. What constitutes an institutional fact is always at least in part a representation. Typically, the representation will represent an object or person *as* such and such. For example, this object is my property; that person is my wife. However, since the whole point of the exercise is to create power relations between human beings, we can sometimes create the relations directly and not have the X term to which the status function is assigned. That is exactly what happens in the case of the free-standing Y terms.

The existence of free-standing Y terms, together with any complex form of institutional structure, requires something in addition to spoken language: there must be written permanent versions of the speech acts that both establish and enable the institutional facts to continue. Pre-literate societies get along just fine without any written documents. However, they do not have corporations, bank accounts, legal systems or disputes about copyright. All of these require writing. They all require some form of documentation, both to establish the institutional facts in question and to sustain it in its existence.

The double direction of fit is characteristic of the creation of the institutional structure and the institutional fact, but it is also characteristic of the continued existence of the status functions. So when a corporation is created, a Declaration is performed. A corporation is declared to exist. Now you might think that any further reference to the corporation requires only the downhill or word-to-world direction of fit. We need say only true or false things about it, but in fact the continued acceptance of its existence is essential to its continued existence, and the use of the vocabulary marks that acceptance. One sees the role of the vocabulary in the activities of revolutionary and reformist movements. They try to get hold of the vocabulary in order to alter the system of status functions. The feminists were right to see that the vocabulary of "Lady" and "Gentleman" already involves a deontology that they wanted to reject. Again, the Communists in Russia wanted people to address each other as "Comrade" as a way of creating new status functions and destroying old ones. Another way to see this phenomenon is to see how words that mark status functions can gradually fall into disuse with a corresponding erosion of the status function itself. The word "spinster" figures prominently in older American laws, but it is not a word that you hear in common speech. When I asked my class, "How many of you are spinsters?" only one fierce middle-aged woman had the courage to raise her hand. So the daily use of the vocabulary with the downhill direction of fit already has a cumulative uphill direction of fit in sustaining the existence of the status functions across time.

The mechanism that creates and sustains institutional facts is always the same. It is that of the Declarational utterance whereby we create an institutional reality by representing it as already existing. Thus we achieve world-to-word direction of fit (we change the world to match the words), but we do that by way of the word-to-world direction of fit (we represent the world as having been so changed).

7   The General Logical Form of Status Functions

I need now to make a precise statement of the relationships between the form of the rule and the power creation operator, so that we can see the distinction between (1) the rule as a universally quantified form, (2) the application of the general rule to individual cases, (3) the *ad hoc* creation of status functions without a preexisting institution, (4) the case of the free-standing Y terms, and (5) the relation of all these to the power creation operator.

The distinction between (1) and (3) is illustrated by the distinction between, for example, the rule that says that anybody who meets certain conditions counts as President of the United States and cases where we simply impose status functions in an *ad hoc* fashion, as in the example of the tribe that just selects somebody as the leader without any general rule for selecting leaders. And then we need to add to those the cases (4) where you can bring a set of status functions into existence by fiat. That is what happens when we create corporations.

Furthermore we have to be very careful about the use of the quantifier notation, because on the standard interpretation the variables range over a previously identified domain of objects. In some of these cases (for example, the creation of corporations) we are actually creating objects.

What is the actual formulation? Well there will be more than one. There will be the general constitutive rule(s), there will be particular applications of the constitutive rule(s), there will be *ad hoc* assignments of status functions, without prior constitutive rules, and there will be creations of status functions that do not require an X term. Furthermore we need to distinguish between the initial creation of the institution, and the continued existence of the institution. The same distinction applies to institutional facts, where we need to distinguish the initial creation from the continued maintenance. The general principle to remember is that:

Institutional facts = status functions → deontic powers.

All institutional facts are status functions and all status functions carry deontic powers.

So here are the complete statements of the forms. We start with the universally quantified form, using the US presidency as an example (I will use small "x" and "y" as variables of quantification and big "X" and "Y" as free variables that may be bound by a neighboring noun phrase):

The constitutive rule. For all x, if x has features f in context C, x counts as the (status function) president of the United States.

Deontic powers: For all x, if x is President, x has the (deontic) powers assigned by constitution and the laws.

Now combine these inside the scope of the collective acceptance operator:

We collectively accept that for all x, if x has features f in C, x counts as the (status function) President of the US and S has the (deontic) powers assigned by the constitution and the laws.

This illustrates the general form of the universally quantified constitutive rule.

Now apply this rule to a particular case.

Barack Obama has f in C.

Therefore

We collectively accept Barack Obama counts as the (status function) president of the US and has the (deontic) powers assigned by the constitution and the laws.

All of this is implicit in the *Construction*. But one point comes out here that is not in the *Construction*: Once you accept the rule and the fact that the X object satisfies the conditions specified by the rule you are logically committed to accepting the particular status function. No further separate acceptance is necessary. This works for all institutions. Once you accept the rules of baseball and you accept that such and such a team scored more runs than the opponent you are committed to accepting that the team with the higher score won. This is why, for example, there was so much dispute about the 2000 presidential election. Did Bush really win a majority of the States' electoral votes? There was never any dispute about the system of constitutive rules, only about whether Bush satisfied the X condition.

In the *ad hoc* case, where for example a tribe just treats Bill as the leader and thus makes him a leader, it looks like this:

We collectively accept that Bill, as X, counts as our leader, as Y, and as leader he has the deontic powers that go with that status.

What about the free-standing cases? Notice that for these cases we cannot have a universally quantified rule that ranges over the domain of preexisting objects, as we did in the example of the US presidency. But more interestingly, we cannot even have an existentially quantified form to the effect that there is some x such that x is a corporation, because by hypothesis there is no preexisting x which is the corporation. On the standard interpretation of the quantifiers, there must be a preexisting domain of objects over which the quantifiers range. But in these cases there is no such range of objects. There are no objects that we are going to turn into corporations, rather we are going to perform a speech act which creates a corporation. The general form of the rule, when codified, would have to be that such-and-such a speech act counts as the *creation* of the corporation. So its logical form of the creation of a particular corporation is not:

For some x, x becomes the corporation, Y

but rather its form is that of a Declaration:

Let it be the case that this speech act X creates the corporation, Y with such-and-such sets of status functions attaching to the officers and shareholders.

The California Code regarding corporations makes this explicit.

Section 200A: "One or more natural persons, partnerships, associations or corporations, domestic or foreign, may *form* a corporation under this division *by executing and filing articles of incorporation.*"

Section C: "*The corporate existence begins upon the filing of the articles and continues perpetually,* unless otherwise expressly provided by law or in the articles." (italics added)

I think it is clear that the logical form of this is not that of existential quantification. It does not say that there is some preexisting x, that is a corporation, rather it says there "begins" a corporation. It says that the performance of these speech acts – "executing and filing articles of incorporation" – counts as the creation of a corporation – "the corporate existence begins upon the filing of the articles and continues perpetually ... ." And the actual statement of the law in California has a universal quantifier ranging over "natural persons, partnerships, associations, corporations, domestic or foreign . . . ."

So the general logical form of the status function in these cases, using the creation of corporations and letting "s" range over natural persons, etc. is: We accept that for all s, if s performs certain sorts of speech acts, (i.e., "executing and filing") these speech acts as X count as the creation of a corporation as Y and with it the Y status functions of President, officers and shareholders are assigned to S1, S2, S3; and with these status functions go the deontic powers assigned to these positions; and we further accept that once created the corporation as Y will last perpetually ("unless otherwise expressly provided") . . . .

Notice that once again, language is crucial to the ontology. The actual sentences from the California Code that I quoted state constitutive rules: the speech acts count as the creation, and once created, the corporation created counts as existing perpetually. It is inconceivable that this could be done without language. But, one might object, couldn't there be a society where corporations, or something just like them, naturally evolved without any law? You can imagine such things, but to do so is to imagine the Declarational speech acts doing their job without benefit of the legislature. My point remains: No language, no corporations.

The creation of money without currency by banks has a different logical structure. Typically the bank creates money by issuing loans of money it does not have. Again these are Declarations. Suppose the Bank of America loans Jones $1,000 dollars. The Declaration has this form: We, the Bank of America, make it the case by Declaration that Jones has $1,000 in his account. So the speech act, as X, makes it the case that Jones has the Y status function. Possessor of $1,000. But there need be no physical reality to the $1,000, other than the representation. Jones now has the deontic power to spend the money as he wishes.

To repeat the crucial point, the logical structure of the creation of all institutional reality is the same as that of the performative. You make something the case by representing it as being the case.

We can now see in a deeper sense why language is the fundamental social institution and why it is not like other institutions. All other institutions require linguistic representation because some new non-semantic fact is created by the representation. Money, government, and private property are created by semantics, but in every case the powers created go beyond semantics. Meanings are used to create a reality that goes beyond meaning. But language itself does not have a meaning that goes beyond meaning. It says on the twenty dollar bill, "This note is legal tender for all debts public and private." Now why don't they add a sentence next to that which says, "This is really a sentence of English and it means what it says"? Would we find that reassuring? The sentence on the bill purports to be a Declaration. It certifies that the note is legal tender by representing it as legal tender. But the sentence I imagined can't certify or otherwise make it the case that the other sentence really is a sentence. Nor does it need to. The language is enough to determine that it is a sentence. The status functions of language (e.g., sentences) are self-identifying for anyone who knows the language. But other status functions require language not only for their identification but for their very existence.

Language itself does not require some further representation in order to be language. The status functions assigned to words and sentences do not require any further representation in order that they exist. Obama can only be President and this property can only be my property if both are represented as such, but the words you see on the page are words with meanings without any further representation of their having these status functions. The way in which the double level applies to all status functions is different in the case of language from what it is in all other institutional facts. It has always bothered me that I say all institutional facts create deontic powers. The "powers" of language are quite different. The double direction of fit can never be part of meaning, except where it is used to create meanings (i.e., linguistic performative speech acts).

8   Conclusion

I think the upshot of this discussion is the following. For standing institutions, we need to distinguish between the creation of the institution and institutional facts within the institution. For the institutional facts themselves, we need also to distinguish between the creation of the institutional fact, and the maintenance of the institutional fact in its continued existence. The logical form of the creation of an institutional fact is always a Declaration. This is concealed from us by the fact that it might evolve gradually over a long period of time. But all the same, the double direction of fit has to be manifest in the creation of any institutional fact. The maintenance of an institutional fact in its continued existence also has the logical form of a Declaration. The reason for this is that the fact can only exist as long as it is represented as existing but in its creation and maintenance the representation does more than represent an independently existing phenomenon, rather it brings about and maintains that phenomenon in its continued existence. So the most general form of the creation of an institutional fact is that of a Declaration, like God creating light: "Let there be light!" For humans it has to be an institutional entity: "Let there be a corporation!"

The "counts as" formula requires a preexisting term to stand as the bearer of the status function which is created by the "counts as" operator. In the creation of a free-standing Y term, the X term is a Declarational speech act that counts as the creation of the Y status function, the free-standing Y term. The sense in which it is free-standing is that after the creation, there is no further physical realization of the Y status function. There are, however, the actual individuals who have the status functions that attach to the corporate or other status. Thus under California law, there must be officers of the corporation, stockholders, etc.

REFERENCES

Searle, J. 1979. *Expression and Meaning: Studies in the Theory of Speech Acts.* Cambridge: Cambridge University Press.
  1995. *The Construction of Social Reality.* New York: The Free Press.
Tsohatzidis, S.L. 2007. *John Searle's Philosophy of Language.* Cambridge: Cambridge University Press.

1 – Comment
De Rerum Natura: Dragons of Obliviousness and the Science of Social Ontology

Mark Turner

In *Parallel Lives*, Plutarch praises the efforts of Theseus, the legendary Athenian king, to establish and settle a Hellenic commonwealth. Theseus, having made secure acquisition of the country about Megara to the territory of Athens, erected a pillar in the Isthmus and inscribed upon it two verses, one on either side, to distinguish the boundary. One said:

This is not Peloponnesus, but Ionia.

Its counterpart said:

This is not Ionia, but Peloponnesus.

These assertions, graven on a pillar, articulate a fundamental social ontology for the Hellenic world and for anyone since then who has completed a Western education. They prompt for something in the mind and thereby create something in the world. The fable of the pillar is so captivating that its verses have been recited since Strabo (9.1.6), probably well before, flying on repetitive wings across millennia, irresistibly perpetuating themselves through the minds of generations of students.

John Searle, in "Language and Social Ontology," proposes to clarify what is going on in cases such as Plutarch's story. Searle's assertions are sweeping and they penetrate to the root of the question posed for this volume, namely, how can scientific practice in the social sciences be improved? His essential claim is that Western social thought would be better, or less vacuous, if it attended to the crucial relationship between language and social ontology. Western social thought on the subject of social ontology is, Searle judges, pretty much mistaken from the start, producing pretty much empty conclusions. The problem, Searle asserts, is that theorists in the social sciences take language for granted.

I have also a polemical aim for wishing to discuss social ontology and that is that I believe we have a long tradition, going back to the ancient Greeks, of misconstruing the role of language in the creation and constitution of social and political reality. It is characteristic of, I believe, all the authors known to me in our tradition, from the Greeks to the present, that they do not accurately see the role of language in the creation, constitution, and maintenance of social reality. Social and political theorists assume that we are already language-speaking animals and then go on to discuss society, without understanding the significance of language for the very existence of human social reality. Every author I have read, from the ancient Greeks right through such contemporary authors as Foucault, Bourdieu, and Habermas, takes language for granted. What I mean when I say they take language for granted is that in their discussions of social reality they are discussing people who already have a language. ... The worst offenders in this regard are the Social Contract theorists who simply assume that we are language-speaking animals and that we all get together in the state of nature and form a social contract. The point I make in this chapter is that once you have a language you already have a social contract. The social contract is built into the very essence of language.

Searle singles out a dragon – *Taking Language for Granted* – as a monster of ruination, a destroyer of legitimate scientific practice in the social sciences. His purpose is to slay the dragon. *Taking Language for Granted* leads us to mistake rehearsing tautologies for doing science. Rehearsing tautologies deludes us into assuming that we are saying something new when we are only restating what we have assumed.

Searle's assertion is important and accurate. I add to it here that *Taking Language for Granted* is not an only child. There is a litter of sibling dragons, hatched from a single clutch, bearing family resemblances and family names. For each aspect X of cognitively modern human performance, there is a dragon "*Taking X for Granted.*" These Xs work together. Evolutionarily and developmentally, they arise together and support each other. Aspects of higher-order human performance – language being only one of them – form a cooperative pack. Insidiously, to take one of them for granted is to take much of what is involved in the others for granted. Accordingly, this way of proceeding is prone to vacuity, just as Searle observes for the case of language and social ontology.

This is discouraging news for the ambition of doing social science. The news is unwelcome and therefore resisted, because social science would be easier if we could treat human behavior as a linear sum of partitioned categories of performance – language being one; social ontology being another. On that model, science could investigate these behaviors separately. Searle sheds light on how this model fails for language and social ontology. The failure, I assert, is general. This news is discouraging, but it appears to be true, and science needs to deal with the bad news if it is to advance.

I will call these powerful, seductive, mostly invisible siblings "dragons of obliviousness." To dispel them requires recognizing their charms. Some of the dragons of obliviousness I will indicate in this comment are:

> *Taking Political Ontology for Granted*
> *Taking Intentionality for Granted*
> *Taking Personal Identity for Granted*
> *Taking Personal Identity as a Stable Chooser for Granted*
> *Taking Counterfactuality for Granted*
> *Taking Material Anchors for Granted*
> *Taking Viewpoint, Perspective, and Focus for Granted*
> *Taking Roles for Granted*
> *Taking Social Memory for Granted*
> *Taking Narrative for Granted*

Cognitively modern human beings – that is, all neurotypical members of *homo sapiens sapiens* during roughly the last 50,000 years or so – possess higher-order cognitive abilities and a consequent suite of performances and artifacts: language, science, mathematics, social pragmatics, art, music, advanced use and invention of tools, personal identity, external representations, fashions of dress, social rank, and institutions and organizations of many kinds: political, cultural, and social. It is a mistake to try to explain any one of these things, or any subgroup of these things, by taking the others for granted. Searle is right that it is bad science to assume that human beings had full language and then tried to create social reality. Social ontology and language derive jointly. It is equally bad science to assume that human beings had any one of these higher-order cognitive abilities and then tried to create from scratch what flows from any of the others.

Consider the pillar and the verses. "This is not Peloponnesus, but Ionia." "This is not Ionia, but Peloponnesus." Consider the social ontology of Ionia. What is involved in a conception like that? To answer this question, we need to focus on the notion of *human scale*. Human beings are built to work with *scenarios at human scale*. For example, we are quite good at parsing our sensory worlds into objects and events, even though the input to our sensory fields shows no such partitioning. The world does not come with labels. But we are good at labeling it in particular ways, built to do so. For example, we are quite good at framing certain kinds of sensory experience as an agent's performing an action on an object that causes it to move in a direction. In parallel, language is built to express such human-scale frames. We can say, "I threw the ball over the fence." This is a caused-motion frame and a caused-motion clausal construction. Similarly, we are quite good at framing sensory experience as an agent's performing an action on an object with a result for that object. In parallel, language is built to express this human-scale frame. We can say, "I painted the wall white." This expresses the human-scale frame for Resultative action, and it does so through a Resultative clausal construction.

How do we grasp what is beyond human scale? All mammals grasp what they are built to grasp – they recognize a predator, a mate, a dominant member of the species, food, damage to their bodies. However wondrous the minds of mammals – even some corvids – seem to be, their limits are clear. No dog is going to follow the rudiments of quantum mechanics. Cats will never paint. Chimpanzees communicate but do not have language. But cognitively modern human beings, unlike other species, have an advanced form of a mental ability that is crucial for social ontology and for the subjects of the social sciences, and this advanced mental ability makes these higher-order human behaviors possible. My central purpose in this comment is to focus on this advanced human mental ability, its operation mostly below the horizon of observation, its fundamental role in enabling higher-order human performances, and the necessity to recognize it if we are not to be lulled to sleep by the dragons of obliviousness.

The mental ability is *blending*, and the advanced human ability is called "double scope" blending (Fauconnier and Turner 1998, 2002; Turner 1996, 2001a). Blending is the mental operation that integrates two or more different conceptual arrays, and double-scope blending is its most advanced form. Double-scope blending typically involves integrating different conceptual frames. Human beings, it turns out, can blend conceptual arrays that are quite incompatible. This is a species-wide ability, but apparently exclusive to human beings. A double-scope network has inputs with different (and often clashing) organizing frames and an organizing frame for the blend that includes parts of each of those organizing frames and has an emergent structure of its own. In such networks, both organizing frames make central contributions to the blend, and their sharp differences offer the possibility of rich clashes. Far from blocking the construction of the network, such clashes offer challenges to the imagination and the resulting blends can turn out to be highly creative.

Very conveniently, *double-scope blending makes it possible for human beings to transform diffuse ranges of information that are not at human scale into useful and congenial human-scale scenes.* As a result, we can say not only "I threw the ball over the fence," for a human-scale caused-motion scenario, but also "Hunk choked the life out of him" and "France moved England toward war." These situations are not in themselves human-scale caused-motion scenarios, but in the blend, they are structured using the human-scale caused-motion scene, and expressed through that language. Through blending, they become human-scale, and we do not need new language to express the emergent content of the blend. Similarly, for the Resultative case, we can say not only "I painted the wall white," for the human-scale scenario, but also "The earthquake shook the building apart" and "Roman imperialism made Latin universal." In the blend, diffuse conceptual arrays are packed to a human-scale scenario with structure projected from a Resultative frame, and expressed through Resultative language. Again, we do not need new language to present the creative content of the conceptual blend.

Through double-scope blending, we can use human-scale conceptual frames such as caused-motion and resultative to bring to human scale conceptual ranges that are not already at human scale. The result is that we can understand them, follow them, manage them, and supervise them dynamically, and, since language is one of the human higher-order performances that derives from double-scope blending and can collaborate with conceptual construction, we can express these understandings in a manner that prompts for such construction of double-scope conceptual integration networks by people listening to us.

Now for Ionia and Peloponnesus. Cognitively modern human beings can activate their human-scale conceptual frame for two entirely distinguishable adjacent physical objects and integrate that conceptual frame with some very diffuse conceptual information about human agency in a certain geographical area during a certain epoch to come up with Ionia and Peloponnesus. To have human language is already to have the ability to conceive of social reality in this way, so *Taking Language for Granted* is *Taking Political Ontology for Granted*.

One might feel inclined to say at this point, but aren't Ionia and Peloponnesus simply in fact clearly different? If so, why do we need to appeal to some grand double-scope achievement? The answer is that, to cognitively modern human beings with language, of course they are clearly different. But that is just the point: cognitively modern human beings are built to be double-scopers. This highly creative form of conceptual integration is running all the time, as a baseline of human conception, and is not cognitively costly at all. I have stood on the Isthmus of Corinth often, and, of course, observed that there isn't anything in the visual field that would tell you they are different, or that one is Ionia and the other Peloponnesus. Ionia and Peloponnesus are social realities, available to us because we can do double-scope blending. Without double-scope blending, there is no human language and no social ontology as we know it.

If we consider further the meaning of "Ionia," we see other consequences of double-scope blending. "Ionia" consisted of all those polities that traced their origin from the city of Athens and kept the festival of the Apaturia. From the Corinthian Isthmus, you can't see the Ionian coast of present-day Turkey (home of Homer), and you can't see the festival of the Apaturia at all, because the festival of the Apaturia is another social ontology – a religious festival held annually in nearly all Ionian towns. In Athens, it took place in the month called Pyanopsion. What is a month? What is Pyanopsion? We can divide the year up into twelve months because we are able to make double-scope conceptual integration networks (Fauconnier and Turner 2008), a mental feat that seems like nothing to us double-scopers but is indescribably beyond the abilities of any other species. Double-scope blending gives us the concept of a month, and the notion – at human scale because of blending – that events can "take place" temporally "in" "the span" of that "month." What's the Pyanopsia? The Pyanopsia was a social religious festival in honor of Apollo. Who is Apollo? And so it goes. Any of these cognitively modern social, political, cultural, and theological realities is connected to very many more.

What was the purpose of the Pyanopsia? It lasted three days, on which occasion the various phratries, or clans, of Attica, met to discuss their affairs. What is a clan? A piece of social ontology. Some think the name "Pyanopsia" means something like "festival of common relationships," a profound social ontology.

We see in the pillar and its phrases another dragon of obliviousness: *Taking Intentionality for Granted*. Theseus set up this pillar. The Caused Motion frame has an intentional agent, as does the Resultative frame. Other species seem to be surprisingly inferior to us in the ability to attribute nuanced states of intentionality. Recent research suggests that chimpanzees, *pan troglodytes*, working at the limits of the ape brain, with scaffolding from their most robust abilities, i.e., abilities having to do with eating and with dominance hierarchies, might have a little more ability to attribute a nuanced intentionality than we had guessed, but not much (Tomasello *et al.* 2003, 2005).

Where do our robust abilities for attributing nuanced intentionality come from? The answer from blending theory is that this ability comes from our ability to blend self and other, so as to attribute a nuanced intentional mind to others. *Taking Language for Granted* automatically invokes *Taking Intentionality for Granted*. Where you find one of these dragons, the others are not far behind.

What is a person? It is important to avoid *Taking Personal Identity for Granted*. Our garden-variety conception of personal identity turns out to be an extremely complicated higher-order human performance, dependent on the construction of impressive double-scope integration networks. To make sense of ourselves, we must do work to manufacture understandings at human scale. Working across great and diffuse ranges of conceptual structure over extended periods of time, we must construct a stable identity. Despite the manifest evidence of discontinuity and variation across our individual lives, we manufacture a sense of stable personal identity. Moreover, despite the swarm of detail in which we are embedded, we manufacture small narratives of ourselves as agents with stable personal identities. This is part of cognitively modern human cognition, and other species do not seem to have it in any substantial measure. It is a result of blending across many complex and detailed experiences, with analogies and disanalogies. There are analogies and disanalogies across all the memories in which we feel our identity plays a role. In the blend, the analogies are compressed to one element, a stable personal identity, and the disanalogies are compressed to change for that personal identity. We make the same kinds of compressions to provide us with notions of personal identity for other people.

The branches of the social sciences that are particularly vulnerable to *Taking Personal Identity for Granted* are those that depend upon the concept of an individual with preferences, who chooses, decides, and judges. Personal identity is far more complicated, more creative as a concept, than our folk theories suggest. This complexity is in keeping with the general news from cognitive science that human thought is remarkably more complicated and diffuse than our folk theories of mind suggest. Consider, as an analogy, vision and language. There is no controversy in vision science and language science that the mechanisms of vision and language are extraordinarily complex, quite unlike commonsense conceptions of how they work, and mostly invisible to human beings, who see and talk and offer folk theories such as "I just open my eyes and the scene comes in" or "Words have meanings so I say what I mean." Great ranges of backstage cognition make vision and language happen. The principal reason that human beings think that sight and language happen in fairly simple ways, with fairly simple principles and with intelligible, human-scale frames, is that vision and language do produce some small, integrated, useful packages and deliver them to consciousness, and these little packages do seem fairly simple, with simple principles and with intelligible, human-scale frames. The cognitive scientist is in a curious situation: human beings are not built to understand actual mental functioning scientifically – doing research in the field is slow, hard work – but human beings are built to grasp the little packages of consciousness, and to blend the frame for the scientific question with the frames of conscious experience, and so to produce, in the blend, human-scale folk theories of who we are and what we do. One result is that citizens who believe that physics and chemistry are very difficult can have a hard time grasping what a cognitive scientist does. Vision and language must be child's play; a three-year-old can see and speak, right, so how hard could it be?

Actual mental functioning is distributed, complex, and not at human scale. We are not designed to look into what we are and how we operate. Obliviousness to the complexity of mental mechanisms, and our species-wide inability to inquire easily into them, are no surprise. What evolutionary benefit would there have been in our environments of evolutionary adaptation for a cognitive power to analyze vision and language? Seeing and speaking are useful; analyzing their mechanisms is not, at least until now.

For centuries, scientific notions of perception depended on the "Cartesian theater," that is, the implicit idea that there is a little perceiver in the head, a kind of attentive homunculus, who pretty much watches a representation of what we are watching in the world, and who figures it out. In this simple conscious frame, each of us is an attentive self looking at the world and making sense of it. To answer the question *what is the mind doing?*, we blend that simple conscious frame with our frame for the scientific question and so create a folk theory of mind as pretty much an attentive agent looking not at the world but at a mental representation of the world. This folk theory of the watchful little perceiving guy in the head is a frame blend. It had influential scientific standing for centuries. But it turns out that perception works nothing like that. Perception is far more complicated, there is no attentive homunculus in the mind, and there is no anatomical spot where sensory data are assembled into a unified representation of the sort we imagine. Indeed, it is a deep scientific problem to explain how something like a coffee cup   with its hue, saturation, reflectance, shape, smell, handle for grasping, temperature, and so on – can seem in consciousness like one unified object. In neuroscience, this problem is called "the binding problem" or "the integration problem." We are built to think that the reason we can see a coffee cup as one unified object is simply that it *is* one unified object whose inherent unity shoots straight through our senses onto the big screen in the mind where the unity is manifest, unmistakable, no problem. It is natural to hold such a belief, but the belief turns out to be just a folk theory, another case in which we make a frame blend of the scientific question with a frame of consciousness. It does not seem to us in consciousness that we are doing any work at all when we parse the world into objects and events and attribute permanence to some of those objects, but these performances constitute major open scientific problems.

In consciousness, typically, we frame experience as consisting of little stories: our basic story frame includes a perceiving self who is an agent interacting with the world and with other agents. Despite the swarm of detail in which we are embedded, and the manifest discontinuities in our lives, we manufacture small conscious narratives of ourselves as agents with stable personal identities, and these small narratives are at human scale and easily intelligible.

In these narratives, we possess straightforward powers of choice, decision, and judgment. Consciousness is equipped for just such little human-scale stories of choice: we encounter two paths, or a few fruits, or a few people, and we evaluate, decide, choose. We act in such a manner as to move in the direction of one of the possibilities. We say, "I'll have an espresso." We are not set up to see the great range of invisible backstage cognition that subtends what we take to be evaluation, decision, and choice, any more than we are set up to see the work of vision or language. But we are set up to make a blend of (1) the human-scale conscious experience of a chooser choosing and (2) the scientific question of how the mind works.

The result is *homo economicus* – a folk theory of a rational actor in the head, with preferences, choices, and actions. *Homo economicus* is a homunculus much like the little mental perceiving guy in the Cartesian theater. The Cartesian homunculus looks at the screen and perceives; *homo economicus* looks at choices and chooses. In the *homo economicus* blend, each of us is a stable chooser with interests, living a narrative moment as an agent with a personal identity, encountering other such agents. This human-scale narrative of the self as a stable identity with preferences that drive choice toward outcomes is marvelously useful, and instrumental in action, motivation, and persuasion. It is a worthy human-scale fiction that helps us grasp ranges of reality that are diffuse and complicated (Turner 2001a). *Taking Language for Granted* involves *Taking Personal Identity for Granted*, and, in passing, *Taking Personal Identity as a Stable Chooser for Granted*.

The next dragon of obliviousness is *Taking Counterfactuality for Granted*. "This is not Peloponnesus, but Ionia." "This is not Ionia, but Peloponnesus." These words assert not just the existence of a social reality, but the exclusion of another social reality: *not* Peloponnesus; *not* Ionia. *Taking Language for Granted* brings *Taking Counterfactuality for Granted* right in. Because of double-scope blending, cognitively modern human beings are able to take two tightly analogous but incompatible conceptual arrays and blend them (Turner and Fauconnier 1998). Here's a simple example. The gazelle reacts really well to an open plain. It also reacts really well to a predator in front of it in a plain. But human beings are able to integrate these two scenes to create a blend in which the open plain is not just an open plain, but rather, a plain with *no predator*. We look in the refrigerator. Sometimes we see goat's cheese, seltzer water, and roast beef. Sometimes we see goat's cheese, seltzer water, milk, and roast beef. We are able to integrate these two so that the first one becomes a refrigerator with *no milk*. We say, "There is no milk in the refrigerator," and we can see it right there. There's the *no milk*. We even say, "I see no milk in the refrigerator." The *no milk* is a real object in the blend. It just happens to be the one with the feature of *absence*. *Taking Language for Granted* involves *Taking Counterfactuality for Granted*, and this is a truly big dragon of obliviousness. The human being in *not Peloponnesus, but Ionia* can sense the great latitude that comes with freedom from the restrictions of Peloponnesus. In *not Peloponnesus*, there is the great feature of *absence of restrictions, privileges, duties, and other social aspects* of Peloponnesus. The counterfactuality is in fact more important for the maintenance of the Greek commonwealth than the positive creation of a social element. It's utterly important for peace that the Ionians recognize that this is *not Ionia*, and for the Peloponnesians to recognize that this is *not Peloponnesus*.

It is not just the propositional content, but also the form of language that is important in these expressions. The form of the language in "This is not Peloponnesus, but Ionia; this is not Ionia, but Peloponnesus" is called a *chiasmus*. A chiasmus is formally symmetric about a midpoint; it is a crossing, a crossing-over, an equilibrium. As Kenneth Burke so famously observed, we integrate (I would say blend) the form and the content. Assenting to the form primes us to assent to the content. The content is an equilibrium, a meeting, a balance about a midpoint: Ionia lies in one direction, Peloponnesus in the other. *Taking Language for Granted* includes taking for granted the blending of form and content.

The linguistic form of chiasmus ushers in the dragon *Taking Material Anchors for Granted*. The pillar is also a form that marks chiasmus. The pillar is not just rock, not even fashioned rock, but rather a social marker, a material anchor for the social ontology. Social ontology has depended throughout the history of cognitively modern human beings on material anchors, rituals, performances, patterns of affiliation and understanding, such as the festival of the Apaturia. Taking them for granted is essentially taking everything (aside from specific elaborations) about the social sciences for granted.

These social realities are imposed upon other realities, just as Searle asserts. The isthmus of Corinth can be seen (by cognitively modern human beings, and by no other species) as a geographical chiasmus. Here is an image from Google Earth:

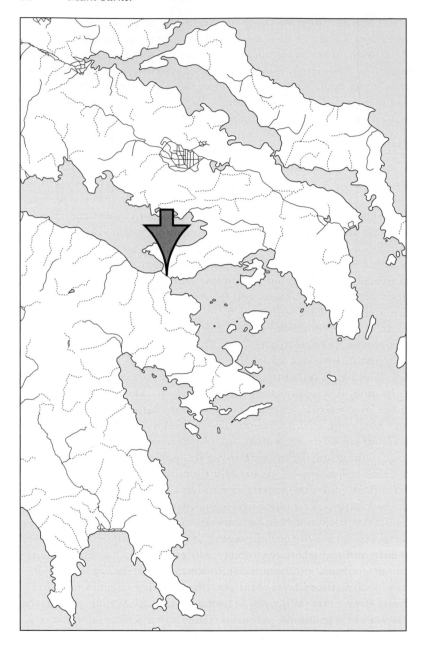

Figure 1.1 An image from Google Earth

Theseus, Strabo, and Plutarch (those three choosing stable personal identities, social entities themselves, with names, created as such through double-scope blending) did not have Google Earth, but they could stand at the Isthmus, or on another isthmus, and integrate the heraldically symmetric geography of meeting at a central point with the chiasmus of language and the idea of a political boundary. These kinds of compounding, reinforcing integrations come with language.

Consider the language on the pillar. On which side of the pillar do you imagine each verse is engraved? There are at least two main possibilities. In the first, the viewpoint resides with the *reader* of the words. So "this" refers to what the reader sees. In this version, we have a blend that draws on our knowledge of actual human conversation, where the speaker is indicating what the hearer sees, perhaps even pointing at it. This conversation is integrated with the factual scene in which someone is looking at the pillar. In the blend, the pillar is the speaker, quoting the political arrangement, and referring to what the reader is seeing. In that case, "This is not Peloponnesus, but Ionia" must be on the western side. It indicates that the reader is *leaving behind* all kinds of social ontology, and stepping into new features of social reality. But now suppose the viewpoint is the *speaker's*. In that case, "this" must refer to what the speaker (i.e., in the blend, the pillar, the material anchor) sees. Accordingly, in that blend, "This is not Peloponnesus, but Ionia" must be on the *eastern* side of the pillar. *Taking Language for Granted* already involves *Taking Viewpoint, Perspective, and Focus for Granted*, not only for visual scenes, but for intellectual, political, social, and cultural content, as in "What we are looking at is fraud" or "The IRS will take the view that you have a reporting responsibility." Viewpoint, Perspective, and Focus can also be prompted for by a great range of other constructions in language that do not in fact use lexical items associated with vision, an example being the difference between "come" and "go," as in phrases like, "The corporation must come into compliance." Viewpoint, perspective, and focus are equally used for temporal ontology, as in the difference between "This *is* Ionia" and "This *was* Ionia."

There are many other dragons of obliviousness, such as *Taking Roles for Granted* (Theseus was king of Athens), *Taking Social Memory for Granted* (both Plutarch and Strabo are handing down what we can no longer see), and *Taking Narrative for Granted* (Once upon a time, there was a great king of Athens who set about to establish and settle a Hellenic commonwealth ...). Limitations of space prevent us from doing more than pointing out and naming some of these marvelous, lethal dragons. They are good at silently, invisibly casting easeful spells that can lull the social sciences into a fantasy of knowledge. This is just a more general restatement of Searle's point.

In Turner (2001a,b), I propose that the social sciences and cognitive sciences are destined to combine, partly because the cognitive neuroscience of human beings is bound up with the fact that we are built for culture, but also partly because explanation in the social sciences usually takes for granted the nature of higher-order human cognition and tries to depart from there, when in fact human social performance is part of the way we think. Social reality and social performance are conditioned by the nature of our basic mental operations and our most characteristic capacities, language prominent among this crew. Our characteristic higher-order human cognitive capacities are a pack, a troupe, a team. They arise together developmentally and evolutionarily, they support each other, and they labor together. Taking any of them for granted in analyzing how the others arose is an error. It is important for social science not to take any of them as given when we attempt to analyze how any of the others came into being.

REFERENCES

Fauconnier, G. and M. Turner 2008. "Rethinking Metaphor." In R. Gibbs (ed.) *The Cambridge Handbook of Metaphor and Thought*. New York, NY: Cambridge University Press, pp. 53–66.

Fauconnier, G. and M. Turner 1998. "Conceptual integration networks," *Cognitive Science* **22**(2): 133–187.

Fauconnier, G. and M. Turner 2002. *The Way We Think: Conceptual Blending and the Mind's Hidden Complexities*. New York, NY: Basic Books.

Tomasello, M., J. Call, and B. Hare 2003. "Chimpanzees understand psychological states: The question is which ones and to what extent," *Trends in Cognitive Sciences* 7(4): 153–156.

Tomasello, M., M. Carpenter, J. Call, T. Behne, and H. Moll 2005. "Understanding and sharing intentions: The origins of cultural cognition," *Behavioral and Brain Sciences* **28**: 675–735.

Turner, M. 1996. *The Literary Mind: The Origins of Language and Thought*. New York, NY: Oxford University Press.

2001a. *Cognitive Dimensions of Social Science: The Way We Think About Politics, Economics, Law, and Society*. New York, NY: Oxford University Press.

2001b. "Toward the founding of cognitive social science," *The Chronicle of Higher Education*. 5 October 2001.

2008. The Blending Website: http://blending.stanford.edu.

Turner, M. and G. Fauconnier 1998. "Conceptual Integration in Counterfactuals." In J.-P. Koenig (ed.) *Discourse and Cognition*. Stanford: Center for the Study of Language and Information, pp. 285–296.

2   Shared Agency

Michael E. Bratman

Human beings act together in characteristic ways.[1] Forms of shared activity matter to us a great deal, both intrinsically – think of friendship and love, singing duets, and the joys of conversation – and instrumentally – think of how we frequently manage to work together to achieve complex goals. My focus will be on activities of small, adult groups in the absence of asymmetric authority relations within those groups. My approach begins with an underlying model of individual planning agency, and then seeks a conceptual and metaphysical bridge[2] from such individual planning agency to modest forms of sociality.

1   Shared Intention

Suppose you and I are painting a house together. What makes this a *shared intentional* activity?[3] We could imagine a contrast case in which

---

[1] This chapter is culled from a larger unpublished manuscript that I hope someday will be a book. That manuscript, in turn, draws in part from earlier work of mine on this subject: "Shared Cooperative Activity," "Shared Intention," "Shared Intention and Mutual Obligation," and "I Intend that We *J*" – all reprinted in my *Faces of Intention* (1999a); "Shared Valuing and Frameworks for Practical Reasoning," as reprinted in my *Structures of Agency: Essays* (2007); and "Dynamics of Sociality" (2006: 1–15). My recent thinking about these matters has benefited from discussions with many people; but let me mention in particular discussions with Scott Shapiro when we both had the privilege of being Fellows at the Center for Advanced Study in the Behavioral Sciences in 2003–2004, and discussions with Facundo Alonso and Jules Coleman. Work on this chapter was supported in part by a Fellowship at the Stanford University Humanities Center.

[2] This bridge will involve, but not be limited to, the "interlocking" intentions I have appealed to in earlier work and that Abraham Roth calls "bridge intentions" (Roth 2003: 76–80). Thanks to Roth for this apt metaphor of a bridge.

[3] I do not say shared *cooperative* activity, since in my view the idea of "cooperative" brings into play a further constraint of non-coercion that I will not address here. See my "Shared Cooperative Activity" in Bratman 1999a: 101–102. In that discussion (1999a: 103–105), I also treat a kind of commitment to helping the other as characteristic of shared cooperative activity, but not strictly necessary for shared intentional activity. In contrast, my present discussion includes a disposition to help within the model of shared intention. See below note 16.

we each intentionally go through the same motions as we do when we paint the house together, and yet there is no shared intentional activity. Perhaps we are each set only on our individual painting project and respond to each other only with an eye to avoiding collisions. Echoing Wittgenstein's question about the difference, in the individual case, between my arm's rising and my raising it, we can ask: what is the difference between such a contrast case and shared intentional activity? In the case of individual human action we can see the difference as involving an explanatory role of relevant intentions of the individual agent. I propose an analogous view of the shared case: the difference in the case of shared agency involves an appropriate explanatory role of relevant *shared* intentions. Our painting together is a shared intentional activity, roughly,[4] when we paint together because we share an intention so to act.

But what is shared intention? And what is an appropriate explanatory relation? Here I focus on the first question, though certain ideas about this explanatory relation (as I call it, the *connection condition*) will feed back into our model of shared intention.

I see shared intention as consisting in relevant intentions of each of the individual participants, in a suitable context and suitably interrelated. Further, the contents of these intentions of each will involve appeal to the group activity. Our shared intention to paint together involves your intention that we paint, my intention that we paint, and relevant further contents, interrelations, and contexts.

This involves violating the *own-action condition* on the content of intention.[5] According to the own-action condition it is always true that the *subject* of an intending is the intended *agent* of the intended activity. And it does seem initially plausible that intentions should respect some such constraint. Since my view involves rejecting the own-action condition, something needs to be said.

One reaction to this is John Searle's (1990: 401–415). Searle focuses on what he calls "we-intention." What he means by this is *not* what I mean in talking about *our* intention. Searle's we-intentions are attitudes in the head of an *individual*, though attitudes that concern the activity of a "we." You could have a we-intention, in Searle's sense, if you were the only person in the world, but thought there were others with whom you might paint. A Searlean we-intention is, then, a candidate for the intentions of individual participants that together help constitute a shared intention, though Searle himself does not say how the we-intentions

---

[4] Putting aside complexities I allude to in "Shared Cooperative Activity."
[5] For this terminology see "I Intend that We *J*" in *Faces of Intention* (Bratman 1999a).

of different participants need to be interrelated for there to be shared intention.

Searle's we-intentions violate the own-action condition. This may be why he claims that we-intentions are not just ordinary intentions with a special *content*, a content that involves the activity of a "we." We-intentions are, rather, a special intending *attitude*, to be distinguished from the ordinary attitude of intending involved in individual agency. If we suppose that the ordinary attitude of intending is subject to the own-action condition, and if we countenance we-intentions, then it will be natural to see we-intentions as distinctive attitudes rather than as ordinary intentions with a special content.

This contrasts with a view that appeals to a special content that involves an activity of a "we," but sees the relevant *attitude* as ordinary intention. Just as I can believe that *I* will do something, and also that *we* will do something, but in both cases what I have is an ordinary belief; so in the case of intention. My approach to shared intention is a view of this kind. Such an approach allows us to draw directly on what we know about the nature of ordinary intention, whereas this is apparently blocked by Searle's strategy, since Searle's we-intentions are not themselves ordinary intentions. But if we do proceed in the way I favor we must explain why it is acceptable to violate the own-action condition. I return to this matter below; but first I outline my overall approach to shared intention.

2   Constructivism about Shared Intention

Begin with a trio of ideas. First: we make progress in understanding aspects of mind by articulating characteristic functions or *roles* together with associated *norms.* This is a central idea behind my approach to the intentions of individuals, an approach I have called *the planning theory* (Bratman 1999b). The planning theory sees intentions as guiding intentional action and as coordinating forms of planning central to our temporally extended agency and to our associated abilities to achieve complex goals across time and interpersonally. The theory appeals to the guiding, coordinating, organizing roles of intentions as elements of larger – and, typically, partial, hierarchical, and future-directed – plans. And the theory appeals to norms associated with these roles, norms that are normally operative in a planning agent's psychic economy. Primary among these norms are norms of consistency, agglomeration, means–end coherence, and stability: intentions are to be internally consistent, and consistent with one's beliefs. It should be possible to agglomerate one's various intentions into a larger intention that is consistent in these ways. Intentions impose demands, roughly, to settle on known necessary means and the like – demands of means–end coherence. And intentions should involve a certain resistance to reconsideration and change.[6]

Second: we apply this methodology directly to *shared* intention. We ask: why do we bother with shared intentions? What roles do they play in our lives, and what norms are associated with those roles? And my initial answer appeals to analogues, in the shared case, of the coordinating, structuring, and guiding roles of intention in the individual case. In particular, the characteristic roles of shared intention include interpersonal coordination of action and planning in the guidance of shared activity, and the structuring of related bargaining and shared deliberation. And relevant norms of social agglomeration and consistency, social coherence and social stability will be associated with these social roles. It should be possible to agglomerate the relevant intentions of the different participants into a larger social plan that is consistent, that adequately specifies relevant means, and that is associated with an appropriately stable social psychological structure.[7] Failure to satisfy these norms will normally undermine the distinctive coordinating, guiding, and structuring roles of shared intention.

So we have structures of roles and associated norms both at the level of individual intention and at the level of shared intention. How are these structures related? My third idea is a conjecture about how to answer this question, at least in the cases of small groups that are my focus here. The basic idea is, roughly, that the norm-assessable social functioning characteristic of shared intention emerges from the norm-assessable and norm-guided functioning of a relevant structure of interrelated intentions of the individuals, as understood by the planning theory. We seek a construction of intentions and related attitudes of the individuals in appropriate contexts that would, when functioning in the norm-guided ways highlighted by the planning theory of individual intention, play the roles characteristic of shared intention. And we try to see conformity to central norms characteristic of shared intention – norms of social consistency, social agglomeration, social coherence and social stability – as primarily emerging from guidance by norms that apply directly to the

---

[6] See my *Intention, Plans, and Practical Reason* (Bratman 1999b). My formulation of the agglomeration principle here has benefited from the discussion in Gideon Yaffe (2004: 510–522). I discuss the nature and ground of these norms further in Bratman "Intention, Belief, Practical, Theoretical" (forthcoming).

[7] These norms apply against a background of assumptions about, roughly, what is possible and what is effective. Here I make the simplifying assumption that these will be beliefs with respect to which the participants are in agreement. A more complete account would appeal here to what I call "context-relative acceptance." See my "Practical Reasoning and Acceptance in a Context" in Bratman 1999a.

relevant interrelated structures at the individual level. If we had such a construction we would have reason to say that this construction *is* shared intention, or at least one important kind of shared intention.[8]

In saying this I am distinguishing between being assessable by a norm, being guided by a norm, and conforming to a norm. To think that certain thought and action is *assessable* by a norm (or, alternatively, that the norm *applies* to that thought and action) is to suppose that the violation of that norm by that thought and action is for that reason criticizable. A norm *guides* relevant thought and action when its acceptance is an appropriate aspect of the actual psychological functioning. Thought and action *conform* to a norm when they do not violate it. And the aim is to provide a construction of intentions of the individuals whose individual-norm-assessable and individual-norm-guided functioning would constitute the social-norm-assessable and social-norm-conforming social functioning of shared intention.

Call this *constructivism* about shared intention. We begin with the idea that shared intentions interpersonally structure and coordinate thought and action, and that these structuring and coordinating roles involve associated norms. We then ask: will these norm-assessable social roles emerge from the norm-assessable and norm-guided functioning of appropriate attitudes of the participants – attitudes with appropriate contents, in appropriate contexts, and appropriately interrelated?[9] We seek to answer this question by constructing a special structure of interrelated attitudes of the individuals, and norms that apply to and guide those attitudes, that would induce the norm-assessable and norm-conforming social roles characteristic of shared intention. We want to show that attitudes of individuals with these special and distinctive contents and interrelations would, insofar as they function properly and in a way that is guided by the norms of individual planning agency, play the roles of shared intention in part by conforming to central norms that apply to shared intention.

Constructivism highlights the idea that the individual participants are guided by norms of individual planning agency, but that given the special contents of their intentions, and their interrelations, this brings with it the applicability of, and conformity to, corresponding social norms on shared intention. In this sense, constructivism posits a kind of normative emergence. When the individuals become aware of this normative

---

[8] I discuss differences between these two ways of putting this point in "I Intend that We *J*" in Bratman 1999a: 144.

[9] Much of our theoretical work will be to say which attitudes, contents, contexts, and interrelations are the "appropriate" ones, and precisely what conceptual resources are needed to do this. That is what the basic thesis, to be sketched below, tries to do.

emergence they may go on explicitly to internalize these social norms and directly appeal to them in their practical reasoning. This would be a further step from the basic kernel of individualistic normativity at the bottom of shared agency.

Constructivism aims at sufficient conditions for shared intention. It allows for the possibility that there are multiple constructions, each of which provides some such basis for the social roles and norms characteristic of shared intention. In the face of purported, alternative constructions, the basic issue is which makes better sense of the complexities of these forms of sociality – though the best thing to say, in the end, may be that shared intention is multiply realizable.

3   Building Blocks

How might we proceed with such a construction? Suppose that you and I share an intention to go to New York City together. What construction of intentions and related attitudes of each would be such that *its* norm-guided functioning (as articulated by the planning theory) constitutes the norm-conforming functioning of the *shared* intention?

3.1   *I Intend that We* J, *and Circularity*

As indicated, I propose that we here appeal to the condition that

(i) we each intend that we go to NYC

where the intentions alluded to in (i) are intentions of the sort characterized by the planning theory of the intentions of individuals.[10] Appeal to these intentions in (i) ensures that an intention-like commitment to *our* activity is at work in the practical thinking of *each*. Once *our* activity is an element in *my* plans, I will face characteristic problems of means with respect to our activity and be constrained by characteristic requirements of plan consistency with respect to our activity. This explains something we need to explain, namely: the responsiveness of each to the end of the shared activity, responsiveness that is an element in the characteristic functioning of shared intention.

(i) appeals to the intention of each in favour of *our* activity. But what concept of our activity is at work in the content of these intentions? On the approach we are taking, shared intentional activity will be activity suitably explainable by shared intention. So if the concept of our activity

---

[10] In my early thinking about these matters some remarks of Philip Cohen helped lead me to (i).

that is at work in (i) is the concept of shared intentional activity, we face a problematic circularity in our construction of shared intention.

Such concerns about circularity might suggest that we see the concept of shared intentionality as primitive. Indeed, that is what Searle supposes.[11] But I propose a different tack, one analogous to a common approach to the corresponding issue about individual intentional action.

In many cases we have available a concept of our activity that, while it draws on ideas of individual intentional action, is neutral with respect to *shared* intentionality (Bratman 1999a: 114–115, 146–148 and Kutz 2000: 86–88). We have, for example, a concept of our painting the house that involves only the idea that, roughly, we are each intentionally painting that house in ways that avoid collisions. We then use such neutral concepts in the contents of the intentions involved in our construction of initial cases of shared intention. We depend on the other elements of the construction, including the connection condition, to ensure that when these intentions connect up in the right way to the group behavior there is shared intentional activity. We then try to use these initial cases to build up to cases involving, in the contents of the relevant intentions of the individuals, concepts of shared activity that are not paired with a corresponding concept of our activity that is neutral in this way.[12] In this way we seek to provide an account of shared intentional action by appealing to the appropriate roles of shared intention, but to explain what shared intention is without using, in the most basic cases, the very idea of shared intentionality in the content of the intentions of each.

3.2   Interlocking Intentions

The next step is to note that in shared intention each participant is committed to treating the other participants not merely as aspects of the world that need to be taken into account, but also as *intentional coparticipants* in the shared activity. We can begin to capture this idea by appealing to the condition that the relevant intentions of the participants *interlock* in the sense that each intends that the shared activity go in part by way of

[11] Searle (1990: 401 and 406) claims that "collective intentional behavior is a primitive phenomenon" and we should not seek "a reductive analysis of collective intentionality." In contrast, Searle does not take a similar non-reductive tack to individual intentional action (Searle 1983, ch. 3). My approach to collective intentionality is more in the spirit of Searle's approach to individual intentional action than is Searle's own approach to collective intentionality.

[12] See again Kutz (2000). One example of a shared-intention-involving concept might be engaging in a conversation.

the relevant intentions of each of the other participants (Bratman 1999a: 124).[13] Further, the relevant intended route from intention to action will need to be a route compatible with the action's being a shared intentional action, and so will need to satisfy the connection condition. In the case of our shared intention to go to New York City together, then we appeal not only to (i) but also to

(ii)   we each intend the following: that we go to NYC in part by way of the intention of the other that we go to NYC (where the route to our joint activity satisfies the connection condition).

I intend that we go to NYC in part by way of your intention that we go to NYC; and vice versa. The content of my intention refers to the role of your intention, and vice versa. So there is a *semantic interconnection* between our intentions: it is part of what each intends that the other's intention be realized in the right way.

This interlocking of our intentions goes beyond the idea in (i) that we each intend that we go to NYC. After all, it might be true in, as it were, the *mafia sense*, that each of us intends that we go to NYC by throwing the other into the trunk of the car and driving to NYC (Bratman 1999a: 100). In such a mafia case we have (i) without (ii); and that is one reason why this is not a case of shared intention.

3.3   Intended Mesh

The next step is to reflect on the attitudes of each toward the various sub-plans for the shared activity. In shared intention there will be a tendency to conform to a norm of compatibility of the relevant sub-plans of each. This is tied to the coordinating role of shared intention. If I intend that we go to NYC by driving, and you intend that we go by train, we have a problem. We will normally try to resolve that problem by making adjustments in one or both of these sub-plans, perhaps by way of bargaining. So we want our construction to account for this standard norm-conforming functioning of the shared intention. And a natural way to do that is to use in the construction the idea that each not only intends the shared activity, but also intends that this shared activity proceed by way of sub-plans of the participants that *mesh* in the sense that they are corealizable. We appeal, that is, to the condition that

(iii) we each intend the following: that we go to NYC in part by way of sub-plans of each that mesh with each other.

---

[13] The classic source of ideas broadly in this spirit is H.P. Grice's "Meaning" (1957: 377–388), though my proposal differs from Grice's in important ways.

In this way we ensure that each is committed to, and appropriately responsive to, the coherent and effective interweaving of the planning agency of one another.

Note that yours and my sub-plans can mesh even if they do not match. Perhaps your sub-plan specifies that we not go during rush hour, whereas mine leaves that issue open; yet our sub-plans are corealizable. Further, what is central to shared intention is that we *intend* that we proceed by way of sub-plans that mesh. This can be true even if our sub-plans do not now mesh, so long as we each intend to solve that problem.

3.4   Disposition to Help If Needed

Suppose that I intend that we go to NYC in part by way of your intention that we so act (and meshing sub-plans). My intention engages norms of means–end coherence and consistency. It would be (pro tanto) irrational of me to continue so to intend while believing that to achieve that end it is necessary now for me to intend a certain necessary means, and yet not intend those means. This rational demand grounds rational pressure on me to track necessary means to this intended end. Again, it would be (pro tanto) irrational of me to continue so to intend while also intending something else that I believe to be incompatible with this intended end. This rational demand grounds rational pressure on me to filter further intentions accordingly.

This is so far just to apply the planning theory of the intentions of individuals to the intentions cited in (i)–(iii). But now we need to address a further issue. As I know, our going together to NYC involves actions both of mine and of yours, in each case actions that are explainable in part by my intentions and your intentions respectively. When I intend that we go to NYC, do I thereby intend both that I go (by way of my intentions) and that you go (by way of your intentions)?

Well, sometimes we intend something that involves a certain pre-condition but do not intend that pre-condition: I might intend to respond to your threat or to your offer, but not intend your threat or your offer. Your threat or offer is only a pre-condition of what I intend, not itself something I intend.

However, as I am understanding it, my intention that we go to NYC does not see your contribution to our joint activity as merely an expected pre-condition of our going, a pre-condition to which I am, as Nicholas Bardsley puts it, simply "adding-on" and "providing the finishing touch" (Bardsley 2007: 145).[14] Your contribution to our going to NYC

---

[14] In his positive view, however, Bardsley rejects the idea, in my next sentence, that what I intend really does include your action. (See Bardsley 2007: 152.)

is, rather, a part of what I intend. In intending, in a way that satisfies my side of (i)–(iii), that we go to NYC, part of what I intend is that we both go, where that involves your going in part by way of your intention that we go.[15]

This means that the demands of means–end coherence and of consistency apply as well to my intention in favor of, *inter alia*, your playing your role in our joint activity: I am under rational pressure in favor of necessary means to that, and in favor of filtering options incompatible with that. I am under rational pressure in the direction of steps needed as means if you are to play your role in our joint activity. And I am under rational pressure not to take steps that would thwart your playing your role. This will normally mean that, insofar as I am rational, I will be to some extent disposed to help you play your role in our going to NYC if my help were to be needed.

A qualification, however, is that I can intend our going, and so your role in our going, and still be willing to bear only a very small cost in helping you. And this raises the question: Suppose I am unwilling to incur any costs at all in helping you if need be, while remaining confident that you will not need my help and so will, in fact, play your role? Would such a complete unwillingness on my part to help you, if need be, be compatible with my intending (and not just expecting) our going to NYC, and so your role in that, given that I expect that you will in fact not need my help?

Well, I think we can make sense of some such case of intending, but just barely. We would still need to insist that, insofar as I am rational, I am set to give up my intention that we go to NYC (and that you play your role) if I were newly to come to believe that you needed my help. And we would need to add that at least I am set to filter out options incompatible with your playing your role. So, at the least, I am set not to thwart you in your role. But the idea of being set not to thwart you and yet not being set to help you if need be, is at best tricky. After all, if you need my help then I thwart you if I do not help. In this respect the filtering role and the means-tracking role of intending tend to go together.

So such cases of my intending our joint action, and your role in it, and yet having no disposition at all to help you if need be are, at best, precious and unusual. An intention that satisfied one participant's side of (i)–(iii) but involved no such disposition to help would be, at most, an attenuated intention. So, since our concern is with robust sufficient conditions for shared intention, I will proceed by assuming that the

---

[15] Can I really intend that? The worry here is, again, the violation of the own-action condition. I turn to this issue in section 4 below.

intentions of the participants that satisfy (i)–(iii) are of the ordinary, non-attenuated, sort. These intentions involve, by way of rational pressures toward means–end coherence and consistency, at least a minimal disposition to track means to the joint action and so, more specifically, to the other's contribution to that joint action. For this reason these intentions involve at least a minimal disposition to help if needed.[16]

3.5   Common Knowledge

Analogues of (i)–(iii) will be basic building blocks in our construction. Given the planning theory, these intentions of each will help ensure modes of norm-assessable functioning that are characteristic of shared intention. These modes of functioning will include intention-like responsiveness of each to the end of the shared action, aspects of treating each other as intentional co-participants, the pursuit of coherent and effective interweaving of sub-plans, and dispositions to help.

The next point is that in shared intention the participants will in some sense know of the fact of the shared intention. Such epistemic access to the shared intention will normally be involved in further thought that is characteristic of shared intention, as when we plan together how to carry out our shared intention. Since such shared planning is part of the normal functioning of the shared intention, we need an element in our construction of shared intention whose functioning involves such thoughts of each about our shared intention.

It is here that something like a common knowledge condition seems apt. It is common knowledge amongst us that p if we each know that p, and both p and the fact that we each know it is out in the open amongst us. So we are each at least in a position to know that the other knows, to know that the other knows that we know, and so on. How more precisely to understand these ideas is a difficult question; here I simply work with the intuitive idea of common knowledge.[17]

What we want is that one constituent of our shared intention to J is a form of common knowledge of that very intention. Further, we do not want to re-introduce problems about circularity by explicitly including

---

[16] This represents a slight revision of my discussion of the case of the unhelpful singers in my "Shared Cooperative Activity" (1999a: 103–105). In the terms of that earlier discussion, I am now building it into the (non-attenuated) intentions that satisfy (i)–(iii) that there is at least minimal cooperative stability. In my thinking about this adjustment I was helped by discussions with Facundo Alonso.

[17] Classic statements are in Lewis (1969: 52–60) and Schiffer (1972). See also Barwise (1988) and Harman (2006: 342). Since I am not providing an account of common knowledge, I am not in a position to claim that it can itself be understood solely in terms of structures of individual agency.

in the content of the relevant attitudes the very idea of shared intention. This suggests that we appeal to common knowledge whose content is that the cited components of the shared intention are in place. We do this by adding as a further building block:

(vi)[18] there is common knowledge among the participants of the conditions cited in the construction.

Let us now return to the own-action condition.

4   I Intend that We *J*, and Further Building Blocks

In intending X I normally suppose that my intention will lead to X and that X would not obtain if I did not so intend. Such supposition need not always be belief; it may rather be what I have called "acceptance in a context" (Bratman 1999a: 32). However, to keep the discussion manageable I will proceed on the simplifying assumption that these constraints on intention go by way of belief.[19] Applied to intentions of the sort cited in (i) this means that, normally, in intending that we *J* I believe that my intention will lead to our *J*-ing, and that we would not *J* if I did not so intend. And, normally, in intending that we *J*, you believe likewise. Further, in a standard non-mafia case of shared intention we will both be right in these suppositions. And this raises the question: how could we *both* be right?[20]

My answer,[21] briefly and roughly, is that we could both be right if it is common knowledge between us that:

(a) we each intend that we *J*,
(b) for each of us the persistence of his own intention that we *J* causally depends on his own continued knowledge that the other also so intends, and
(c) if, but only if, we do both so intend then as a result (and in accordance with the connection condition) we will *J*.

When my intention that we *J* would lead in this anticipated way to our *J*-ing, in part by way of your intention that we *J* – an intention that depends on your knowledge of my intention – the control my intention

---

[18] I skip to (vi) to leave room for further conditions, to be discussed below.
[19] Facundo Alonso argues that appeal here to belief is stronger than we need and that what is needed is, rather, a kind of reliance. See Alonso (forthcoming).
[20] As J. David Velleman forcefully puts the worry: "how can I continue to regard the matter as being partly up to you, if I have already decided that we really are going to act?" (2000: 205).
[21] "I Intend that We *J*," in Bratman (1999a), where I also discuss further complexities.

has on our J-ing is a case of what I call "other-agent conditional media-tion" (Bratman 1999a: 152). And such other-agent conditional media-tion need not baffle intention. In believing (a)–(c) each believes that *his* intention that we J appropriately leads to *our* J-ing in part by way of its support of the persistence of the other's intention that we J (and thereby the other's relevant actions). Each can then coherently intend that we J even though this intention violates the own-action condition.

This points to further elements of shared intention. In light of (c), and returning to our shared intention to go to NYC, we will want to add to our construction:

(iv) we each believe the following: if and only if each of us continues to intend that we go to NYC then, as a result (in a way that satisfies the connection condition), we will go to NYC.

And in light of (b) we will want to add to our construction:

(v) we each believe the following: the persistence of his own intention that we go to NYC causally depends on his own continued knowl-edge that the other also so intends; and vice versa.

In (v) each believes the following condition of *interpersonal intention-interdependence*:

(DEP) each continues to intend that we go to NYC if and only if the other continues to intend that we go to NYC.

And we can expect that normally, when there are the beliefs in (v), (DEP) will indeed be true.

Now, the condition of interlocking, in (ii), and the beliefs about effi-cacy, in (iv), build the connection condition into the contents of relevant intentions and beliefs. I think, roughly, that the basic element of this connection condition is that the route from the complex of my inten-tion and your intention to our joint action involves mutual responsive-ness of each to each in a way that aims at the joint action. There will be responsiveness of each to each in relevant subsidiary intentions about means and preliminary steps. This is responsiveness in intention; and a tendency toward this is ensured by the condition that each intends that the activity proceed by way of sub-plans that mesh. There will, further, be responsiveness of each to each in relevant actions in pursuit of the intended joint activity. This is responsiveness in action.

5   The Basic Thesis

Our construction sees analogues of (i)–(vi) and (DEP) as building blocks of shared intention. Bracketing some subtleties, and using **boldface** to indicate generalizations of the conditions cited, the idea is that shared intention, in a basic case, involves:

  (i) intentions on the part of each in favor of the joint activity,
  (ii) interlocking intentions,
 (iii) intentions in favor of meshing sub-plans,[22]
 (iv) beliefs about the joint efficacy (in conformity with the connection condition) of the relevant intentions,
  (v) beliefs about interpersonal intention-interdependence,
(DEP) interpersonal intention-interdependence, and
 (vi) common knowledge of (i)–(vi) and **(DEP)**.

And conditions (ii) and (iv) involve appeal to the **connection condition** and thereby to conditions of **mutual responsiveness**.

The **basic thesis** is that these interlocking and interdependent intentions of the individual participants, and relevant beliefs of those participants, in a context of common knowledge will, in responding to the rational pressures specified by the planning theory of individual agency, function together in ways characteristic of shared intention. This structure will, when functioning properly, normally support and guide coordinated social action and planning, and frame relevant bargaining and shared deliberation, in support of the intended shared activity. And conformity to social norms of social agglomeration, consistency, coherence, and stability that are central to shared agency will emerge from the norm-guided functioning of these interrelated attitudes of the individuals.

Recall the observation that Searle does not say how various "we-intentions" need to be interconnected for there to be shared intention. In contrast, the basic thesis helps characterize this social glue. This social glue is not solely a cognitive glue of belief and common knowledge. This social glue also includes the forms of *intentional* interconnection specified in **(ii)–(iii)** as well as the interdependence cited in **(DEP)**.

While I cannot here provide a full defense of this basic thesis,[23] let me note three points that would bear on such a defense. First, in our shared intention I intend that we *J in part by way of your analogous intention and meshing sub-plans*. This complex content of my intention imposes rational pressure on me, as time goes by, to fill in my sub-plans in ways that, in particular, fit with and support yours as you fill in your sub-plans. This pressure derives from the rational demand on me to make my own plans

---

[22] Where the intentions in **(i)–(iii)** are non-attenuated in the sense discussed in section 3.4.
[23] For further discussion see Bratman (1999a, sec. V).

means–end coherent and consistent, given the ways in which your intentions enter into the content of my intentions. By requiring that my intentions both interlock with yours, and involve a commitment to mesh with yours, the theory ensures that rational pressures on me to be responsive to and to coordinate with *you* – rational pressures characteristic of shared intention – are built right into my *own* plans, given their special content and given demands of consistency and coherence on my own plans. And similarly with you. So there will normally be the kind of mutual, rational responsiveness in intention – in the direction of social agglomeration, consistency, and coherence – that is characteristic of shared agency.

A second point concerns the way in which a shared intention can frame bargaining about means. Recall our shared intention to paint together. Given this shared intention we might, for example, bargain about what color to use, and about who is to scrape and who is to paint; and this bargaining will be framed by our shared intention. How does this work? Well, on the theory, we each intend the shared activity in part by way of the intentions of the other and by way of meshing sub-plans: each of our intentions includes, in its content, the condition that the other's intentions be effective by way of sub-plans that mesh. So we are each under rational pressure to seek to ensure that our sub-plans, agglomerated together, both are adequate to the shared task and do indeed mesh. And that is why our shared intention will tend to structure our bargaining in the pursuit of such adequate and meshing sub-plans.

The third point is that the sharing of intention need not require commonality in each agent's reasons for participating in the sharing. You and I can have a shared intention to paint the house together even though I participate because I want to change the color whereas you participate because you want to remove the mildew. Though we participate for different reasons, our shared intention nevertheless establishes a shared framework of commitments. Granted, extreme divergence in background reasons might undermine the organizing roles of shared intention. Nevertheless, much of our sociality is *partial* in the sense that it involves sharing in the face of divergence of background reasons for the sharing. Our construction of shared intention can recognize and help us understand this partiality.

6   Putting Obligation in its Place

All of this faces the objection that there is an important aspect of shared intention for which we have not accounted. Some have argued that if, for example, you and I share the intention to paint together then we each have distinctive, corresponding obligations to the other, obligations that include obligations not to opt out without the other's permission. This idea is behind much of Margaret Gilbert's work.[24]

Now, when each of us intends that we *J* we each have intentions that are, according to the planning theory, subject to norms of stability. Insofar as the reconsideration and change of either of these intentions violates these norms, some at least *pro tanto* criticism is in the offing. And the interdependence characteristic of shared intention will help extend this stability of the intentions of each.[25] So our constructivism can explain how criticism of opting out of a shared intention can be grounded in the intentions of each and the norms of the planning theory. But this criticism does not by itself ground an obligation of each *to* the other, or an associated *entitlement* of each to the performance of the other; it appeals only to norms of stability on the intentions of each, not to what each owes the other. And the objection is that this is a fundamental lacuna in the theory.

What to say? Well, I agree that shared intention frequently involves obligations of each to another – though I doubt that this is always the case.[26] However, it seems to me that this is because shared intentions normally involve, both in their etiology and in their execution, associated assurances, intentionally induced reliance, and/or promises. Such assurances (and the like) typically induce relevant moral obligations of one to the other.[27] We can recognize this and still see shared intention as consisting in the social-psychological complex articulated in our construction, a social-psychological complex that – taken together with its typical etiology and execution – will typically induce, given relevant principles of assurance-based obligation and the like, obligations of each to another. And once these obligations are on board they can provide further support for the stability of the shared intention.

Let us try to take a larger view of this dispute. I seek to understand

1. characteristic forms of functioning and associated norms of individual intention,

and

---

[24] See for example her essays (2000, 2003, and 2005). Abraham Sesshu Roth (2004) develops an importantly different version of this idea, but here I focus on Gilbert's version.

[25] Facundo Alonso develops this last point in detail in *Shared Intention, Reliance, and Interpersonal Obligation* (2008, ch. 7).

[26] For a defense of this qualification, and aspects of the approach to the connection between shared intention and mutual obligation that I go on to sketch, see "Shared Intention" (Bratman 1999a), "Dynamics of Sociality" (2006), and "Shared Intention and Mutual Obligation."

[27] In contrast, both Gilbert and Roth see the relevant obligations as not specifically moral.

2. characteristic forms of functioning and associated norms of shared intention.

And I want to see to what extent

3. 2. is constituted by special versions of 1.

The idea in 3. is that the norm-conforming roles in 2. emerge primarily from the underlying norm-guided roles in 1., as they are understood within the planning theory of intention, given appropriate, special contents, contexts, and interrelations.

I agree with Gilbert that the functioning in 2. is normally – even if, as I see it, not universally – supported in part by forms of mutual obligation. But when we turn to 3. I do not think we should require that these obligations emerge solely from the underlying roles and norms in 1. Instead, I think we need at this point to draw on further normative principles, in particular moral principles of assurance-based obligation and the like.[28] These resources in hand, we can acknowledge important roles of such obligations of each to another in normally providing support for 2., without supposing that those obligations need to be embedded entirely in the basic social psychology of sharing. To understand these mutual obligations we need also to appeal to further principles of moral obligation.

My constructivism about shared intention begins with an underlying model of individual planning agency. This model – the planning theory – highlights roles and norms characteristic of individual intending and planning. My constructivism then seeks a conceptual and metaphysical bridge from such individual planning agency to modest forms of sociality. This conceptual and metaphysical bridge draws on special and distinctive contents and contexts, and on ideas of intention, interlocking, mesh, mutual responsiveness, interdependence, and common knowledge. It applies the norms of the planning theory of individuals to the distinctive intentions and plans of individuals that are constructed using these further ideas. In these ways this conceptual and metaphysical bridge tries to articulate infrastructures that, when functioning properly, realize salient forms of social-norm-assessable and social-norm-conforming social functioning. Once we have these articulated infrastructures on hand we can ask how their functioning will normally engage moral norms of interpersonal obligation. And we should expect that the ways that such moral norms are characteristically engaged will feed back into the social functioning and support these forms of sociality.

Gilbert's alternative strategy is to see non-moral obligations of each to another as partly constitutive of shared agency. The move from

---

[28] As I note in "Shared Intention and Mutual Obligation" (1999a), other moral principles may also be relevant here.

individual to shared agency involves, *inter alia*, a move to such forms of mutual obligation – where this involves, in her view, the introduction of a non-reducible social concept of "joint commitment" (Gilbert 2000: 25–26). In my judgment, this runs the risk of too quickly passing over complex bridging structures of the sort that I have been highlighting. Granted, our sociality normally involves forms of mutual obligation. These are, as I see it, familiar forms of moral obligation: human sociality and morality are, after all, ineluctably intertwined. But I think that when we examine modest forms of sociality under our philosophical microscopes we can discern, at bottom, even more fundamental intentional, causal, semantic, epistemic, and rational structures. This chapter has been an effort to say, albeit roughly, what those structures are, and to show how they can constitute such sociality. We can then see relevant forms of moral obligation of each to another as grounded in part in those structures.

The concepts I have used, at bottom, to specify these aspects of sociality – concepts of interlock, mesh, mutual responsiveness, interdependence, and the like – are ones that are broadly available to our theory of individual planning agents (which is not to say that planning agency by itself ensures the capacity for modest sociality). Or anyway, this is true with the possible exception of the appeal to common knowledge. So we are in a position to argue that we do not need, for our theory of these modest forms of sociality, a further practical concept that goes beyond the resources of our theory of individual planning agents. The social glue involved in these forms of sociality does go considerably beyond the cognitive glue of common knowledge – here Gilbert and I agree. But I think we can say what else is involved in this special social glue – including its distinctive normativity – without appeal to yet a further, conceptually primitive, and non-reducible practical social relation of the sort that Gilbert has in mind in her talk of "joint commitment." There is a distinctive social-norm-assessable and social-norm-conforming functioning associated with shared intention and shared agency. However, in its most basic form this social functioning is constituted by special versions of the individual-norm-assessable and individual-norm-guided functioning characteristic of individual planning agency. And we can say what these special versions are while staying within the conceptual and metaphysical resources of the planning theory.

REFERENCES

Alonso, F. (forthcoming). "Shared intention, reliance, and interpersonal obligation," *Ethics*.
    2008. *Shared Intention, Reliance, and Interpersonal Obligation*. PhD Thesis, Stanford University.

Bardsley, N. 2007. "On collective intentions: Collective action in economics and philosophy," *Synthese*: 141–159.

Barwise, J. 1988. "Three Views of Common Knowledge." In M. Vardi (ed.), *Proceedings of the Second Conference on Theoretical Aspects of Reasoning About Knowledge*. Los Altos, CA: Morgan Kaufman, pp. 365–379.

Bratman, M. E. (forthcoming). "Intention, Belief, Practical, Theoretical." In Simon Robertson (ed.) *Spheres of Reason: New Essays in the Philosophy of Normativity*. Oxford: Oxford University Press.

1999a. *Faces of Intention*. New York, NY: Cambridge University Press.

1999b. *Intention, Plans, and Practical Reason*. Cambridge, MA: Harvard University Press (1987); reissued by CSLI Publications.

2006. "Dynamics of sociality," *Midwest Studies in Philosophy: Shared Intentions and Collective Responsibility* **15**: 1–15.

2007. *Structures of Agency: Essays*. New York, NY: Oxford University Press.

Gilbert, M. 2000. "What Is It for Us to Intend?" In M. Gilbert (ed.) *Sociality and Responsibility*. New York, NY: Rowman & Littlefield, pp. 14–36.

2003. "The Structure of the Social Atom: Joint Commitment as the Foundation of Human Social Behavior." In Frederick F. Schmitt (ed.) *Socializing Metaphysics*. New York, NY: Rowman & Littlefield Inc., pp. 39–64.

2005. "A Theoretical Framework for the Understanding of Teams." In N. Gold (ed.) *Teamwork: Multi-disciplinary Perspectives*. New York, NY: Palgrave Macmillan, pp. 22–32.

Grice, H. P. 1957. "Meaning," *Philosophical Review* **66**: 377–388.

Harman, G. 2006. "Self-reflexive thoughts," *Philosophical Issues* **16**: 334–345.

Kutz, C. 2000. *Complicity: Ethics and Law for a Collective Age*. Cambridge: Cambridge University Press.

Lewis, D. 1969. *Convention: A Philosophical Study*. Cambridge, MA: Harvard University Press.

Roth, A. 2003. "Practical Inter-subjectivity." In Frederick F. Schmitt (ed.) *Socializing Metaphysics*. New York, NY: Rowman & Littlefield Inc., pp. 65–92.

Roth, A. S. 2004. "Shared agency and contralateral commitments," *The Philosophical Review*: 359–410.

Schiffer, S. 1972. *Meaning*. Oxford: Oxford University Press.

Searle, J. 1983. *Intentionality*. Cambridge: Cambridge University Press.

1990. "Collective Intentions and Actions." In P. Cohen, J. Morgan, and M. E. Pollack (eds.) *Intentions in Communication*. Cambridge, MA: MIT Press, pp. 401–415.

Velleman, J. D. 2000. "How to Share an Intention." In J. D. Velleman (ed.) *The Possibility of Practical Reason*. Oxford: Oxford University Press.

Yaffe, G. 2004. "Trying, intending and attempted crimes," *Philosophical Topics* **32**: 505–532.

2 – Comment
Where Is the Social?

Pierre Demeulenaere

Michael Bratman's thesis builds up shared intentions on the basis of individual intentions, refusing the specificity of Searles's "we-intention" on the one hand, and Gilbert's "joint commitment" on the other. It has a strong "individualistic" commitment, since the aim is to proceed from individual intentions to shared intentions, without introducing any kind of irreducible social level that would not depend on properties of individual intentions. Indeed, the social level is said to "emerge" from an individual level, which is "at the bottom" of "modest forms of sociality." Let us stress a point that is not, naturally, the focus of Bratman's chapter: the argument rests on a series of metaphors like "bottom," "level," "emergence" and others of similar ilk that oppose the locus of the "individual" and the locus of the "social." Since he intends to examine the nature of both individual and shared intention "under a philosophical microscope," it should be noted straightforwardly that such a microscope has very conceptual lenses that should themselves be examined under a sociological microscope. If you refer to the social, should it be part of the microscope or of the object seen under it? It seems to us that all those metaphors used to design the place of the social are not very clear. Social scientists often build up the social on the basis of the individual; they equally often refuse to construct things this way when they accept a specific "social level," the notion of emergence being here ambiguous, since it can refer either to an irreducible social level or conversely to a reducible phenomenon dependent on antecedent individual elements. I will not enter in this venerable debate here. I just want to mention one ambiguity in Bratman's chapter: the fact that shared intentions rest on individual intentions does not seem to me equivalent to the fact that the social level would "emerge" from an individual level. Those are two different things that should not be confused.

In the social sciences the basic "methodological individualism" thesis I myself endorse entails two basic propositions, a negative one and a positive one:

60

- the negative one rejects any kind of "social" agency, that is, it rejects that "collective entities" or institutions are, as such, actors.
- this leads to the positive assertion: only individuals (that is, human beings) are actors as such, and any institution or group (like, for instance, the British government) depends on individuals to be "active."

This thesis is not ontological, since it does not say anything about the status of institutions or of the "role" they exert on individual agents. It is a thesis about identifiable agents: institutions do not act by themselves; only individuals do so. This thesis seems both inescapable and rather trivial, since it does not say anything about the actual behavior of individuals and the type of links they have with institutions. Neither does it specify the way institutions play a "role" in individual behavior. Clearly, for instance, the fact that there are borders between two different countries has major effects on the inhabitants of these countries, even if frontiers do not act by themselves.

Yet the Methodological Individualism thesis does not necessarily proclaim that the social "emerges" from the individual. It seems important to separate and not to conflate the idea that only individuals act and the idea that the social is built up on individuals. The two things do not have at all the same meaning; and the first point is rather obvious, whereas the second is intrinsically controversial, because of problems of defining and localising the "social."

Let us start by a consideration of what is social in Bratman's chapter. It seems that three different uses of the notion can be found.

The first refers to shared intentions that are linked to "modest forms of sociality"; I understand that this modesty corresponds to a small number of participants in joined intentions as opposed to larger forms of sociality that would include political or legal systems (that is, a much larger number of actors).

The second opposes norms that are said to be individual to those that are social: Bratman's thesis, once again, is to build up shared social norms on the basis of individual norms.

The third entails an opposition of "levels": the individual one and the social one, which does not depend, as such, either on action or on the number of participants, and which, accordingly, exceeds the problem of norms.

It seems to me that those various uses should be reconsidered on the basis of one major phenomenon that I want to describe now: when an individual acts, he does so, most frequently, on a social basis. What does this mean? It means that intentions are submitted to social norms, that is, to rules (either prescriptive or proscriptive, formal or informal) that delineate the sphere of acceptable intentions; this is equivalent to saying that there is no such thing as an individual intention that would be pure of any kind of social influence; naturally, this is even truer when a relationship between two or more actors is involved.

Let us take the two simplest examples that are provided by Bratman, going to NYC and painting a house: both examples can be related to one individual or to two or more individuals. Considering individual intentions themselves, social norms are already at play: First, to go to NYC, I have to be allowed to do so; and we know that most inhabitants in our world would not get a visa to go there. Second, the act of going to NYC is linked to the existence of a reason for going there, and such a reason is always social. One goes there to visit the city, to attend a wedding or a professional meeting, to vote, and so on. There are no such things as solipsistic reasons that would lead one to NYC: generally such an intention is more or less linked to social norms (the idea that NYC should be visited, for instance). Indeed, individual intentions have to take into account these norms.

The same can be said of painting the house: How often should a house be painted (regarding aesthetics, health or status criteria) and in which colour? Here again, there are strong social norms that delineate the legitimate sphere of preferences: in Paris, for example, buildings are to be repainted at least every ten years, and only a few colours (white, light grey, beige) are allowed when a building is not made of stone (in which case it should not be painted).

It is certain that, in the end, intentions have to be decided as a personal responsibility: Social norms do not strictly determine intentions; but intentions cannot be understood apart from all social dimensions, which, for their part would intervene only afterwards in joined intentions. The same can be said of two persons walking together: First, are they allowed to walk together? How should they walk (any traveller to Japan realizes that girls do not walk there in the same fashion as European girls)? Should one person, because of his or her status, lead the group? If not, should they talk together when they walk, or are they supposed to remain silent (like monks in a cloister)? If I write a paper in English, should I write "he" or "she" to name any individual?

Naturally, whenever we depart from such extremely simple forms of interaction, we encounter even more important social norms: Let us think of "friendship, love, singing duets, and the joys of conversation," mentioned by Bratman, which are laden with social norms that need not be described here.

When I say that any individual intention is linked to social norms, I do not give up the standard "individualistic" position, in particular the one that is represented by mainstream economic analysis. It is well known that a utility function, the way standard analysis describes it, depends on the ordering of individual preferences. Where do these preferences come from? It is hard not to acknowledge that preferences are always rooted in a social context and that they are linked to social norms. This is obvious for all goods that a consumer tends to choose (like food, clothes, houses, cars, and so on); they always include a social-norms dimension. Therefore, it is not possible strictly to oppose an "individualistic" approach to a "social" one; individual preferences (and intentions) are "individual" in the sense that they are under an individual's responsibility, and at the same time social, since in any circumstance, there is a more or less narrow array of legitimate options an individual has to face in taking his decisions.

Three things should be said at the same time:

First, any intention or any preference depends on a social context that delineates the limits of acceptable, legitimate, and desirable preferences and intentions.

Second, within this social context, individual intentions and individual preferences are to be decided by individuals (this decision making can include negotiations among relevant actors).

Third, as I will discuss later on, it is necessary to understand how those social norms are constituted and do "emerge."

I agree with Bratman's description of the way individual intentions develop and are linked to others' intentions; but I do not agree with the sudden appearance of the "social" after individuals have met to share their intentions, since the social is always already there when individuals act on their own or when they consider their own intentions, regardless of their coordination with others.

I will not take into account all the elements involved in Bratman's description of individual intentions or how they lead to shared intentions, for instance, the possibility of "bargaining" between the parties. I will now concentrate on one major aspect involved, namely norms of rationality. Bratman's strategy is to describe norms that apply to the individual level, and then to derive from them the norms that apply to the social level.

The basic idea is, roughly, that the norm-assessable social functioning characteristic of shared intention emerges from the norm-assessable and norm-guided functioning of a relevant structure of interrelated intentions of the individuals, as understood by the planning theory. We seek a construction of intentions and related attitudes of the individuals in appropriate contexts that would, when functioning in the norm-guided ways highlighted by the planning theory of individual intention, play the roles characteristic of shared intention. And we try to see conformity to central norms characteristic of shared intention – norms of social consistency, social agglomeration, social coherence and social stability – as primarily emerging from guidance by norms that apply directly to the relevant interrelated structures at the individual level. If we had such a construction we would have reason to say that this construction is shared intention, or at least one important kind of shared intention.

Such a stance can be understood in two different ways – one weak, and one strong. The weak interpretation would contrast norms that apply to an individual and norms that apply to a small group, naming the first "individual" and the second "social"; this is done solely on the basis of the number of participants, without regard to a theoretical decision about the origin of the norms that apply to the individuals and their action, and specifically their relationship with the "social" in general. The stronger version – it seems to me that this is Bratman's thesis, but I am not sure of it – would view the norms that apply to the individual level as themselves strictly individual; and view these simultaneously as the basis of emergent norms that apply to the social level and that thus are said to be social.

The first version would focus on the fact that there is a continuity between the norms that apply to a single action and the norms that apply to small-group shared intentions, without making any claims about the intrinsic properties of such norms (i.e., without saying whether they are "individual" or "social").

The stronger thesis would confirm this continuity and similarity, yet add that, clearly, the social derives from the individual. This implies that initial norms of rationality that apply to individual action cannot be considered to be social in any sense, but that they are the basis of the social norms that apply to shared intention.

I would reject this stronger version, but with caution, as I will now briefly show. The problem is to understand where norms of rationality come from and whether they are "social" or "individual." In fact, in my view this either–or alternative is misleading. I will maintain that rationality norms are at the same time individual and social. A full argument should include an analysis of the meaning of the idea of rationality, its significance for human agency, its normative aspect, and, finally, the way it applies to human action. I will naturally not develop such an argument here. I will modestly consider two important things.

First, norms of rationality can be said to be individual in the sense that they should apply to any distinct human action when it cannot be thought to be free from any concern of rationality. This naturally entails importing a philosophical and normative assessment to the social sciences. The individual aspect of those norms is equivalent to the idea that norms of rationality are not cultural in the sense of local conventions that prescribe specific attitudes in specific social settings. Rationality is considered universal, which means that no one can escape its requirements. In this sense it is not social, if "social" corresponds only to a particular social context; since it is not social, it can be said to be "individual."

But in my view this argument is incomplete, even when one supports the idea that norms of rationality are not just variable cultural norms that need not apply to large domains of human behavior. Norms of rationality have to be considered both universal and social in three different ways.

First, there is a long-standing debate within the philosophical and the social sciences tradition regarding what precisely the norms of rationality are. Even if everyone agrees there is a non-cultural (or non-local) dimension of rationality, there is still a wide range of discussions about the exact meaning and scope of this notion. Bratman introduces four dimensions here: consistency, agglomeration, means–end coherence and stability; all this should be contrasted with the way the notion of rationality is understood in various economic or sociological traditions.

Second, it is clear that particular conceptions of rationality, or historical developments of notions of rationality, can reinforce specific types of behavior: for instance, the historical development of capitalism and the concomitant analysis of economic rational behavior can reinforce the attention paid to various means leading to specific ends and pose the problem of the acceptability of these means for these ends. The definitions of rational behavior can have a direct influence on individual behavior. For instance, when I intend to go to NYC, how much attention should I devote to alternative prices offered by rival airlines? Should I take the externalities into account in the means – for instance, the fact that planes pollute more than trains?

Finally, norms of rationality are socially supported: even when it concerns only individual action, it is accomplished within the purview of witnesses that either support or disavow (in some specific circumstances) rational behavior. It cannot be said that rational behavior is at the bottom of social norms: rational behavior is itself, also, a social norm that is reinforced in specific social environments and that leads an individual to conform to rational attitudes.

Certainly our argument should include three dimensions that cannot be developed here: First, it should reflect on the way rationality is at the same time a universal dimension of human behavior and yet is socially defined and reinforced in specific social settings. Second, it should analyse the different types of norms and the way rationality norms can be located among more general social norms, including those that are more local and cultural than rationality norms. Third, it should reflect on the way those local and cultural norms can be explained on the basis of elements that are not considered themselves to be explained.

I will conclude only by expressing dissatisfaction with a common way of approaching these matters, namely with the approach that views the social as emerging from the individual. The social is everywhere, pervading individual activities. Can we properly understand a notion of "individuals" who would be separated from any social dimension, something like individuals in Rousseau's "state of nature," who have not yet met each other? It seems to me that this is not possible. We can neither support the view that the social emerges from the individual nor conversely criticize it the way Durkheim used to (i.e., by maintaining that the social level should not be confused with the individual level); this stance too presupposes the autonomy of something like the individual. Both views presuppose the idea that there is something like the individual, which is to be opposed to the social (this is different from the idea that only individuals act, whereas collective entities do not). But in fact all individual actions take place within social settings. We do not have a strong basis for opposing the individual and the social. What we can do is contrast stable intercultural types of intentions that individuals tend to endorse, according to specific norms, and, on the opposite pole, various local cultural norms that should be explained on the basis of the previous ones. It is also possible, on the basis of typical motives that are characteristic of disconnected individuals (in specific social settings), to derive norms that establish new types of links among those individuals; the best example are the norms that tend to emerge from a prisoner's dilemma situation. But a prisoner's dilemma situation is a social situation.

Nevertheless, in neither case do we reach a primitive non-social dimension on which we would build up a social life: Social life is first given in our experience, knowing that only individuals are in the position to act. On the basis of this experience, we analytically separate different types of intentions (more or less universal); we describe the way they coordinate; and, contrasting different social situations (for instance, a state of anarchy and a state supported by a police force), we try to understand how, logically and/or historically, individual actors can be driven from one situation to another.

3   The Reality of Group Agents

Philip Pettit

Introduction

Human beings form many sorts of groups but only some of those groups are candidates for the name of agent. These are groups that operate in a manner that parallels the way that individual agents behave. They purport to endorse purposes, to form representations and to act for the satisfaction of those purposes according to those representations. And, building on those purported capacities, they make commitments and incur obligations, they rely on the commitments of others and claim rights against them. As candidates for group agents of this kind we might cite the partnership or the corporation, the church or the political party, the university or the state.

But are such entities truly agents? Or are they mere simulacra of agents? Do they replicate the agency of individual human beings? Or do they merely simulate it? That is the question I address in this chapter.[1]

The chapter is in three sections. In the first section I set out the requirements that systems of any kind must fulfill if they are to count as agents. In the second I look at the way in which individuals might seek, on the basis of shared intention, to form a group agent. And then in the final section I show how the sort of entity they construct in that way can meet the requirements given and count as a genuine agent.

---

[1] The question is the third of three questions that I take to be crucial in social ontology. The first is the question between individualism and non-individualism and bears on how far social regularities undermine the agential autonomy that we ascribe in folk psychology to individual human beings. The second is the question between atomism and non-atomism and bears on how far the psychology of individual human beings non-causally or superveniently depends for some of its important features on those individuals having social relations with one another. Those two questions are addressed in Pettit 1993, where I argue for individualism but against atomism. The question considered here divides singularism from non-singularism, as we might call the rival approaches, and bears on how far groups of individuals can constitute agents on a par with individuals.

1   The Requirements of Agency

1.1   *Agential Behavior*

It is possible, on the face of it, for something that is not strictly an agent to display agential behavior. We can imagine finding evidence in the behavior of a system that it is an agent, but then overruling that evidence on the basis of further information. So what is it that we should expect in a system's behavior if that behavior is perfectly agential: if it is to do as well as possible in constituting evidence, however defeasible, that the system is an agent?

This is probably the easiest question in the theory of agency, for almost all sides are agreed that behavior manifests agency to the extent that it instantiates what we may describe as a purposive–representational pattern. Let the behavior of a system be understood, not just as the behavior it actually manifests, but as the behavior that it displays across the fullest range of possible scenarios, actual and counterfactual: that is, as the behavior it is disposed to display in such scenarios. That behavior might consist in an entirely random collection of behavioral pieces, without any rhyme or reason to them. But if it is agential in character, then it will be patterned in a way that links it to certain purposes and certain representations (Dennett 1991).[2]

The candidate purposes of the behavior will be revealed by the outcomes that it reliably achieves. And given an assignment of purposes, the candidate representations of the system will be revealed by the adjustments it makes in pursuit of those purposes, as it registers the nature of the different situations it confronts. The behavior of a system will display a purposive–representational pattern, and exemplify agential behavior, to the extent that there is a suitable set of purposes and representations – ideally, a single set – such that the behavior promotes those purposes according to those representations (Stalnaker 1984). The behavior involves the adoption of means for realizing those purposes that will tend to be effective if the representations, being suitably responsive to situations, are correct.

In explaining the notion of a purposive–representational pattern, I abstract from the extent to which the pattern is enriched or impoverished. It should be clear that agents may differ in how far the purposes they pursue, and the representations they form, relate to the here and now as distinct from the spatially and temporally distant; refer to the

---

[2] The spirit of this chapter is broadly congenial to the views with which Dennett is associated. For a more explicit connection between those views and a realistic model of group agency, see Tollefsen 2002.

presumptively actual as distinct from the counterfactual and possible; engage with a limited, as distinct from an open, range of particulars and properties; quantify over abstract entities like numbers as well as more concrete items; and bear on particular matters of fact as distinct from generalities and laws. In the remainder of the discussion I shall continue to abstract from this issue, since the argument goes through independently of how rich the domain of agential behavior happens to be.

While abstracting from the richness of the purposive–representational character of agential behavior, I focus, without using the word, on the rationality of the pattern. A pattern of behavior will be agential to the extent that, in ordinary terms, it is rational. The purposes and representations must make sense in an attitude-to-evidence dimension, being responsive to the different features of the situations it confronts; they must make sense in an attitude-to-action dimension, being organized to generate whatever interventions are instrumentally required by the purposes of the system according to its representations; and for a mix of evidential and instrumental reasons, they must make sense in an attitude-to-attitude dimension, being more or less consistent with one another, for example, and even perhaps mutually supporting.

Given a conception of agential behavior we can now ask after what we should expect of a system that is to count as an agent. Presumably a system will count as an agent just to the extent that it relates in a certain way to an agential pattern of behavior. But what precisely is the relationship required?

1.2   Agential Behavior and Agency

The simplest theory of agency would say that a system is an agent just to the extent that it instantiates an agential pattern in its behavior. There are purposes and representations that it is independently plausible to ascribe to the system – this constraint may be variously interpreted[3] – and the behavior of the system generally promotes those purposes according to those representations. This is classical functionalism or dispositionalism. Let a system be disposed, no matter on what basis, to display a plausible, agential pattern of behavior. Or, to be more realistic, let it be disposed in general to display such a pattern; naturalistic limitations are bound to make for occasional failure. To the extent that the

---

[3] Not only are there different views as to what is required to remove mystery on this front; the views also differ on how restrictive the requirements are. Those in the "teleosemantic" camp, for example, hold that ascriptions of representations have to satisfy requirements of an evolutionary kind and that these are quite demanding. See Millikan (1984).

agent displays such a disposition, it will count as a center of agency. To be an agent, on this approach, is simply to function as an agent: to pass as an agent on the behavioral front.

There are three broad alternatives that compete in the current literature with a purely functionalist theory of agency. Each would add a further clause to the behavioral condition. And, typically, each would downgrade that first condition in the process: it would allow that in the presence of the further clause, a system may count as an agent without fully satisfying the behavioral condition.

The first of these alternatives would stipulate that in order to count as an agent proper, a system has to be composed of the sort of stuff or substance or material out of which paradigmatic agents – perhaps human beings, perhaps humans and other animals – are composed. It would suggest that only systems that are made up of that same stuff, or perhaps stuff of a broadly similar sort, can constitute agents. A good example is the Cartesian account that takes human beings to be composed of a non-physical kind of thinking substance, and that makes the presence into a prerequisite of agency.

The second alternative would stipulate that in order to count as an agent, a system does not just have to instantiate the dispositions that constitute relevant purposes and representations. Those dispositions have to be realized within the system in a certain psychological pattern: according to a certain architectural design, for example, or with a certain conscious, qualitative feel. The requirement is not just that the dispositions have to evolve in interaction with the environment, and not be rigged in advance (Block 1981); that is already guaranteed by the assumption that representations should be responsive to situational features. It is the much more problematic assumption that any genuine agent has to display something like the architecture of classical computing, or the conscious life of a biological organism like one of us (Searle 1983, Fodor 1975).

Finally, the third alternative would stipulate that in order to count as an agent a system has to have the capacity, not just to conform broadly to a pattern of agential behavior, but to achieve a critical, ratiocinative perspective on that pattern (Davidson 1980). The system has to be able to identify some of the demands imposed by the pattern as regulative or normative requirements, and to let the identification of those demands reinforce conformity and underpin the recognition of non-conformity as a failure. It has to be able to do the sort of thing that we do when we reason.

When we reason theoretically we don't just form representations in accordance with *modus ponens* or any such rule; we don't just come to believe that q, on coming to believe that p and that if p, then q. We also recognize that "p" and "if p, then q" are true and that their truth ensures the truth of "q". And then this further recognition reinforces an independently triggered belief that q, or makes up for the lack of an independent trigger, prompting the belief to form for the first time. Again, when we reason practically we don't just form intentions in accordance with the means–end principle; we don't just come to intend that X, on coming to form the intention to G and the belief that the only way to G is to X. We also recognize that G is a goal to be realized, and that realizing G entails realizing X. And then this further recognition reinforces the independently triggered intention to X or makes up for the absence of an independent trigger. Let a system have the capacity to reason and it will be able to recognize some of the demands imposed by agential pattern – say, the demand to intend any necessary means toward an intended goal – and this recognition will play a role in supporting compliance with that demand.

If we impose the first or second of our three extra conditions on agency, then we cannot admit the reality of group agents. Group members may act and speak as if a single representational and purposive mind lies at the origin of the group's actions and utterances. They may even manage to display a more or less perfect form of agential behavior. But on either of the first two alternatives, what the members achieve together can only be a show of agency, not its substance. The behavior will not be produced by the appropriate stuff or according to the appropriate sort of processing.

I do not think that this need concern us unduly, since both of these alternatives to pure functionalism seem dubious. We ascribe agency to one another in light of our behavior and without giving any obvious thought to the basis on which that behavior is produced. Thus the conception of agency that we deploy in mutual interpretation – and, plausibly, in the interpretation of many other animals – does not necessarily presuppose anything about the stuff or the process in which agency materializes (Jackson and Pettit 1990a). It is more purely functional than either of those approaches suggests.

What should we say about the divergence between pure functionalism and the third alternative? Here it is possible to be ecumenical (Pettit 1993, chs. 1–2). The system that instantiates an agential pattern of behavior, at least in the absence of perturbation, can count as a regular or non-ratiocinative or non-critical agent. The system that is capable, in addition, of recognizing and responding to the demands of agential behavior can count as a special, ratiocinative or critical agent. This ecumenism is attractive because it enables us to countenance many non-human animals as agents, which we surely have reason to do, and yet at the same time to acknowledge an important gap between such animals and human beings like you and me. Most of the time we human beings may operate as agents in the unthinking manner of other animals. But we sometimes adopt the ratiocinative pose exemplified by Rodin's sculpture of *Le Penseur*. Moreover, we are always ready to resort to such a perspective when the red lights go on. And we can rely on the possibility of such resort to help keep us in line with the demands of agential pattern: critical reflection can guard us against incautious or sloppy processing at the more spontaneous level.

The groups whose claim to agency we will be exploring, as we shall see in the next section, are groups that adopt a critical perspective on agential pattern, like individual human beings and unlike other animals. These groups are capable of going through something like a process of reasoning and of using that process to guard against failure. This means that even those who refuse the title of agent to systems that are not capable of a critical perspective on agential pattern can find the argument that follows congenial; nothing in the argument presupposes the truth of the view they oppose.

The upshot of this discussion is that if a system is to count as an agent, then it must display a purposive–representational pattern of behavior, whether on a critically informed or uninformed basis; it must function appropriately, however that functionally is supported. But there are three aspects of the functioning required that are easy to miss and I conclude with a brief discussion of each; they will be relevant again in the final section. I describe these respectively as the issues of systematic perturbability; contextual resilience; and variable realization.

1.3   Systematic Perturbability

There are two quite different ways in which a system might be perturbed in the display of an agential pattern. There might be systematic factors such that in their absence, there is little or no noise; most perturbation derives from those factors and only a little materializes as random disturbance. On the other hand more or less all of the perturbation to which the system is subject might appear as random disturbance: as an unpredictable breakup of the pattern, like the spasmodic trembling of an otherwise steady hand.

Systematic perturbability is easy to square with agency, unsystematic perturbability not. If an entity behaves like an agent, subject to random, unsystematic departures from form, it is easy to think that the appearance of agency may be an illusion. But if there are established, independently intelligible sources of perturbation and in their absence the system behaves like an agent, with only a very little failure, then it is harder not to take the appearance of agency seriously. There will be no temptation to think that the system is an aleatoric device that just happens to project a broken image of agency in a sequence of chance events. Given the systematic character of the perturbers, it will be natural to take them as constraints under which the system was designed or selected for fidelity to a purposive–representational pattern.

1.4   Contextual Resilience

A pattern may be very reliable under certain boundary or contextual conditions. But those conditions can be very demanding, so that even if the pattern is almost certain to obtain so long as the conditions are fulfilled, it will break up under even a slight variation in those conditions. Consider the sequential pattern generated under John Conway's game of life by the initial figure in which, roughly, there are four squares placed close to one another to form a larger square on a grid; this is called the exploder pattern.[4] The kaleidoscopic, explosion–implosion sequence that is generated from that starting figure is entirely reliable. But it is fragile or non-resilient, in the sense that it will fail if even a single box in the grid, adjacent to the starting figure, is also filled in.

We cannot count a system as an agent if it is inflexible and fragile in this manner. A system will be an agent insofar as it is disposed, on an uncritical or critical basis, to display a purposive–representational pattern of behavior. But as this disposition is tied more closely to a suitable context – as it becomes the disposition to display that pattern in context x and only context x – it becomes more and more implausible to describe the pattern generated in terms that abstract from that context. The disposition will amount to nothing more than a reactive habit that is tailored to cues provided in that particular situation.

Consider the Sphex wasp that Daniel Dennett (1979) discusses. This wasp brings its eggs to the edge of a hole that it has found or dug, enters the hole to make sure that it is still provisioned with the paralysed prey that it has previously deposited there, then comes up and takes the eggs back into the hole. But it turns out that if the eggs are moved even a little bit away from the edge while the wasp is in the hole, then the wasp goes through the whole routine again and that it can be forced by this intervention to repeat the exercise an indefinite number of times. The failure here prompts us to recognize that the wasp is not displaying the

---

[4] See www.bitstorm.org/gameoflife/

pattern of ensuring that its eggs are placed in a suitable hole, as if it were focused in an agential way on that abstract purpose. The pattern that it is displaying is tied inflexibly to the context in a way that makes that ascription of purpose unwarranted.[5]

1.5   *Variable Realization*

Suppose that a system displays fairly systematic perturbability and a high degree of contextual resilience in the generation of a purposive–representational pattern. Will it tend to count, then, as an agent? Not necessarily. For one further condition that we expect agents to satisfy is that they generate a purposive–representational pattern, not just over variations in surrounding context, but also over variations in how precisely they are compositionally organized and in how, therefore, the generative dispositions are realized within them.

The relevance of this issue is illustrated by the fact that a simple system for controlling the temperature in a room or building might otherwise count as an agent. This system generates a purposive–representational pattern of behavior, albeit one that is impoverished to the point where only one purpose is in play – to keep the temperature in a certain range – and only one sort of representation: that which registers the ambient temperature in the relevant space. The system might generate this behavior with only systematic perturbability and with a high degree of contextual resilience: no matter how the space is cooled or heated beyond the set range, for example, the system will restore the temperature to that range. But we would not for a moment think of the system as an agent. It would be quite extravagant to do so.

The reason why this is so, I suggest, is that while a heating–cooling system might be designed on this or that basis, any given system manifestly operates on the basis of a single, simple mechanism; this is going to be manifest even where it is unclear which particular mechanism is in operation. The agential pattern will be realized without exception or variation in that mechanism, then, and the causal relevance of any given agential configuration – say, the system's registering a drop in temperature – is going to be put in question by the causal relevance of the simple mechanism. There will be no information given about the genesis of the ensuing adjustment by tracing it to that configuration over and beyond the information given by tracing it to the state of the mechanism that uniquely realizes that configuration.

---

[5] The requirement of contextual resilience is close to John Searle's (1983) requirement that an agent satisfy "the background" condition of having sufficient skills to be able to adjust appropriately under situational variation.

Things would be very different if there were a variable pattern of realization for the agential configuration. In that case the appearance of that configuration would be causally relevant to the ensuing adjustment, since it would ensure the appearance of the adjustment, regardless of the state of the mechanism by which it is realized. If we think of the state of the mechanism as producing the adjustment, we might think in this case of the agential configuration as programming for that production: ensuring, regardless of its mode of realization, that the adjustment occurs (Jackson and Pettit 1988; Jackson and Pettit 1990b; and Jackson and Pettit 1992).[6] In this case there will be extra information given about the genesis of the ensuing adjustment by tracing it to that configuration over and beyond the information given by tracing it to the state of the mechanism that uniquely realizes that configuration. The extra information will be that the adjustment is ensured, not just in the world where the actual realizing state is present, but also in those worlds where different mechanical states play the realizing role.

2   Candidate Group Agents

2.1   The Transparent Group

If agency requires just the display of purposive–representational pattern – specifically, in a way that satisfies systematic perturbability, contextual resilience and variable realization – then it is at least logically possible that a group of people might be an agent without the members of that group recognizing the fact; they might mediate the agency of the group in the unthinking, zombie-like manner in which, on a naturalistic picture, my neurons mediate my agency. Equally, it is at least logically possible that a group of people might be engineered into constituting an agent, while only one or two members recognize the fact; those in the know might recruit others to suitable roles, without revealing the agent-constituting point of the roles. And furthermore, it is certainly possible that a group of people might constitute an agent under a procedure that gives them differential roles and that makes the workings of the group relatively opaque to those in lesser offices, if not opaque in quite the same measure as in the other possibilities.

---

[6] The language of producing and programming should not suggest that there is a difference of kind between the way causality is exercised at the two levels. Considered in relation to the subatomic states that realize it in turn, the mechanical state can also be described as programming for the adjustment. For all that need be presupposed, there may be an infinite number of levels of this kind, and no bottom level at which causality is exercised in a different fashion.

In considering groups from the viewpoint of the question about real agency, I shall concentrate on more transparent possibilities of group formation. Specifically, I shall consider only groups in which there is full and equal awareness of the aspiration to agency among members, and full and equal participation in the attempt to realize that aspiration. This strategy makes sense. The existence of such transparent groups is not open to empirical doubt, so that it will be a significant result if we can establish that they are real agents. And if we can establish that such transparent groups are real agents, then there can be little hesitation about ascribing agency to variants in which the crucial factors remain fixed, but transparency is reduced.

Let us consider the case of groups, then, in which the members have a shared intention or commitment that they form a group agent; they jointly intend that together they operate in a way that parallels the manner in which an individual agent might behave (Searle 1995; Tuomela 1995; Bratman 1999; Gilbert 2001; and Miller 2001). There are many analyses of what shared intention requires but for our purposes here, we need not endorse any one of those analyses rather than others; all we have to assume is that there is good sense in the idea of shared intention.[7]

In order to emphasize the transparent character of the groups envisaged, let us assume in addition that the content of the shared intention is this:

- that the purposes and representations of the group be formed on the basis of member views – in effect, votes – as to which attitudes ought to be adopted;
- that the deputies who enact such purposes and representations on behalf of the group are selected on the basis of member views about selectional procedures;
- that in this formation and enactment of attitudes members are treated equally, having the same group roles, or the same chance of playing group roles, as others.

2.2   The Majoritarian Assembly

The sort of group to which these stipulations saliently direct us is the association or partnership or assembly in which members gather to discuss and vote on decisions about matters that engage them collectively and agree to be bound by those decisions, authorizing suitably chosen

---

[7] Pettit and Schweikard (2006) argue that an analysis that is broadly in the spirit of Bratman works well for a theory of group agency.

deputies to enact the decisions in their name. The standard image of such a body is classically associated with the image of the democratic assembly presented by Hobbes (1994, ch. 16) and Rousseau (1973, Bk 4, ch. 2) and, in less explicit mode, by Locke (1960, Bk 2, ch. 8.96). In this image, members unanimously agree to be bound by the majority vote of the assembly. They authorize the assembly, and those who are chosen to act for the assembly, as figures by whose agreed words and actions they are bound or committed.

Despite the distinguished heritage, however, this tradition is mistaken in suggesting that an assembly might operate as an agent on the basis of majority voting. Suppose that the assembly has to resolve logically connected issues, whether at the same time or over a period of time. No matter how deliberative and democratic the assembly is, and no matter how consistent the individual members are, majority voting may generate an inconsistent set of resolutions on such issues. And no assembly can be expected to function properly as an agent if the representations and purposes it endorses are inconsistent and incapable of being realized together. Assuming that the assembly will only vote on matters that are near the coal-face of action, and not for example, on abstruse issues of metaphysics or theology, any inconsistency in the representations or purposes is liable to affect its capacity to act; the attitudes will guide it at once in different directions. And in any case the endorsement of inconsistent representations or purposes will mean that other agents, including its own members, cannot think of it as a potential partner in reasoned exchange; no one can take seriously the commitments of an agent that does not care about the inconsistency of the positions it endorses.

The unreliability of majority voting is revealed by the discursive dilemma (Pettit 2001a, ch. 5), a problem that generalizes the doctrinal paradox in juridical theory (Kornhauser and Sager 1993). For an illustration of the dilemma consider the way a group of three members of a political party, assuming they endorse a balanced budget, might vote on whether to increase taxation, increase defense spending and increase other government spending. The members, A, B and C might each vote in a consistent pattern on these issues, yet the group view of A-B-C, as determined by majority voting, might involve an inconsistency. The possibility is registered in Table 1.

There are many variations in which the discursive dilemma appears, all suggesting that no group can expect to function as a proper agent if it insists on forming its representations or purposes on the basis of majority voting (List 2006). But the problem is not restricted to majority voting. It turns out that making a group responsive to its individual members in the manner that is exemplified by majority voting, but not only by

Table 1

	Increase taxation	Increase defense spending	Increase other spending
A	Yes	Yes	Yes
B	No	Yes	No (reduce)
C	No	No (reduce)	Yes
A-B-C	No	Yes	Yes

majority voting, rules out an assurance that the group displays collective rationality. Specifically, it rules out an assurance that, if the group faces logically connected issues, then it can resolve them completely and consistently. This is a significant result. Every group will tend to confront connected issues, at least over time. And since these issues will typically be restricted to questions that the group needs to resolve in order to pursue its purposes, a failure to resolve them completely or consistently will undermine its agential capacity. A failure of consistency will leave the group unable to decide between rival courses of action; a failure of completeness will leave it without any purpose or representation to act on.

There are broadly three respects in which we might expect a paradigmatically transparent group agent to be responsive to its membership. First, it should be robustly responsive to its members, not just contingently so; the group judgments should be determined by the judgments of members, independently of how the members judge. Second, the group should be inclusively responsive, not just responsive to a particular member – a dictator – and not just responsive to named individuals; otherwise it would fail to use its members as its eyes and ears, as epistemic considerations suggest it should do, as well as failing on a democratic count. Third, the group should be issue-by-issue responsive – if you like, proposition-wise responsive (List and Pettit 2006) – with its judgment on any question being determined by the judgments of its members on that very question, and with its attitude to any proposed goal being determined by the attitudes of the members to that goal.

There are a number of possible voting procedures under which a group would be responsive to members in a robust, inclusive and issue-by-issue way, and majority voting is only one example. It can be shown that under a variety of interpretations, however, any suitably responsive group will tend to fail the requirement of collective rationality. If it is faced with logically connected issues that it is required to resolve, then it is liable to endorse resolutions that are inconsistent with one another. Thus there will be a hard choice for the group, as in the discursive dilemma, between endorsing individual responsiveness and aspiring to collective rationality. One example of a result that demonstrates this general problem is proved in List and Pettit 2002, and others have since followed.[8]

2.3   The Sequential-Priority Assembly

But if the majoritarian version of the assembly is not going to give us an example of a presumptive group agent, there are variants that certainly can do so. One obvious variant would be to have the assembly follow the sequential priority rule (List 2004). This would order issues so that whenever a group faces an issue on which its prior judgments dictate a resolution, voting is suspended or ignored and the judgment recorded on that issue is the one that its existing judgments dictate. The most salient ordering might be a temporal one.[9] Consider our A-B-C group in the earlier example and imagine that it took votes on the taxation issue first and then on the issue of defense spending. Under a sequential priority rule, the group would suspend or ignore the voting on the issue of other spending, for the first two votes would have mandated a decrease in such spending. This rule might be followed on the basis of reflection about what existing resolutions require. But equally, at least in formal domains, it might be followed on a mechanical basis, with a computing device registering the entailments from existing resolutions that dictate the response to new issues.

The reason why the sequential priority rule would enable the group to be consistent is that while it forces the group to be robustly and inclusively responsive to its members, on intuitive interpretations of those conditions, it allows failures of issue-by-issue responsiveness. On any question where prior judgments dictate a certain line, the group may adopt a position that goes against the views of a majority of members on that particular issue. The position taken will be driven by the positions

---

[8] See for example Pauly and Van Hees (2006) and Dietrich and List (2007). Notice that the three dimensions of responsiveness are not always reflected in a one-to-one fashion by three exactly corresponding conditions. The List–Pettit result demonstrates the problem under the following interpretation of the three responsiveness conditions:

- robust responsiveness: the procedure works for every profile of votes among individuals (universal domain);
- inclusive responsiveness: the procedure treats individuals as equal and permutable (anonymity);
- issue-by-issue responsiveness: the group judgment on each issue is fixed in the same way by member judgments on that very issue (systematicity).

[9] A variant on this procedure would divide issues into basic, mutually independent premise-issues and derived issues – this will be possible with some sets of issues, though not with all – and treat those judgments as prior, letting them determine the group's judgments on derived issues. (See Pettit 2001b and List 2004.)

that members take on other issues, but not by their positions on that issue itself (List and Pettit 2006).

It should be clear that a group might avoid inconsistency by having all of its attitudes formed under the sequential priority rule, or suitable variants. But such a group could scarcely count as a rationally satisfactory agent. It would be entirely inflexible in its responses and potentially insensitive to the overall requirements of evidence. When I realize that some propositions that I believe entail a further proposition, the rational response may well be to reject one of the previously accepted propositions rather than to endorse the proposition entailed. Those are the undisputed lessons of any coherence-based methodology and the group that operates under a sequential priority rule will be unable to abide by them; it will not be reliably sensitive to the demands of evidence.

The evidential insensitivity of the sequential priority rule is apparent from the path-dependence it would induce.[10] One and the same agent, with access to one and the same body of evidence, may be led to form quite different views, depending on the order in which issues present themselves for adjudication. The group agent that follows the rule will be required to respond to essentially conflicting bodies of testimony – conflicting majority judgments among its members – without any consideration as to which judgment it seems best to reject. It will be forced by the order in which issues are presented not to give any credence to the judgment its members may be disposed to support on the most recent issue before it. And this, regardless of the fact that often it will be best to reject instead a judgment that was endorsed at an earlier stage.

2.4   *The Straw-Vote Assembly*

Happily, however, there is a variation on the sequential priority rule that would enable a group to escape the conflict between individual responsiveness and collective rationality, without forcing it to be evidentially inflexible. Under this variation, the assembly would consider different issues in turn, depending on the order in which they arise, and with every issue that arises the assembly would determine whether existing resolutions dictate a resolution of that issue. But at that point it would take a different approach from the sequential priority rule.

Rather than automatically endorsing the resolution dictated in such a case – rather than suspending or ignoring a vote on the most recent issue – the assembly would adopt the following sequence of steps: take a straw vote on the new issue; if the vote gives rise to an inconsistency,

---

[10] For ways of mitigating the effects of path-dependence see List (2004).

identify the other resolutions that combine with that straw resolution to generate the problem; and then vote on which of the problematic resolutions to reject in order to restore consistency. Under this straw-vote procedure the upshot might be the same as under the sequential priority rule: a rejection of the straw vote. But equally it might be the revision of some of the group's existing commitments. Both possibilities are open and so the group that follows this line can avoid the path-dependency, and the associated inflexibility, of the sequential priority rule.

Consider, then, how the A-B-C group might operate under the straw-vote constitution. They will register that their existing commitments on taxation and defense spending require them not to increase — in fact to reduce – other spending. And then, if their vote goes in favor of increasing other spending, they will register the inconsistency and reflect on which of their existing and proposed resolutions to drop. They may decide to reduce other spending, as the sequential priority rule would require them to do. But equally they may decide to revise their commitment not to increase taxation or to increase defense spending. All three options are open.

The assembly that follows the straw-vote procedure is as standard a candidate for group agency as the assembly that cleaves to the strategy of determining every issue by majority vote or that follows the sequential priority rule. And it has the virtue, in similar measure, of being wholly transparent. If we can show that this sort of group should count as a real agent, then we will have shown that a very plausible sort of group can display real agency. Moreover, we will have shown that real agency is a prospective feature for other sorts of groups that depart from it in ways that are not crucial to the argument provided in support of real agency.

3   The Real Agency of the Straw-Vote Assembly

Does the straw-vote assembly introduced in the previous section count as a real agent, by the criteria of agency emerging from our discussion in the first section? The general question is whether it can display a broadly agential pattern of behavior, however subject to perturbations. The more specific questions are whether the perturbability of the pattern is systematic rather than unsystematic; whether the pattern is relatively resilient across different contexts; and – most important, as we shall see – whether it is variably realized across the contributions of individual members.

3.1   *The General Question*

There can be little doubt about the capacity of a straw-vote assembly in general to display a broadly agential pattern of behavior. The members of such an assembly will collectively endorse certain purposes, perhaps revising them from time to time, and they will ensure that the path taken to the realization of any endorsed purpose is determined by the representations that they also collectively endorse, and no doubt revise as occasion demands. The guiding representations will bear on a variety of matters such as the opportunities available for satisfying their purposes, the relative importance or urgency of those purposes, and the best means at their disposal for realizing one or another purpose.

Why does it seem so natural to ascribe a purposive–representational pattern of behavior to a straw-vote assembly? The reason may be that the members are required to reason as a group and to develop a critical perspective on the demands associated with that pattern. For it is hard to think that a group which reasons about what is demanded under a certain pattern of purposes and representations, and which regulates itself for fidelity to those demands, might not actually display such a pattern, at least in broad outline. This feature of the straw-vote assembly marks it off dramatically from any simpler arrangement, such as the assembly that operates by majority vote (Pettit 2007a).

In the majoritarian assembly the members share an intention that they together perform as an agent but under the majority rule they need never reflect on what is demanded of the group agent in view of its existing commitments. All they each have to do, at least in the formation of attitude, is to play their local part, voting as required on the different issues posed and trusting in the majoritarian constitution to assemble their individual contributions into a collective, sensible whole. Under such a mode of organization the members as a group would never conduct anything akin to reasoning in sustaining the performance of the group. They would each follow a personal rule of voting: say, that of voting according to their individual judgment on any issue of collective purpose or representation. And blind adherence to that rule is all that the group would require of them; in sustaining the group they might each be as unreflective as the ants that sustain a colony or indeed the neurons that sustain an individual agent.

But the members of an assembly cannot rely blindly on a majoritarian constitution to ensure that their individual contributions are assembled into a sensible group profile, whether in the space of purposes or representations; as we saw, majority voting might lead the group to endorse an inconsistent set of attitudes. Nor can the assembly members rely blindly on a sequential priority rule – say, one applied by a computer – that would automatically discount any vote that generates inconsistency with prior resolutions; such a rule might lead them, path-dependently, to endorse an evidentially unsupported set of attitudes. That is why we resorted to the straw-vote assembly in order to identify a plausible candidate for group agency.

In the straw-vote assembly, however, members are expected to reason. They do not let the group attitudes be generated blindly, as under the majoritarian constitution. Nor do they conform blindly to *modus ponens*, as the sequential vote procedure would have them do. They have to consider those propositions that have already been endorsed as purposes and representations; they have to determine what those propositions imply, if anything, for the answer to any issue that is currently up for voting; and in the event of the vote going contrary to such implications, they have to decide on which of the conflicting resolutions to drop. This means adopting precisely the sort of critical stance on the demands of a purposive–representational pattern that individual human beings embrace when they question their spontaneous processes of attitude formation.

The fact that the members of the straw-vote assembly can recognize and respond to the demands associated with purposive–representational behavior should give us confidence that they will generally display that sort of behavioral pattern. It means, after all, that while their behavior may sometimes drift away from that pattern, there are correctives available that should serve to guard against this. But even when those correctives fail to work in a given case, the critical character of the group may give us grounds for continuing to ascribe agency; it may mean that the actual display of suitable behavior is less important than it would have been with a non critical agent. For if a straw vote assembly is truly sensitive to purposive–representational demands, then in a case where its behavior drifted away from the required pattern it can presumably be made to recognize the fact and to acknowledge it as a failure. To the extent that it can do this, and can use the recognition of the failure to guard in some measure against repeat failures, we can be much more confident that this is a system that should count as an agent. Whatever past lapses from suitable behavior, it apparently has the capacity to guard against similar lapses in the future.

But now I should turn to more specific questions that may be raised about the claim of a straw-vote assembly to count as a real agent. Those questions bear on the systematic perturbabilty, the contextual resilience and the variable realization that we should expect an agent to display, as we saw in the first section.

3.2   *Systematic Perturbability*

Every natural agent, human or animal, individual or collective, is bound to fall away from the demands of the purposive–representational pattern, whether on the attitude-to-evidence, attitude-to-action, or attitude-to-attitude front. There is a great difference, however, between two sorts of perturbability. A system may be subject to random perturbations whose origin remains opaque, or it may be subject to perturbations that derive from identifiable factors. It may be subject to unsystematic or systematic perturbability.

In the unsystematic case, we may certainly say that the system approximates the profile of an agent. While many of the responses it makes do not have any agential sense, they are few enough in number to count as noise in a system that is otherwise constructed to display agential pattern. But approximating the profile of an agent is not necessarily being an agent; it may just mean simulating agency in a more or less imperfect manner. In the systematic case, things are different. Being able to identify the sources of perturbation, we may be able to see them as constraints on the operation of the system that its history of selection or design had to take as given.

We have no hesitation in identifying systematic sources of perturbation in our own performance as individual agents, given that we reason with one another about the demands imposed by purposive–representational pattern. We assume that we can target the same purposes, the same representations, and the same requirements of rationality. When we have access to the same information, therefore, but diverge from one another in relevant judgments – say, judgments on what the evidence supports, on whether certain purposes or representations are inconsistent, on what means are required for a given purpose – we assume that at least one of us has been subject to perturbation. And so we have a heuristic available for identifying perturbers: we look for factors such that their presence tends to generate judgments – and, presumably, corresponding behaviors – that are discrepant from those of others. This heuristic has produced a wealth of folk knowledge on the perturbing effects of perceptual obstacles and interpersonal pressures, of bias and passion and inattention, and of paranoia and compulsion and other pathologies. And this body of knowledge has been greatly expanded with psychological studies of cool and hot irrationality.

What is true of individual human beings, performing as individuals, will presumably carry over to human beings in assemblies that reason in the manner of the straw-vote group. While every such assembly is going to fail as an agent in various respects, there is surely ground for expecting that the failures will be traceable to systematic sources of perturbation of the kind with which we are familiar with individual human beings. There may also be sources of perturbation that operate on assemblies and other groups but do not affect human beings in isolation. But again these will tend to be more or less familiar or identifiable, such as the contagion effect whereby panic or pack-behavior can be generated among people in crowds, or the hierarchy effect whereby some group members may mindlessly defer to others.[11]

3.3   Contextual Resilience

The first lesson of our earlier discussion was that the perturbability of a would-be agent should be systematic rather than unsystematic and we have found that it is borne out with the straw-vote assembly. The second lesson was that equally, the purposive–representational pattern displayed by the system should be contextually resilient. It should not be tied to such a specific context of performance that it is misleading to think of the system as aiming at a more abstract goal; the lesson was illustrated by Dennett's Sphex wasp.

As the critical, reasoning character of the straw-vote assembly ensures that its perturbability is as systematic as that of ordinary human subjects, so that character makes it natural to ascribe a high degree of contextual resilience to the pattern of behavior that it displays. Let the group behave in a given context after a certain purposive–representational pattern. Should we expect it to be able to adjust so as to continue to realize the same purposes, as the context changes? Or should we expect it, like the Sphex wasp, to be capable of displaying the pattern only in a stereotypical, context-bound way?

If the straw-vote assembly is capable of reasoning then it has to be capable of recognizing the abstract goals it targets, and the different means that may be appropriate for realizing them in different situations. And if it is capable of reasoning then it has to be capable, equally, of responding to that recognition and choosing the appropriate means. But the presence of those capacities means, then, that the purposive–representational pattern it displays is more or less bound to enjoy a high degree of contextual resilience. The group will be robustly disposed to

---

[11] Our earlier discussion shows why we should be loath to ascribe agency when the perturbability is unsystematic. The majoritarian assembly might pass as an agent that is unsystematically perturbable, since it acts in a purposive–representational way, subject to the random, ramifying disturbance introduced by discursive dilemmas and the like. But it would be wrong to think of such an assembly as an agent, since it lacks a crucial, agential resource: the capacity to register and remedy attitudinal inconsistency.

pursue relevant purposes according to appropriate representations, not disposed to do so only under fixed situational parameters.

3.4   Variable Realization

But now we come to a third, more problematic issue. When a straw-vote assembly operates as an agent, does the realization of that pattern by individual members vary in such a way that we have to see the agential configuration at any point – say, the group's endorsing such and such a means of achieving such and such a goal – as causally relevant to what ensues? Or is the variation so limited that that configuration has no causal relevance over and beyond the relevance of the realizing set of attitudes that is present in individual members? Does the group configuration program for the ensuing behavior cover a range of possible ways in which it might be realized at the individual level? Or is there no significant variation at the individual level and no ground for assigning a programming role to the group attitudes?

This issue may seem more problematic than the other two, precisely because the straw-vote assembly is a critical, reasoning body in which individuals do not play their parts blindly, like ants in a colony or neurons in the brain. Individual members monitor where the group is going, and rely on issue-by-issue voting only when there is no issue of consistency between the different purposes or representations they endorse. And so it may well seem that while that reasoning character made it easy to give appropriate answers to the questions of systematic perturbability and contextual resilience, it makes it difficult to defend an appropriate answer to the question of variable realization. If individuals play the critical part required under the straw-vote model, why not think of the configuration that programs at any point for action as the configuration of individual attitudes? Why ignore individual attitudes and invoke the group-level configuration as the causally relevant antecedent?

When an assembly operates with the straw-vote procedure, it is certainly true that what the group does it does with the full endorsement of members; nothing happens behind their backs. But that does not undermine the possibility that group attitudes are variably realizable and that they program for group responses independently of how they may be realized in the dispositions of members. On the contrary, the reasoning character of the straw-vote assembly actually increases the variability with which the group attitude on any issue is likely to be realized at the individual level. And so here, as with the other two issues, it argues for an answer that supports taking that sort of group as a real agent.

Consider the majoritarian group that constructs its purposes and representations on the basis of majority vote. In this case there will be one obvious source of variation in the way the constructed attitude is realized at the individual level. The majority may be any of those sets of members that have one more member than the set of other members; any of those sets of members that have two more members than that other set; and so on. And so to know that the group is committed to a given purpose or representation is not to know much about how in particular the individual members are disposed. It is only to have knowledge of the statistical, aggregative fact that a majority supports that purpose or representation.

This source of variation as between individual and group levels is expanded in the case of the sequential priority group by a further factor. If such a group holds by a certain purpose or representation, that may be because a majority of members support it, as in the majoritarian case. But it may also be because, while a majority reject that purpose or representation, the group is required in consistency to endorse it, given prior majority support for certain other purposes or representations. Thus the ways in which individuals may be disposed, consistently with the group endorsing that purpose or representation, are greater in number than the ways in which members may be disposed in the counterpart case with the majoritarian assembly.

The two sources of variation that are relevant with the sequential-priority group remain in place with the straw-vote assembly. But at this stage, a third source of variation enters as well, for there is a further way in which individuals may adjust so as to ensure that the assembly endorses a certain purpose or representation. This is the sort of adjustment that members make when they revise an earlier commitment and endorse a certain group attitude, because that adjustment is taken by them to be the best way of responding overall to the demands of evidence.

When we take all of these sources of variation into account, it becomes hard to see any reason why we should balk at treating a straw-vote assembly as a real agent. Not only can such a body display a purposive–representational pattern of behavior under a systematic mode of perturbability and with a high degree of contextual resilience. It is more or less bound to display this pattern in a way that is radically discontinuous with the attitudes of the individual members who constitute it. We should have no hesitation in looking to the group attitudes at any point as causally relevant factors that program for what the group goes on to do. For that complex of attitudes will program for the group response over an indefinite range of variations in how it is realized in the dispositions of individual members. The contrast with the simple heating–cooling system could not assume a starker profile.

4   Conclusion

The criteria I proposed as tests of agency are hard to question and the straw-vote assembly that I identified as a candidate for group agency is hard to dismiss as an institutional possibility. Yet by those criteria it is demonstrable that that candidate does indeed count as an agent. Thus there is every reason to conclude that groups can be real agents. Nothing but prejudice can stand in the way.

But prejudice on this matter is in no short supply. It comes in two major forms, one associated with an epistemological presupposition, the other derived from a complex of metaphysical and normative fears.

The epistemological presupposition that may block the admission of group agency is the assumption that if group agents are real, then they must have a mental life of their own, in particular a mental life that is not accessible to other agents. Thus they may have to draw on the conceptual and ratiocinative abilities of their members in order to operate properly; but they must have a consciousness that individuals as such do not access. They must operate, in some sense, behind the backs of their members.

Group agents in the straw-vote mold certainly do not operate behind the backs of their members; they exist by virtue of the monitoring and management that those individuals exercise. But agents are distinguished by the sets of attitudes that they embody, and by the principles of development to which those attitudes are subject, not by the extent to which their attitudes are inaccessible to others (Rovane 1997). And even though the attitudes of a straw-vote assembly are fully accessible to its members, being intentionally formed and enacted by the membership, they constitute a developing set that bears no systematic relationship to the attitudes by which the individual members are each characterized. Group attitudes have to satisfy criteria of rationality in order to support a unified pattern of agency and their being robustly rational means, as we saw, that they cannot be systematically responsive to the attitudes held by individual members.

There is also a complex of metaphysical and normative fears that may stand in the way of admitting the reality of group agency. Thus, people will balk at the admission on the grounds that it makes groups mysteriously emergent, that it would raise again the totalitarian specter that was banished in the last century by Popper's attack on holism, or that it is liable to introduce group rights as restrictions on the rights of individuals.

These sorts of fears, however, are baseless. The straw-vote assembly has an agential profile that may go with any of a variety of individual profiles but it is nothing and it does nothing except on the basis of the contributions of its members; it is superveniently dependent on how those members are disposed to behave (List and Pettit 2006). The state may well count as a real agent, by extension from the case of the straw-vote assembly, but that does not mean that the state has to be given the totalitarian role that it assumed amongst fascists and communists (Pettit 2003). And while group agents of different kinds – associational, commercial, and political – may be given certain rights under the law, the rights they are given should surely be constrained by how well and how far they serve individuals; they are the tail, we are the dog.

This is not to say that ascribing reality to group agents has no important normative and explanatory implications. On the normative side it means, as I have argued elsewhere, that group agents should be held responsible for programming for certain actions, even though it may also be appropriate to hold members responsible for enacting their programmed roles (Pettit 2007b). And on the explanatory side it means that there is good reason to seek explanations at a level where group agents are treated as agents in their own right without always exploring the nuts and bolts of individual contribution; the refusal to go to the fine grain of causal mechanism may be crucial for the pursuit of certain explanatory purposes (Pettit 1993, ch. 5). More generally, recognizing the reality of group agents opens up an enormous range of questions as to how such entities can and should be designed, both in general and in certain political or other contexts. It places an important research program on the agenda of explanatory and normative social theory.[12]

REFERENCES

Block, N. 1981. "Psychologism and behaviorism," *Philosophical Review* **90**: 5–43.
Bratman, M. 1999. *Faces of Intention: Selected Essays on Intention and Agency*. Cambridge: Cambridge University Press.
Davidson, D. 1980. *Essays on Actions and Events*. Oxford: Oxford University Press.
Dennett, D. 1979. *Brainstorms*. Brighton: Harvester Press.
1991. "Real patterns," *Journal of Philosophy* **88**: 27–51.
Dietrich, F. and C. List (2007). "Arrow's theorem in judgment aggregation," *Social Choice and Welfare*, **29**: 19–33.

---

[12] This chapter owes an enormous debt to my collaboration with Christian List on a book about group agents; in that book we detail and illustrate the sort of research program at which I only gesture here. I benefited greatly from discussion at three presentations of the material: one, in June 2007, at Witten/Herdecke University; a second, "Collective Intentionality VI" in July 2008 at the University of California, Berkeley; and a third in October 2008 at the University of Pittsburgh.

Fodor, J. 1975. *The Language of Thought*. Cambridge: Cambridge University Press.

Gilbert, M. 2001. "Collective preferences, obligations, and rational choice," *Economics and Philosophy* **17**: 109–120.

Hobbes, T. 1994. *Leviathan*. E. Curley (ed.) Indianapolis, IN: Hackett.

Jackson, F. and P. Pettit 1988. "Functionalism and broad content," *Mind* **97**: 381–400; reprinted in F. Jackson, P. Pettit, and M. Smith, 2004. *Mind, Morality and Explanation*. Oxford: Oxford University Press.

   1990a. "In defence of folk psychology," *Philosophical Studies* **57**: 7–30; reprinted in F. Jackson, P. Pettit, and M. Smith, 2004. *Mind, Morality and Explanation*. Oxford: Oxford University Press.

   1990b. "Program explanation: A general perspective," *Analysis* **50**: 107–17; reprinted in F. Jackson, P. Pettit, and M. Smith, 2004. *Mind, Morality and Explanation*. Oxford: Oxford University Press.

   1992. "In defence of explanatory ecumenism," *Economics and Philosophy* **8**; reprinted in F. Jackson, P. Pettit, and M. Smith, 2004. *Mind, Morality and Explanation*. Oxford: Oxford University Press.

Kornhauser, L. A. and L. G. Sager 1993. "The one and the many: Adjudication in collegial courts," *California Law Review* **81**: 1–59.

List, C. 2004. "A model of path-dependence in decisions over multiple propositions," *American Political Science Review* **98**: 495–513.

   2006. "The discursive dilemma and public reason," *Ethics* **116**: 362–402.

List, C. and P. Pettit 2002. "Aggregating sets of judgments: An impossibility result," *Economics and Philosophy* **18**: 89–110.

   2006. "Group agency and supervenience," *Southern Journal of Philosophy* **4**: 85–105.

Locke, J. 1960. *Two Treatises of Government*. Cambridge: Cambridge University Press.

Miller, S. 2001. *Social Action: A Teleological Account*. Cambridge: Cambridge University Press.

Millikan, R. 1984. *Language, Thought, and Other Biological Categories*. Cambridge, MA: MIT Press.

Pauly, M. and M. Van Hees (2006). "Logical constraints on judgment aggregation," *Journal of Philosophical Logic* **35**: 569–585.

Pettit, P. 1993. *The Common Mind: An Essay on Psychology, Society and Politics*. Paperback edition 1996. New York, NY: Oxford University Press.

   2001a. *A Theory of Freedom: From the Psychology to the Politics of Agency*. Cambridge, New York, NY: Polity and Oxford University Press.

   2001b. "Deliberative Democracy and the Discursive Dilemma," *Philosophical Issues supplement to Nous* **11**: 268–299.

   2003. "Deliberative Democracy, the Discursive Dilemma, and Republican Theory." In J. Fishkin and P. Laslett (eds.) *Philosophy, Politics and Society Vol 7: Debating Deliberative Democracy*. Cambridge: Cambridge University Press, pp. 138–162.

   2007a. "Rationality, reasoning and group agency," *Dialectica* **61**: 495–519.

   2007b. "Responsibility incorporated," *Ethics* **117**: 171–201.

Pettit, P. and D. Schweikard 2006. "Joint action and group agency," *Philosophy of the Social Sciences* **36**: 18–39.

Rousseau, J.-J. 1973. *The Social Contract and Discourses.* London: J.M. Dent & Sons Ltd.

Rovane, C. 1997. *The Bounds of Agency: An Essay in Revisionary Metaphysics.* Princeton, NJ: Princeton University Press.

Searle, J. R. 1995. *The Construction of Social Reality.* New York, NY: Free Press. 1983. *Intentionality.* Cambridge: Cambridge University Press.

Stalnaker, R. C. 1984. *Inquiry.* Cambridge, MA: MIT Press.

Tollefsen, D. 2002. "Organizations as true believers," *Journal of Social Philosophy* **33**: 395–410.

Tuomela, R. 1995. *The Importance of Us.* Stanford, CA: Stanford University Press.

3 – Comment
A Note on Group Agents

Diego Rios

Pettit provides a general framework detailing the conditions for ascribing agential status to groups (such as political parties, assemblies, churches, states, etc.) in a way that parallels the attribution of agency to individuals. We normally use an intentional vocabulary to refer to the behavior of complex collective organizations, and we implicitly assume that these organizations behave as true agents. We say, for instance, that the aim of Parliament at the moment of a vote on new legislation is to reduce poverty by increasing welfare allocations; or, we say that the objective of the government with this or that measure is to reduce unemployment. In both cases, parliaments and governments are conceived as agents having specific goals and objectives that they attempt to promote. These ascriptions of intentional and purposive behavior have both normative and explanatory consequences. From the normative point of view, they are used to create obligations and other commitments: we say, for instance, that such and such an assembly has promised to do this or that, and we can criticize it for failing to honor its self-imposed obligations. From the explanatory point of view, ascribing purposes and goals to organizations and groups is a way to account for their behavior: we explain, for instance, the behavior of governments and states by ascribing goals to them and assuming that they attempt to satisfy – with different degrees of success – such goals. I take as uncontroversial the *existence* of this kind of agential talk about groups and organizations. The problem is how literally we are ready to interpret this agential vocabulary.

Group agency has sometimes been looked at with some scepticism within the social sciences. There are two major sources for this scepticism. One source is associated with social choice theory. Social choice literature has isolated systematic failures at the moment of aggregating individual preferences, generating an array of well-known social paradoxes (Arrow 1951). A paradigmatic example is the case of majority voting, that could generate – at the group level – an incompatible set of social preferences. The prospects for granting agential status to collective bodies governed by an inconsistent set of preferences looks rather dubious. The group or assembly in question could, after all, endorse incompatible courses of action and push through resolutions that violate elementary rationality constraints. This analysis has been enlarged to cover a wide range of paradoxes (List 2004 and 2006; List and Pettit 2002 and 2004; Kornhauser and Sager 1993; Pettit 2001a and 2001b).

The other source of scepticism about group agency has its roots in the old philosophical tradition associated with individualism. According to Hayek and Popper, groups are not true agents. They would probably concede that we sometimes *speak* as if groups were intentional agents, ascribing them intentional features; nevertheless, this is just a *façon de parler*. There might be different ways to state what these authors understood by individualism, but they seem to have amalgamated individualism with singularism – the idea that only individuals are agents (Gilbert 1989). Early individualist literature seems to have rejected both the emergent status of groups and their agential standing, maybe assuming – without much discussion – that granting agential status to groups inevitably leads to conceiving them as emergent. Independently of how appropriate this amalgamation is, it seems to be true that Popper and Hayek were emphatically committed to the idea that only individuals are agents, and that talk about group agency is purely metaphorical.

These are then two possible sources of scepticism for the project of granting agency status to groups. Pettit provides a general framework to dispel some of the assumptions underlying this scepticism. His strategy is developed in three steps. First, he sets the conditions for ascribing agential status to a given system. Pettit's reasoning on this issue has a strong Dennettian flavor: a system will count as properly agential when it exhibits the disposition to display, in a wide range of contexts – actual and counterfactual – a purposive representational pattern. The second step consists in showing how this framework could be applied to fully transparent groups – groups where all their members have full and equal awareness of the collective goals. The third – and last – step consists in generalizing what has been said about fully transparent groups to less than fully transparent ones: the main line of the argument is that once the agential status of fully transparent groups has been granted, it is natural to extend the claim to cases where transparency is reduced.

The first part of the chapter discusses the conditions that must be met for a system to count as an agent. One common objection to the instrumental theory of agency is that it is too generous at the moment of granting agential status: too many systems count as agents. Some of these difficulties are avoided by introducing further conditions – systematic perturbability, contextual resilience, variable realization – restricting the set of systems that could eventually be granted agential status. Nevertheless, I am not sure that all our intuitive judgments are captured by this apparatus. Compare Dennett's Sphex wasp and a Coke machine: they will both exhibit similar scores when submitted to the perturbability, resilience and variable realization tests. Intuitively I would say however that Dennett's Sphex wasp might count as an agent, while the Coke machine cannot be one.

In the second part of the chapter, Pettit dispels some of the reasons for being sceptical about group agency. Although majoritarian rule could give rise to inconsistent sets of preferences, more demanding voting procedures – like the straw-vote procedure – might help the members of the group to rule out manifest inconsistencies in group resolutions, avoiding at least one important source of trouble. Note that Pettit's rehabilitation of group agency does not imply that groups have a life of their own. The behavior of the group *supervenes* on the behavior of its members: the group, even if conceptualized as a true agent, is still dependent on individual behavior. Every variation at the group level will be accompanied by at least one change or variation at the individual level. Individuals are always the underlying causally efficacious elements responsible for group behavior. This line of thought makes good sense when looked at with other of the author's major contributions to the field – the idea of program explanations (Jackson and Pettit 1990, 1992). Group agents could be conceived as *programming* individual behavior: in a way they contribute to *canalizing* individual behaviors along specific lines. Although not causally efficacious, group agents are nevertheless causally *relevant* in the production of social outcomes: they raise the frequency of certain types of outcomes and contribute to directing the behavior of individuals (Pettit 1993: 258). They are then an important part of the causal history of a given social event.

The primary aim of Pettit's paper is to reconsider fully transparent and quasi-transparent groups as potential agents. I find his analysis on this topic very convincing. Pettit focuses his analysis on fully transparent and quasi-transparent groups. His framework opens a fresh research agenda not only for the philosophy of the social sciences, but also for institutional design. I would like to speculate about the possibility of extending this analysis to fully opaque groups. Although this is certainly *not* the primary aim of Pettit's essay, I find the topic of fully opaque groups intriguing enough to deserve some discussion. Pettit makes a brief mention of this issue:

If agency requires just the display of purposive–representational pattern – specifically in a way that satisfies systematic perturbability, contextual resilience and variable realization – then it is at least logically possible that a group of people might be an agent without the members of that group recognizing the fact; they might mediate the agency of the group in the unthinking, zombie-like manner in which, on a naturalistic picture, my neurons mediate my agency.

This paragraph can be interpreted in different ways. The parallel to zombie-like neurons suggests that there is room at the group level for true agency, even when *all* the individual realizers are unaware of the high-level agential goals and purposes of the group. In the case of fully opaque agents, the parallelism between the individual and the collective agency would be strong: the constituent elements – neurons, in the case of an individual agency; individuals, in the case of group agents – produce high-level purposive outcomes via purely blind, mechanical interactions. None of the individuals – as in the case of the neurons – is aware of the global outcomes that its own behavior is helping to promote.

How plausible is this option? It could be argued that fully opaque groups are exactly the examples that critics like Popper and Hayek had in mind when they criticized the reification of groups as agents. Note that transparent or semi-transparent groups are easier to tackle: it is not impossible to imagine a group designed by some of its members in such a way as to behave as a group agent; the designers need just to *canalize* the behavior of the other individuals. Although unaware of the global consequences of their actions, these individuals nevertheless contribute by their own behavior to produce the collective purposive outcome. None of the members of the group except the designers – need know about the general purpose of the group. In the case of partially transparent groups, the explanation of the purposive outcome will be given in terms of those constituents of the groups that are transparent – the designers. This move however cannot be made when the group is fully opaque, because, by definition, fully opaque groups lack individual designers: *all* the constituents of such a group are blind about the ultimate goals of the group.

The most serious problem connected to granting agential status to fully opaque groups is that the ultimate global outcomes of the system are left totally *unexplained*. Many times in the social sciences, fully opaque groups have been described as agents having their own goals. The problem with these accounts is that they normally lack the mechanism generating the purposive outcome. This is a powerful reason for skepticism. But maybe selectional considerations could be introduced to explain the emergence of group purposive outcomes, even when all the members of the group are unaware of the collective goals of the organization to which they belong. I am unsure about the prospects of this strategy .

Two further notes. The first concerns the scope of the theory of agency when applied to fully opaque groups. Pettit argued in favor of a functionalist interpretation of agency that could – eventually – be enriched with thicker conditions requiring the existence of a ratiocinative capacity. The straw-vote model left room for this ratiocinative process to take place: the members of the group were "forced" to reason as a group and to take a critical attitude to the potential intentional profile of the group. Obviously this enriched conception of agency as involving sensitivity to rational requirements can only be applied to fully transparent or semi-transparent group agents. Fully opaque group agents cannot be critical agents in the same way: they lack the internal resources to be so. In order to be to able to take an evaluative stance toward the intentional profile of the group, the member must be aware of the purposes of the group as such – this condition is absent in fully opaque group agents. In a way, fully opaque group agents can only be part of the *thin* functional theory of agency. This makes fully opaque group agents much less interesting, especially when considered from the perspective of institutional design.

The second point generalizes what has been said before concerning the relationship between singularism and individualism. One of the important points of the chapter is that granting agential status to groups need not violate individualist constraints. A critic might think that this would not be an option when dealing with fully opaque groups. Nevertheless this objection is not correct. All social outcomes will be traceable to individual behavior: social outcomes supervene on individual behaviors. This is so, independently of the degree of opacity of group agents. The two issues are conceptually different. Granting fully opaque groups agential status need not imply a commitment to a reified conception of groups. Even fully opaque group agents would be the result of the interaction of individual constituents: granting agency to them does not necessarily entail conceiving them as mysteriously emergent players in the social world.

To sum up: Pettit's analysis is a powerful contribution to one of the fundamental problems in the philosophy of the social sciences – to wit, the possibility of enlarging the set of social actors to include not only singular individuals but also groups. This opens a rich philosophical agenda, raising new questions and challenges for the entire domain. Apart from quite obvious implications for social theory in general, the issues raised in this paper have the potential to generate a fresh look at old practical problems, concerning how to design institutions in order to help the members of a group to police and filter the consistency of the organization's resolutions and attitudes. There might also be important normative consequences for the way we assess the responsibility of individuals for the actions of the group they belong to. The claim that some types of groups, but not others, contribute to making individuals *think* about their goals as a group – a process that amounts to properly collectivizing reason – promises to be of paramount importance in the near future.

REFERENCES

Arrow, K. 1951. *Social Choice and Individual Values*. New York: Wiley.
Gilbert, M. 1989. *On Social Facts*. Princeton, NJ: Princeton University Press.
Jackson, F. and P. Pettit 1990. "Program explanation: A general perspective," *Analysis* 50: 107–117.
    1992. "In defense of explanatory ecumenism," *Economics and Philosophy* 8: 1–21.
Kornhauser, L. and L. Sager 1993. "The one and the many: Adjudication in collegial courts," *California Law Review* 81: 1–59.
List, C. 2004. "A model of path-dependence in decisions over multiple propositions," *American Political Science Review* 98(3): 495–513.
    2006. "The discursive dilemma and public reason." *Ethics* 116(2): 362–402.
List, C. and P. Pettit 2002. "Aggregating sets of judgments: An impossibility result," *Economics and Philosophy* 18: 89–110.
    2004. "Aggregating sets of judgements: Two impossibility results compared," *Synthese* 140(1): 2007–235.
Pettit, P. 1993. *The Common Mind: An Essay on Psychology, Society and Politics*. Oxford: Oxford University Press.
    2001a. *A Theory of Freedom*. Cambridge & New York: Polity Press 2001.
    2001b: "Deliberative democracy and the discursive dilemma," *Philosophical Issues*, 11: 268–299.

Laws and Explanation in the Social Sciences

Part II of the book, which is devoted to the methodology of the social sciences, starts with an exchange of arguments about reductionism an issue originally debated by the logical positivists, then abandoned for many decades. Fodor's paper on "Special sciences: or the disunity of science as a working hypothesis" again shifted attention to the issue. *David Papineau* starts from the proposition that we are all physicalists now and explores its methodological implications for the human sciences. Even if one maintains physicalism, i.e. the position that human entities must be physical if they are to make a difference in the real world, one does not need to be a classical reductionist. Fodor has respected the requirements of physicalism, while maintaining the autonomy of the human sciences, a claim that Papineau questions, arguing that the human sciences will only contain a rich body of laws to the extent their categories have a uniform physical nature. If he is right, does it matter? He thinks it does matter a great deal, since the prospects for successful human sciences will consequently depend on the extent to which their categories can be viewed as in principle equitable with physical categories. *Shulman and Shapiro* are very critical of this stance and regard Papineau's view as reflecting an overestimation of the reductive achievements of the social sciences and a flawed conception of what is required for science. As practicing scientists, they find that Papineau's plea for the yet-to-be-found reducing theories is irrelevant and even dangerous for their fields, since if it were to be successful, it would serve to bind valuable creative energies of the working scientists to a futile reductive enterprise.

The issue of the existence of laws in the social sciences also concerns the interaction between *Sandra Mitchell* and *James Alt*; but the question they address is not whether there are or are not laws in the social sciences, but how we can successfully replace the standard conception of laws with a more spacious conceptual framework. Mitchell rejects what she regards as a spurious dilemma arising from representing the knowledge of our world either as laws or non-laws. She also argues against the common way of modifying the standard account of strict laws by means of adding a *ceteris paribus* clause. Starting from the observation that the actual products of scientific practice hardly fit the ideal image of laws, she endorses a pragmatic strategy by defining a law functionally as a claim that permits explanation, prediction, and intervention; consequently less-than-universal, exception-laden generalizations can qualify as laws or in any case function as laws in our explanations of social phenomena. In Jim Alt's comment, he takes up the suggestions of Mitchell and shows how they can be fruitfully applied in empirical research. His general position is that the model of the natural scientist who provides causal explanations of phenomena is an appropriate one for the budding social scientist and that reliance on statistical techniques and far-from-perfect correlations still enables the social scientist, and in particular, the political scientist, to offer causal explanations. By examining a series of examples, Alt shows what good practices in political science look like – when they bring together theory and data.

In the third exchange, *Daniel Little* starts with the premise that the construction of good social explanations depends on having a more realistic understanding of the ontology of the social world than we currently possess. Little argues for "methodological localism" and against the availability of strong social regularities, and he advocates the discovery of concrete social-causal mechanisms as a basis for social explanation. Social-causal mechanisms are constituted by socially situated individuals and the social pathways through which their actions are aggregated, constrained, and transmitted. Social institutions and structures are plastic across time and place, embodied in and transformed by choices of socially situated actors and leaders. In his comment *Jack Knight*, first questions Little's approach that casts doubt on the search for generalizable theories across numerous societies. Knight contends that many scholars in the social sciences would admit that universal laws should not be the basis of social scientific explanations, but he thinks that many of these same scholars would likely question efforts to undermine the pursuit of generalizations in their research. In a second step he shows that the methodological approach that Little offers as an alternative, what he calls "methodological localism" based on the pursuit of causal mechanisms, is really consistent with a significant role for generalizations and he asks: Do we really need to de-emphasize generalization in order to reap the considerable benefits his proposed focus on causal mechanisms and methodological localism would produce? And similarly, would we be unwise to do so?

The final interaction of Part II between *Nancy Cartwright* and *Gerd Gigerenzer* turns our attention to another paradigmatic problem of the social sciences: How policy measures should be designed that respect some minimal conditions of scientific rationality or, formulated in more heroic terms, how science and politics are linked. The discussion is focused around the value and limitations of Randomized Controlled Trials (RCTs) for suggesting policy measures. Nancy Cartwright asks the question, "What is this thing called 'efficacy'?" Efficacy is what is established about causes in RCTs. For their part, RCTs are often described as the "gold standard" in medical science and, increasingly, in the evaluation of social policies as well. If an RCT demonstrates that a cause has an effect (efficacy), this does not imply that the cause has the same effect in the field (effectiveness). Thus, without some theory about the capacities of the causes we are studying, the evidentiary value of an RCT is not just weakened, but, Cartwright argues, made empty, i.e., zero. Gerd Gigerenzer applies her philosophical analysis to the "field" of health care. He does agree with Cartwright about the existence of the efficacy–effectiveness gap due to the problem of induction and he highlights the way that this gap is dealt with in practice by briefly presenting two case studies of cancer screening in Germany and the United States. His case studies highlight the political economy of screening for breast cancer with mammography: Since the government agencies that are in charge of the screening programs are the same as those in charge of the dissemination of the information about cancer screening, they have a vested interest to present this information in a biased way. In other words, policy making is shaped by conflicting interests rather than evidence. Gigerenzer concludes that in practice the problem is not, as Cartwright thinks, that the results of RCTs are often quickly assumed to hold true in the field, without any further input from theory. The case is rather the opposite: They are mostly willingly ignored by policymakers because of the conflict of interests inimical to democratic politics.

The four interactions of this part address the four core issues of the methodology of the social sciences: the issue of reductionism; the issue of laws; the issue of causal mechanisms and explanations; and the issue of how empirical findings should be used for policy interventions. Any research agenda on *Laws and Explanation in the Social Sciences* must include these four issues.

4   Physicalism and the Human Sciences

David Papineau

1   Introduction[1]

We are all physicalists now. It was not always so. One hundred years ago most educated thinkers had no doubt that non-physical processes occurred within living bodies and intelligent minds. Nor was this an anti-scientific stance: the point would have been happily agreed by most practicing scientists of the time. Yet nowadays anybody who says that minds and bodies involve non-physical processes is regarded as a crank. This is a profound intellectual shift. In this chapter I want to explore its methodological implications for the human sciences. I do not think that these have been adequately appreciated.

It is sometimes suggested that the modern enthusiasm for physicalism is some kind of intellectual fad, fanned by the great successes of physical science during the twentieth century. But this underestimates the underpinnings of contemporary physicalism. The reason that scientists one hundred years ago were happy to countenance non-physical processes is that nothing in the basic principles of mechanics ruled them out. Mechanics tells us how material bodies respond to forces, but says little about what forces exist. Prior to the twentieth century, orthodox scientists countenanced a far wider range of independent forces than are admitted today; these included not only separate chemical, cohesive, and frictional forces, but also special vital and nervous forces. Consider the term "nervous energy." This was originally a nineteenth-century term for the potential energy of the nervous force field. Nervous energy was supposed to be stored up during cognition and then converted into the kinetic energy during action.

The verdict of the twentieth century, however, has been that there are no such special forces. A great deal of detailed experimental research,

---

[1] Much of the material in this chapter is also in my "Can Any Sciences be Special?" in Macdonald, C. and Macdonald, G. (eds.) *Emergence* (forthcoming). Thanks are due to the editors of this volume and to Oxford University Press for permission to reproduce this material.

including detailed physiological research into the internal working of living cells, has failed to uncover any evidence of material processes that cannot be accounted for by a few fundamental forces (gravity, electro-magnetism, the strong and weak nuclear forces). Because of this, special vital or mental forces are now discredited, along with chemical, cohe-sive, and frictional forces. The basic physical forces are almost univer-sally regarded as adequate to account for all material processes.[2]

Where does this leave thoughts, feelings, relationships, institutions and the other familiar human entities that form the subject matter of the human sciences? At first sight it might seem that they must be dismissed as illusory. If all material effects are due to purely physical influences, then doesn't this show that the putative components of human reality don't make a differ-ence to anything? But this would be too easy. Perhaps these human com-ponents are themselves part of the physical world, and so perfectly able to influence material processes. This is the reductionist option. We don't take the advances of physical science to show there are no thoughts or institu-tions. Rather, we conclude that thoughts and institutions are themselves physical entities, and so perfectly real. (Compare the way that heat was reduced by the kinetic theory of gases, rather than eliminated. The kinetic theory showed that all supposed effects of heat can be explained by the motion of molecules. But science didn't conclude that therefore there is no heat. Rather it said that heat is nothing more than molecular motion.)

This reductionist option promises to save the subject matter of the human sciences. But at the same time it threatens their autonomy. Before the rise of physicalism, the human sciences could regard themselves as identifying mental, behavioral, and social patterns that were separate from any physical principles. Of course, such human processes could have effects in the material world, just as Descarte's immaterial mind could have effects on the body. But these human processes would not themselves be part of the physical world, and so would not be governed by physical principles. Mental, behavioral, and social patterns would be quite independent of the laws of physics.

However, this autonomy is threatened by physicalism. According to Ernest Nagel's classic model of reduction (1961), any patterns displayed at the level of a "reduced" science are special cases of the laws of the "reducing" sciences. On Nagel's conception, reduction requires the categories of the reduced science to be identified with categories of the reducing science, via "bridge laws." In consequence, any regularities of the reduced science can in principle be rewritten as regularities of the

---

[2] For the history of physicalism, see Papineau 2002, Appendix.

reducing science. In the kind of case we are interested in, this would mean psychological, economic, and other human categories must be specifiable in purely physical terms, and that any laws involving these categories must be expressible as purely physical laws.

In due course we shall consider further how far this classical reductionist model really does impugn the autonomy of the human sciences. But first we need to consider whether classical reduction is really forced on us by physicalism. This would be denied by many philosophers today. Over the past fifty years, philosophers have devoted a great deal of energy to developing varieties of "non-reductive physicalism." The idea here is to go along with the basic physicalist thought that human entities must be physical if they are to make a difference in the real world, but to deny that the specific requirements of classic Nagelian reduction follow. (The terminology can be a bit confusing here. By ordinary standards, 'non-reductive physicalism' would be counted as a species of reductionism, since it rejects any ontological pluralism and collapses all reality, including human reality, into the physical realm. But in this chapter I shall adhere to contemporary philosophical jargon, reserving 'reductionism' for the stronger requirements of Nagel's classic model, and using 'physicalism' for the more general denial of ontological pluralism.)

2   Laws Without Reduction

Non-reductive physicalism promises to restore the possibility of autonomous laws in the human sciences by allowing for human patterns that are not special cases of physical laws. The classic explanation of how this might work is Fodor's "Special sciences: Or the disunity of science as a working hypothesis" (1974). Fodor made his analysis graphic in what must be the most-reproduced diagram in philosophy (Figure 4.1).

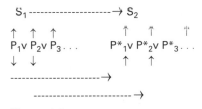

Figure 4.1

Here S_1 and S_2 are special kinds and $S_1 \rightarrow S_2$ is a special law. Fodor gives the example of Gresham's Law – "bad money drives out good."

If there are two kinds of money in circulation, the money that people trust more will be hoarded, and the less trusted money will be used for exchanges. So in this case S_1 would stand for the presence of two kinds of money, and S_2 for the disappearance of the good money from circulation.

Now, if Nagel's classic reductionist model applied here, we should be able to equate S_1 with some specific physical category P, say, and S_2 with some specific physical category P*, and thus reduce the special Gresham's Law S_1 -> S_2 to the physical law P -> P* (Figure 4.2).

Figure 4.2.

But on Fodor's picture this will no longer be possible, because Fodor does not require that S_1 or S_2 be identified as *types* with physical categories. Rather he holds that these special categories will be *variably realized* at the physical level, by $P_1, P_2, P_{3\ldots}$, and $P^\star_1, P^\star_2, P^\star_{3\ldots}$ respectively. For example, in some cases of Gresham's Law the two kinds of money will be two species of cowrie shell, in other cases they will be coins and notes, and in yet others they will be values in electronic registers. Fodor is a physicalist all right, in that he supposes that in each such case S_1 and S_2 will be realized by nothing but physical facts. But he resists classical reduction by denying that there is any common physical nature to all the different cases of S_1 and S_2, so *a fortiori* denying that the law S_1 -> S_2 can be expressed in purely physical terms.

When Fodor talks of "variable realization," this should be understood as the converse of metaphysically necessary determination: S_1 is realized by P_1 if and only if P_1 metaphysically necessitates S_1. This is what ensures Fodor is a physicalist. Nothing more than P_1 is needed to ensure S_1. Not even God could make something that is P_1 without S_1. At the same time, not everything that has S_1 will have P_1, or have any other physical kind, since there are always other physical ways ($P_2, \ldots$) in which S_1 can be realized. This is why S_1 is not type-reducible to any physical kind, and why laws involving S_1 will not be expressible in physical terminology.

Let us look a bit more closely at the way the S_1 -> S_2 law is consistent with physicalism without itself being a physical law. At the physical level, the various physical Ps which realize S_1 will generally give rise to P*s which realize S_2. Thus, when S_1 is realized by some P_1, this will instigate physical processes that give rise to a $P^\star_1$, which in turn then determines S_2. These physical processes are thus consonant with the special law S_1 -> S_2.

According to Fodor, such a P_1 -> P^*_1 link needn't hold in every single case. Some of the P_1s that realize S_1 will fail to give rise to a P^*_1 that determines S_2. This is why, says Fodor, the laws of the human sciences only hold *ceteris paribus*. The relevant physical processes won't always fit with the S_1 -> S_2 law, and so the law will sometimes have exceptions.

3   Reduction Required

Fodor thus promises to respect the requirements of physicalism while maintaining the autonomy of the human sciences. All particular human facts are realized by physical facts. But the general patterns that appear at the human level have no counterpart at the physical level. There is no physical pattern corresponding to Gresham's Law, for lack of any common physical categories to cover all the different instances of this law.

But this only saves the autonomy of the human sciences if Fodor's picture is coherent. I have always had my doubts (Papineau 1985, 1992, 1993). Here is the obvious worry. If the realizations of S_1 are all so physically different, then how come they all give rise to a similar result, namely, some physical state that determines S_2? Won't it be an unexplained coincidence that they should all display this common result? Unless more can be said about what ties the P_1s together at the physical level – as would be provided by a traditional reduction – won't the variability of the P_1s undermine the idea that S_1 is regularly followed by S_2?

Here is an example that will illustrate the point (Papineau 1993, ch. 2). Suppose we find some initial evidence that people who eat reheated brussels sprouts (S_1) come to suffer from inflamed knees (S_2). However, when we investigate this phenomenon, we find that there is no common feature that accounts for this syndrome. Rather, in one case the sprouts harbour a virus (P_1) that infects the knees (P^*_1). In another the sprouts contain a high level of uric acid (P_2) that leads to gouty attacks (P^*_2). In a third the sprouts involve some toxin (P_3) that deplete the cartilage that protects the knee joints (P^*_3). And so on.

This story doesn't hang together. It beggars belief that reheated brussels sprouts should always give rise to inflamed knees, yet the physical process that mediates this should be different in every case. Surely either there is some further feature of the sprouts that can explain why they all yield the same result, or we were mistaken in thinking that there was a genuine pattern in the first place, as opposed to a curious coincidence in our initial sample of cases.

Yet this looks just like the picture that Fodor is inviting us to accept for human scientific laws. So I am inclined to say just the same about Fodor's picture. Either there is something more to say about why S_1

should always give rise to S_2, or it can't be a genuine pattern to start with.

Does it help that Fodor's human science laws are only supposed to be *ceteribus paribus* and not strict? Not really. Note that the puzzle about the reheated brussels sprouts leading to inflamed knees doesn't depend on this being an invariable pattern. In the absence of a uniform explanation, it would be just as puzzling if *most* people who eat reheated brussels sprouts get inflamed knees – or even if reheated brussels sprouts merely *raises the probability* of inflamed knees. Any such correlation would seem to call for a uniform explanation. It would be mysterious that reheated brussels sprouts should so much as increase the probability of inflamed knees, if the mechanism were different each time it did so.

Some readers may wonder whether an analytic functionalist account of human science concepts can resolve the puzzle. Analytic functionalism defines concepts in terms of causal structures. Thus it might be definitionally required that something only counts as an "S_1" if it gives rise to an S_2. For example: something might only count as a "pain" if it leads to efforts to avoid the source of the pain; something might only count as "inflationary pressure" if it generates a fall in the value of money; and so on. Given this kind of definition, it will scarcely be a surprise if many different physical kinds P_1 realize S_1 and yet all give rise to a $P^\star_1$ that determines S_2. After all, if they didn't do this, then they wouldn't count as realizations of S_1 in the first place. Something that doesn't generate avoidance behavior just isn't a "pain"; something that doesn't lead to a fall in the value of money isn't an "inflationary pressure"; . . . so, given this, it will be inevitable that all S_1s will lead to S_2s, notwithstanding their variable realization, for that's what it takes to count as an "S_1".

Unfortunately, nothing in this line of thought helps explain variably realized human science *laws*. It may explain how definitional truths can be variably realized, but that is a different matter. Genuine laws can be expressed by synthetic statements with the antecedent definitionally independent of the consequent, as opposed to the analytic truths that result when "S_1" is defined as a precursor of S_2. And that is precisely why there is a puzzle about their variable realization. Given that the antecedent circumstance S_1 in a genuine law can be identified independently of whether it produces the consequent S_2, we expect there to be some further account of why such S_1s are always (or at least unusually often) followed by S_2s – and that is what the variable realization seems to preclude (Millikan 1999).

4   Kinds of Kinds

Despite the points made so far, it may seem that there can't really be a problem about variable realized laws as such. After all, surely there are plenty of familiar examples of such laws. What about the law that a temperature of 100°C will make water boil? Aren't there many different molecular movements that can realize a water temperature of 100°C? Yet there clearly isn't any puzzle about why we find the boiling in all these cases.

But this is a different kind of set up. To see why, we need to be a bit more explicit about the idea of "variable realization." For a category S to be variably physically realized, it isn't enough that the instances of S display *some* differences at the physical level. We wouldn't want to say that being square, say, is variably physically realized just because different square things have different masses. Nor should we say that being in pain is variably physically realized just because different people have different-sized C-fibres. For a category S to be genuinely variably realized, the requirement is not the weak demand that there be some physical differences between the Ss, but rather that there should be *no* physical property that is peculiar to them. The members of a genuinely variably realized kind will share no physical property that is not also shared with non-members.

With temperatures, there is of course a common physical property of the right kind. All samples of water at a given temperature have the same mean molecular kinetic energy, notwithstanding any further differences between the specific motions of their constituent molecules. And that is why there is no puzzle about why water boils at 100°C. Despite the different molecular motions involved, all water at 100°C shares the same mean molecular kinetic energy, and this allows a uniform physical explanation of the boiling. By contrast, if there is no common physical feature to a category, then there is no room for such a traditional type–type reduction of any patterns it enters into.

Might Fodor just be saying that human science categories are like temperature? That is, might he simply be pointing out that there can be physical differences between different instances of some human type, like circulating money, just as there are differences between different samples of water at 100°C, and that this is consistent with their having some physical commonality that will explain why they fit into some uniform pattern?

But this suggestion is not consistent with other claims Fodor makes. Thus consider his original response to the obvious query raised by his diagram: why isn't the disjunction $P_1 \vee P_2 \vee P_3 \ldots$ a physical property with which S_1 can be type-identified, thereby yielding a traditional physical reduction of S_1? Fodor's response is that even if we can formulate this disjunction, it won't represent a genuine physical *kind*, as opposed to a heterogeneous collection of different physical kinds. Correspondingly, even if we can write down the generalization $P_1 \vee P_2 \vee P_3 \ldots \rightarrow P^\star_1 \vee P^\star_2 \vee P^\star_3 \ldots$, this won't constitute a genuine physical *law*, as opposed to a representation of a bunch of different physical processes. There is of course an element of circularity here, in that the standard explications of kinds is that they are categories that figure in genuine laws, while the standard explications of laws is that they are patterns that involve genuine kinds. But any such circularity doesn't affect the point currently at issue, which is that Fodor is explicit that there is no single physical kind that characterizes all instances of his human Ss.

5   Dilemma for Fodor

Given the points just made, the challenge facing Fodor can be put in the form of a simple dilemma. If the realizations of human S_1 and S_2 are genuinely variable and don't form kinds, then doesn't this immediately imply that the empirical generalization $S_1 \rightarrow S_2$ won't be a law, but rather a collection of heterogeneous processes? Alternatively, if the realizations of S_1 and S_2 do form kinds, doesn't this mean that $P_1 \vee P_2 \vee P_3 \ldots \rightarrow P^\star_1 \vee P^\star_2 \vee P^\star_3 \ldots$ will be a genuine law that constitutes a traditional reduction of $S_1 \rightarrow S_2$ (Kim 1992)?

Fodor responds to this putative dilemma in his splendidly-named "Special Sciences: Still Autonomous After All These Years" (1997). He argues that the dilemma begs the question. True, he allows, human categories aren't identical to *physical* kinds, and so any generalizations involving them won't be *physical* laws. But that's not decisive, he insists. For it is still possible that these categories constitute *human* kinds, in virtue of entering into *sui generis human* laws. Fodor takes it to be a datum that psychology, economics, and the other human sciences contain genuine laws covering categories that can't be type-reduced to physics. Given this, he concludes that the categories of such sciences are *kinds* all right, in virtue of entering into these human laws. From this perspective, the brussels sprouts example is misleading: it appeals to our intuitive knowledge that there is no real law in the case and that *reheated brussels sprouts* is thus not a medical kind. By contrast, Fodor suggests, in areas where there are real laws covering physically heterogeneous categories, like psychology and economics, we have every reason to ascribe kindhood to those variably realized categories.

At first pass, this response may seem reasonable enough. There is no immediate reason why the only laws of nature should be physical laws.

After all, it is clearly consistent with supervenience physicalism that there should be a finite few cases in which, say, eating reheated brussels sprouts lead to inflamed knees via disparate physical processes. So there can scarcely be any outright contradiction in supposing that such a variably realized pattern should be repeated indefinitely.

However, note that human categories don't just enter into laws connecting them with other human categories. They are also systematically related to *physical* categories. For example, a *drought* in cocoa-producing areas will raise the price of chocolate. Economic growth without environmental regulation will lead to an *increase in atmospheric CO_2*. And so on. (Indeed this kind of interaction was implicit in our original rationale for physicalism: it was precisely in order to explain how human facts *can influence the material world* that we needed to credit them with a physical nature in the first place (Papineau 2002, ch. 1).)

But this now reinstates the dilemma once more. If human categories are going to feature in physical laws, then doesn't this mean that the disjunction of their physical realizations will itself need to be a physical kind? If physical kinds are just those categories that feature in physical laws, then it follows that the human categories that feature in physical laws must be type-identical with physical kinds after all.

We can make the point graphic by considering situations where a supposedly variably realized category has some uniform physical cause and physical effect. For example, if someone *ingests alcohol*, this will engender inebriation, and this will lead to *slower reactions*. But now suppose, for the sake of the argument, that the category of inebriation is not physically reducible. Then there will be quite different physical processes mediating between the initial physical cause and the final physical effect. What then ensures that all these different intermediary processes converge on the same final effect? It is not as if the inebriation can exert some independent causal influence to bring this about – that would require interactive dualism and "causal gaps" in the physical realm. Rather, the causal influence of the inebriation in each instance is exhausted by the causal influence of its physical realization. But then we seem to be left with a mystery. We are supposing that the initial cause, ingesting alcohol, generates a divergent range of intermediary neurological effects. But why then should these inexplicably converge on the same physical result, the slower reaction times?

6   Methodological Consequences

In the last four sections I have been arguing, contra Fodor, that any laws in the human sciences must be reducible to physical laws. Let us suppose for the moment that I am right about this. Does it matter?

One reason why it might be thought to matter is that it would under-mine the independent authority of the human sciences. Fodor's ter-minology of "autonomy" suggests that the human sciences will be threatened as independent academic disciplines if their categories are reducible to those of physics. The worry is that type-reduction would mean that any human laws would simply be special cases of the physi-cal laws that reduce them, and the human sciences therefore little more than sub-departments of physics.

Still, is this a serious worry? There is of course a sense in which the reducibility of some human science means that it is not independent of physics – in principle its laws will follow from physical laws. But this in principle possibility need have no practical implications. For the in principle derivability may be practically unfeasible, in which case the reducibility of the human science will make no methodological differ-ence to its practitioners. They will still proceed to investigate the rel-evant special laws using direct empirical evidence. This is surely how it goes in many science departments. Nobody doubts, I take it, that chemical, meteorological or geological laws have uniform physical explanations. But at the same time nobody tries to derive these laws from basic physics, at least once we are dealing with systems more com-plex than the hydrogen atom. Instead special scientists investigate the relevant complex systems directly, using observation and experiment to ascertain the laws they obey – which is why we have separate chemis-try, meteorology and geology departments in universities. By the same coin, a purely in principle requirement of reducibility to physics would seem to leave plenty of room for human sciences that in practice owe nothing to physical theory.

However, a rather different methodological worry can also be occa-sioned by a requirement of in principle reducibility to physics. This isn't Fodor's worry that the human sciences won't be able to call their laws their own, if they are reducible to physical laws. Rather it is the converse worry that the human sciences won't have any laws in the first place, if human laws need to be reducible to physical laws.

To understand this threat more fully, it is helpful to see Fodor as mak-ing two claims. First, the categories of the human sciences are variably realized at the physical level. Second, even though physicalism is true, this variable realization doesn't preclude serious laws in the human sci-ences. I have disputed the second claim, arguing that that variable reali-zation is indeed incompatible with human laws. But I haven't queried the first premise. And this may well stand on its own. Even if we disagree with the rest of Fodor's position, it can seem plausible that the categories used in the human sciences are variably realized at the physical level. But put this together with my claim that variable realization is incompatible with human laws. It will follow that there are no human laws.

Why should we suppose that the categories of the human sciences are variably realized at the physical level? The standard argument is that such categories are normally functional types, constituted by a structure of causes and effects,[3] and that different physical states will fill this causal role in different instances. Thus people who share the belief that Britain is an island, say, will all have some internal state that is derived from relevant information and will cause appropriate behavior – but there is no reason to suppose that the same *physical* state will carry this information in all those individuals. Again, to take Fodor's example, something counts as money if it plays a certain economic role – but this doesn't require that all examples of money have a common physical nature.

Some thinkers will be unperturbed at the suggestion that there are no laws in the human sciences. After all, a whole tradition holds that the "Geisteswissenschaften" are distinguished from the natural sciences in just this way: where the natural sciences seek to *explain* events by bringing them under empirical laws, the Geisteswissenschaften do not deal in laws, but yield *understanding* by discerning the meaning of actions. However, it would be premature to conclude at this stage that the advocates of Verstehen over Erklären have been vindicated.

For one thing, the above argument against the possibility of human laws is by no means conclusive. Let us formulate this argument explicitly.

Premise (I) The categories of the human sciences are variably realized.

Premise (II) Physicalism implies that all laws must be reducible to physical laws.

Conclusion (III) There are no laws in the human sciences.

If the two premises are given, then it will indeed follow that there are no human laws. However, neither of the premises is incontrovertible. As we shall see below, both of them are subject to important qualifications, and these qualifications will mean that there is room for laws in the human sciences after all.

Moreover, there are of course more immediate grounds for thinking that the above argument against human laws must be too strong – namely, that the human sciences can offer many plausible examples

---

[3] Note that this functionalist view of the *nature* of human states is different from the "analytic functionalist" view discussed earlier about the *definition of terms* for human states. The two views are quite independent of each other.

of actual laws. If the above argument were sound, it would not only rule out strict exceptionless laws in the human sciences, but also any kind of systematic correlation. (As I pointed out earlier, any kind of positive correlation between brussels sprouts and inflamed knees, and not only a perfect association, would seem to call for a uniform physical explanation.) And it is very hard to deny that the human sciences can sometimes give us systematic correlations. Psychologists, economists, sociologists, and political scientists will all have their favourite examples. To take just one that has already featured in this chapter, what about Gresham's Law and the tendency for bad money to drive out good?

It will be worth pausing here to say a bit more about this particular example, as this will help to distinguish the two different ways in which the above argument against human laws can be evaded. At first sight it might seem as if Gresham's Law is a counter-example to premise (II): isn't it a clear case of a human law that is variably realized at the physical level, so cannot be reduced to a physical law? However, this would be too quick. As we shall see below, there are indeed cases of genuine human laws that are variably realized and so contra-exemplify premise (II). But it is doubtful that Gresham's Law is itself such a counter-example.

I am not here querying whether Gresham's Law is a genuine law. Rather, I doubt that it is variably realized (i.e., maybe Gresham's Law violates premise (I) rather than (II)). At first sight this might seem an odd suggestion. Fodor takes it to be obvious that Gresham's Law is variably realized at the physical level, on the grounds that *money* can have quite different kinds of physical embodiment. But this is a relatively superficial feature of the example. For note that even so there is a uniform *psychological* reduction of the category of money: X is money if and only if the population in question expects to be able to exchange X for goods and services. Moreover, this kind of psychological reduction allows an obvious uniform explanation of Gresham's Law in terms of the psychological principles of decision theory: people would rather hold on to the form of currency about which they have higher such expectations. From this perspective, the variable physical make-up of different forms of money becomes a superfluous detail that can be ignored in the psychological explanation of Gresham's Law. True, this only gives us a uniform reduction of Gresham's Law to *psychological* decision theory, and not yet to *physical* theory. Still, as we shall see below, there is reason to suppose that the relevant parts of human psychology will themselves have a uniform physical explanation. If this is right, then Gresham's Law will in principle be reducible to physics, thus undermining premise

(I) above, rather than a variably realized counter-example to (II), of the kind Fodor has in mind.

7   Selectional Laws

In section 9 I shall come back to Gresham's Law and the idea that some human laws may be in principle reducible to physics. But first let me explore the other way in which the argument against human laws might break down. Even if Gresham's Law is not a good example, there are other cases of genuine laws that are variably realized at the physical level and so do violate premise (II).

As a number of writers have observed (Macdonald 1992; Block 1997; and Papineau 1985, 1992), one possible way in which variably realized laws might arise is as the result of *selection processes*. Consider this example. In all electrical hot water heaters, the current is switched off at some temperature below boiling point. But when we look at the physical process that mediates between the high temperature and the switching off, we find that it is different in each case. Each heater contains a thermostat, but there are many different kinds of thermostat, each using different physical components in different combinations (including bi-metallic strips, expansion gases, mercury bulbs, and thermocouples).

Given this, we can imagine someone asking why so many different physical processes should all have the same effect – namely breaking the circuit. If there is no uniform physical explanation for this commonality, is it not a mystery that all the divergent effects of temperature increases should converge on this single effect?

But here of course there is an obvious answer. All these different physical processes were *designed* to produce the same effect. The people who construct heating systems make sure they contain a thermostat. They want a device that will shut off the current when the temperature gets too high, and any of the different thermostats on the market will serve for this purpose. That's why we can have a genuine law with physical antecedent and consequent even though the intermediate process is variably realized. Designers want the antecedent to produce the consequent and there are different ways of achieving this.

I have illustrated the point with an example of human design, but the point generalizes. There are other selection processes in nature apart from conscious design by intelligent agents, such as the intergenerational selection of genes, or the selection of cognitive and behavioral elements in the course of individual and social learning. These selection processes can also give rise to variably realized laws.

Take the paradigm of a putatively variably realized category – cross-species *pain*. It is widely supposed that pain is variably physically realized across different life forms: pain involves quite different physical processes in octopuses and humans, say. Yet the category of pain nevertheless enters into cross-species laws mediating between physical causes and effects, such as the law that bodily damage gives rise to pain and the law that pain in turn leads to avoidance of the source of the damage. Here too there would be an obvious answer if someone asked why all the disparate physical processes caused by bodily damage have the same effect. Natural selection favours organisms that have *some* mechanism that mediates between bodily damage and the avoidance thereof. It doesn't care too much about how this is done. Or, to speak less metaphorically, natural selection will foster any mechanism that plays the pain role within a given species. This is why pain mechanisms can be different across different species, yet all underpin the same damage-avoidance law.

Here is another example. Animals who maintain individual territories will respond to the presence of conspecifics with some territorial display that makes the invaders retreat. Here there is a regular antecedent–consequent pattern – invasion followed by retreat – but the displays that play the intermediary role in this pattern will vary widely from species to species. But once more the explanation is clear enough – natural selection will encourage any display that plays this role, even if it is different from species to species.

We can expect something similar in the human realm. Grown-up human beings respond to untied shoelaces by tying them. Yet they have different ways of doing this – the only common feature is that they get the shoelaces tied. How come all these different responses to untied shoelaces produce the same effect? Again the answer is obvious enough. Humans learn in large part by trial and error. If they light by chance on some behavior that produces a successful result, then they will persist in this behavior. That's why different humans end up with different ways of tying shoelaces. Learning ensures that they will find some way of doing the job, but it doesn't mind how exactly they do it.

Many other examples offer themselves. Most mature humans will have some way of recognizing and thinking about common objects (cats, dogs, telephones, bicycles) but there is no reason to suppose they use the same brain states to achieve this. Most mature humans will have some technique for solving common intellectual problems (numerical addition, planning tomorrow's activities, balancing their budgets) but these will vary across individuals. Most mature humans will have some way of putting others at ease, but they won't all do this the same way. And in general people with shared ends will generally figure out some way of achieving their common aim, but will light on different means of doing this (Millikan 1999).

In all these cases, the variability of the means which lead to some given result can be explained by selection processes operating during individual and social development. Humans and other complex animals are learning machines. They embody a hierarchy of processes that operate at many different levels to preserve items that produce such-and-such effects. These items may well be physically different in different individuals, but this won't matter to the selection mechanisms, provided they produce the reinforcing effects. So the means by which the effects are produced will be variably realized at the physical level across different individuals.

8   The Limits of Selection

How seriously should we take the kinds of variably realized patterns that can result from common selective pressures? Do they have the same standing as normal scientific laws?

One possible worry is that selection-based patterns are not precise enough. After all, pains don't always lead to avoidance of the source of damage, territorial displays don't always succeed in repelling invaders, and untied shoelaces don't always get tied. These regularities look more like rules of thumb that anything worth dignifying with the name of "law."

I don't think that this is a decisive reason for downgrading selection-based patterns. There are surely plenty of sciences in good standing whose laws need to be understood probabilistically or as *ceteris paribus* claims. This was why the problem I originally posed for Fodor's picture was not how variably realized kinds can enter into exceptionless laws, but rather how they can so much as figure in projectible correlations. And the selection-based patterns from the last section certainly amount to projectible correlations. They carry information about as-yet unobserved cases, and they support counterfactuals. (Any damaged animal will respond by avoiding the source of the damage; if some animal were damaged, it would avoid the source of the damage ...) These projectible patterns may be a lot less precise than the fundamental laws of physics, but they still display the characteristic properties that distinguish genuinely projectible patterns from merely accidental regularities.

However, selection-based patterns might be argued to be deficient in a rather different respect. The laws that arise from selection processes tend to stand on their own, rather than fitting into networks of interrelated laws. In this respect they contrast with laws involving paradigm examples of natural kinds, which enter into *lots* of laws, not just single ones.

For paradigm natural kinds, we can project a wide range of properties. Thus, chemists can study many properties of gold: its density, color, melting point, electrical conductivity, and so on. And this hinges on the fact that all samples of gold have a uniform physical realization. It is precisely because all gold has the same atomic structure that there are many different further features that all samples of gold have in common.

The point isn't restricted to basic chemical kinds, but applies to any kind with a uniform physical realization. For example, there are many general truths about chicken pox: its gestation period, characteristic symptoms, ease of transmission, susceptibility to various drug treatments, and so on. Again, it is because of a common structure at the physical level that we are able to assume that all these different features will hold good across different instances of chicken pox.

This kind of multiple projectibility will not apply to the variably realized kinds that enter into selection-based patterns. Take cross-species pain, considered as a category that is variably realized in different species. This enters into the law that pain leads to damage avoidance, as this is part of the role for which pain mechanisms are selected. But there is no reason to expect that the category of pain will enter into any further laws. Thus there won't be any cross-species laws about the sensitivity of pain mechanisms to stimuli, their susceptibility to analgesics, or the time it takes pains to abate. Precisely because the physical basis is different, such things will vary across different species.

The same point applies to other variably realized categories. There is no cross-species science of territorial behavior, nor any cross-person science of shoelace tying or bicycle recognition. And this is precisely because these categories are variably realized. We can say that in general territorial behavior will tend to repel invaders, but the fact that different species repel invaders in different ways blocks any other generalizations about territorial behavior as such. The same goes for shoelace tying and bicycle recognition. We know that all normal people can do these things, but there are no further general facts about the means they adopt, precisely because the means vary across individuals.

We can emphasize the point by comparing variably realized categories with some of their more specific instantiations. Take human pain, as opposed to cross-species pain. It seems highly plausible that pain is underpinned by the same physical realization in all humans, even if it is variably realized across species. (Remember that this doesn't require that there are *no* physical differences between individuals' pain mechanisms – just that there is enough physical commonality to yield uniform physical explanations of patterns involving pain.) Given this uniform realization, it makes perfect sense to investigate the many properties of human pain as such (sensitivity to stimuli, effective analgesics, and so on). Again, there would seem to be no barrier to a complex of laws about the territorial displays of some particular bird species – goldfinches, say – covering triggers to aggressive behavior, song patterns, seasonal variation, and many other things. Here too there are many laws because the physiological basis of the behavior is presumably constant across goldfinches. There could even be a range of general truths about a particular individual's shoelace tying or bicycle recognitions, given that there is likely to be a uniform physical basis for these abilities within any given individual.

Biologists distinguish between *analogous* and *homologous* traits. Analogues are independently derived products of convergent evolution that serve a common purpose, like the wings of insects and birds. Homologues are traits that share a common descent, even if they now serve divergent functions, like the flippers of seals and the hands of humans. The last few paragraphs explain why homologous categorizations are standardly taken more seriously by biologists than analogous ones (Brigandt and Griffiths 2007). Analogues do enter into common patterns, but they are once-off selection-based patterns. Both insect and bird wings lead to flight, but beyond that there is not much they have in common, because they have no common underlying physical basis. Homologues, by contrast, will be physically similar, even if they serve divergent functions, and because of that they will share a wide range of further developmental, structural, and other similarities.

9   Human Sciences

Where does this leave the human sciences like psychology, economics, and political science? Does the fact that variably realized categories fail to figure in multiple laws impugn the status of these disciplines as sciences?

If these disciplines could uncover no other laws apart from the once-off patterns that arise from selection mechanisms, then they would be sciences in only a weak sense. They wouldn't be able to boast the rich patterns of interconnected laws that are characteristic of other sciences.

Still, it is by no means obvious that the human sciences are stuck with this status. This would be their fate if all the categories that they work with were variably realized at the physical level. But there is no good reason to assume this. On the contrary, it seems highly likely that many of the categories that matter to these sciences are uniformly realized at the physical level within humans, even if they are variably realized across other species.

I have already made the point in connection with human pain. There is every reason to suppose that the pain mechanism is uniformly realized across humans, and that as a result there will be a rich nexus of laws about human pain. The same applies to many other cognitive abilities. Sensory mechanisms in general are uniformly realized across humans, which is why there is a substantial set of laws about human perception. The basic mechanisms that underpin human learning are physically similar across humans, which is why we have a wide range of generalizations about it. Again, it seems plausible that the basic mechanisms of reasoning – the processes that govern interactions between learned and other cognitive states – will be uniformly realized in all humans, and that here again we can expect a serious collection of generalizations about human reasoning.

It should not be supposed that the only attributes that are uniformly realized in humans are those that are genetically determined. Many of the physically uniform processes that occur in human ontogeny will hinge on interaction with environments as well as on common genetic endowment. (This may well include interaction with other humans as well as with the physical environment.) The question at issue is whether the overall developmental process produces a uniform physical structure, not whether this structure is determined by the human genome on its own.[4]

To the extent that human categories are uniformly physically realized, they will function as scientific kinds in a full sense. There will be a wide range of projectible general truths about various facets of human pain, human vision, and human learning. Moreover, to the extent that subjects such as economics and sociology formulate generalizations that depend only on the basic structure of human reasoning, rather than on variably realized learned states, we can expect them to deal with complexes of interrelated generalizations too. It is plausible that many of the principles of economics, political science, and social choice theory will fit this bill. As with Gresham's Law, the relevant generalizations will depend on the fact that people value certain things to certain degrees,

---

[4] Some philosophers explicate "innate" as "a product of normal development that is not due to learning" (Samuels 2002). If we assume that the products of learning are generally not uniformly physically realized, for reasons indicated in previous sections, then anything that is physically uniform across humans will need to be "innate" in the suggested sense, since it is not due to learning. However, it is highly controversial whether "due to normal development but not learning" is a legitimate reading of "innate" (Mameli and Papineau 2006).

and that they have certain expectations, but these will be independent of what exactly it is that they value, be they cowrie shells or silver coins.

Still, the human sciences often aim to go beyond matters that are uniformly realized within humans. They don't just study the upshots of sensory and other basic cognitive mechanisms. They also aim to generalize about the varied products of these mechanisms, including the many different things that people learn about.

For instance, sociologists will generalize about the way that dispersed empires keep bureaucratic records, political theorists about the way that democracies avoid famines, social psychologists about the way that people recognize and defer to authority. And here things will work differently. The patterns observed in such cases will not be the manifestation of common physical structures, but of similar selective pressures operating in different contexts. The humans involved will have been shaped to achieve the same results, but they will often have different ways of doing so. There are different ways of keeping bureaucratic records, of avoiding famines, of identifying people who wield authority, and so on. And this will limit the range of general truths we can expect to find in such cases. We might be confident that certain categories of people will all have some way of achieving some end, but there will characteristically be little to say about the many idiosyncratic ways in which they achieve this.

Does all this mean that the human sciences are not really *sciences* in the full sense? I don't think that this is a particularly fruitful question to press. As we have seen, the subject matter of the human sciences contains both physically uniform cognitive mechanisms and variably realized selectional categories. Correspondingly, some human kinds will enter into a thick nexus of projectible laws and others into a few thin selection-based laws. Once we are aware of this, there seems little point in continuing to ask whether economics as a whole, say, is a "science." The answer is that it resembles a paradigm science like chemistry in some respects, but not others.

The more interesting issue is to figure out how much of the human sciences can be grounded in uniform physical mechanisms and how much depends on common selectional pressures. I have been writing as if the dividing line is reasonably clear-cut. But on reflection it is by no means obvious where it lies. This is because the subject matter of the human sciences is largely constituted by human cognition, and the role of learning and other selective processes in the ontogeny of human cognition is a highly disputed matter. I would say that this should be a central issue for those thinking about the methodology of the human sciences. If we want to know about the kind of general truths we can hope to find in the human sciences, it is crucial that we figure out which might rest on uniform physical mechanisms and which are the products of selection.

At first sight, the abstract metaphysics of physicalism may seem unlikely to have any concrete methodological implications for the practitioners of the human sciences. And it would certainly be a mistake to take the fact that human reality is physically constituted as an argument for trying to infer facts about humans directly from physical theory. Still, I hope I have done something to show that there are other ways in which the metaphysics of physicalism can matter to human scientific practice. Given any human scientific category, we mightn't need to *know* its specific physical make up, but it can still be very fruitful to ask whether it *has* such a uniform physical make up. In some cases, as with human pain, we can be confident that there is a uniform physical reduction, and therefore that it is sensible to seek a nexus of interconnected laws about human pain. In other cases, as with avoiding famines, it will be clear that there isn't any such uniform physical realization, and that any generalizations will at best be of the thin selection-based variety. Either way, judgments about the physical reducibility of human kinds can be a crucial guide to the prospects for further research.[5]

REFERENCES

Block, N. 1997. "Anti-Reductionism Strikes Back." In J. Tomberlin (ed.) *Philosophical Perspectives* 11: 107–132.
Brigandt, I. and P. Griffiths 2007. "The importance of homology for biology and philosophy," *Biology and Philosophy* 22: 633–641.
Charles, D. and K. Lennon 1992. *Reduction, Explanation and Realism*. Oxford: Oxford University Press.
Fodor, J. 1974. "Special sciences: Or the disunity of science as a working hypothesis," *Synthese* 28: 77–115.
    1997. "Special Sciences: Still Autonomous After All These Years." In J. Tomberlin (ed.) *Philosophical Perspectives* 11: 149–164.
Kim, J. 1992. "Multiple realizability and the metaphysics of reduction," *Philosophy and Phenomenological Research* 52: 1–26.
Macdonald, C. and G. Macdonald (eds.) (forthcoming). *Emergence*. Oxford: Oxford University Press.
Macdonald, G. 1992. "Reduction and Evolutionary Biology", In Charles and Lennon 1992.
Mameli, M. and D. Papineau 2006. "The new nativism: A commentary on Gary Marcus's *The Birth of the Mind*," *Biology and Philosophy* 21: 559–573.

---

[5] I would like to thank all those who responded to my talk at the Witten/Herdecke conference on Explanation in the Social Sciences in 2007, and Gabriel Segal, Mark Textor, and Nils Kurbis for later comments on this written version.

Millikan, R. 1999. "Historical kinds and the 'special sciences'," *Philosophical Studies* **95**: 45–65.

Nagel, E. 1961. *The Structure of Science*. New York, NY: Harcourt.

Papineau, D. 1985. "Social Facts and Psychological Facts." In G. Currie and A. Musgrave (eds.) *Popper and the Human Sciences*. Dordrecht: Nijhoff, pp. 57–71.

1992. "Irreducibility and Teleology." In Charles and Lennon 1992.

1993. *Philosophical Naturalism*. Oxford: Blackwell.

2002. *Thinking about Consciousness*. Oxford: Oxford University Press.

Samuels, R. 2002. "Nativism in cognitive science," *Mind and Language* **17**: 233–265.

4 – Comment
Reductionism in the Human Sciences:
A Philosopher's Game

Robert G. Shulman and Ian Shapiro

David Papineau contends that the possibility of science depends on there being uniformly realized phenomena that are reducible to physical laws, not merely the variably realized selection mechanisms that are characteristic of much social science. Some who share Papineau's view regard it as fatal to the possibility of social science on the grounds that their subject matter lacks the uniformly realized phenomena that he regards as necessary for science.

Papineau disagrees with this pessimism, asserting that, like "pain mechanisms," many cognitive abilities are uniformly realized across humans. As a result, there can be a "rich nexus of laws" about them – though he says nothing about what these laws might be. Not everything social scientists study exhibits what Papineau regards as the necessary reductive feature, but although he sidesteps any attempt to demarcate what he takes to be the scientific zone of the social sciences, he is confident that enough falls within its ambit to make the game worth the candle.

To the extent that human categories are uniformly physically realized, they will function as scientific kinds in a full sense. There will be a wide range of projectible general truths about various facets of human pain, human vision, and human learning. Moreover, to the extent that subjects such as economics and sociology formulate generalizations that depend only on the basic structure of human reasoning, rather than on variably realized learned states, we can expect them to deal with complexes of interrelated generalizations too. It is plausible that many of the principles of economics, political science, and social choice theory will fit this bill. (Papineau, this book, p. 120)

We regard Papineau's view as reflecting an overestimation of the reductive achievements of the social sciences and a flawed conception of what is required for science. There is a lot less out there in the social world that can be shown to meet his reductive criterion than Papineau seems to realize, but the criterion itself embodies a view of the scientific enterprise that should be rejected – and it is rejected by the great majority of practicing scientists.

1   Papineau's Overestimation of Reductive Achievements

Papineau is right that "sensory mechanisms in general are uniformly realized across humans" (Papineau, this book, p. 120). However, efforts by cognitive neuroscience have failed to show that cognitive, intentional, and emotional issues are similarly realized. Cognitive concepts like memory and attention are not realized in the same brain response even in the same individual. Responses to such concepts have been observed by brain imaging to depend upon their context (i.e., on whether the memory is of shapes, words, numbers, or faces, and whether presented aurally, visually, or by touch).[1] This means that sensory responses are not a suitable model for other mental processes despite the hopes that generated many imaging experiments.

One can, of course, always insist that the pot of gold is just behind the next tree. Perhaps the future will fulfill Papineau's hopes for a logical processing of information, but for the present we must agree with Jerry Fodor who, in response to results, changed his views and concluded that "the mind doesn't work that way" (Fodor 2001).

To relate the fact that the theory doesn't work as claimed in enthusiastic publications[2] one has to realize that the hopes for identifying rigorously specific brain responses with innate psychological modules have been replaced by fuzzy concepts showing up in broad brain regions. Imaging results, for example, claimed that a consort of registered Republicans voted "emotionally" because their amygdala was activated while a "Democratic" consort voted reasonably because of an activated pre-frontal cortex (Kaplan *et al.* 2007). These unspecific localizations raise the question whether the brain imaging results are as specific or as reliable as the standard assessment of the motivations of Republican and Democratic voters. It would not be the first time that political scientists have been awed by the apparent rigor practiced in other disciplines, which turn out on inspection to involve hopelessly vague operationalization and measurement of explanatory variables (Green and Shapiro, 1994).

2   Papineau's Misguided Reductive Expectations

Papineau prefaces his argument about the reductive requirements for science with the claim that the "basic physical forces are almost universally

---

[1] Interview with R. G. Shulman in *Journal of Cognitive Neuroscience*. See Schulman 1996.
[2] For a review see Van Eijsden *et al.* (forthcoming).

regarded as adequate to account for all material processes" (Papineau, this book, p. 104).

Most practicing scientists and almost all social scientists would be agnostic about that assertion. Moreover, they would recognize that, even if true, the connections between physical forces and social and political outcomes are so complex that few, if any, are likely to be identified any time soon – if ever. This is only partly for the reason we have already identified, to wit, that within even one individual the physiological bases of cognitive processes vary enormously. It is also because the great majority of what social scientists study concerns the interactions among individuals and groups of individuals. The observable behavior of voters, consumers, politicians, students, parents, criminals, soldiers, armies, parties, and nations, often depends vitally on what others do, or are expected to do in relation to them. If there are laws governing these interactions, trying to reduce them to physical processes would be like trying to link changes in stock prices to changes in the weather.

Papineau conflates his physicalist metaphysics with the epistemological claim that there should be no limit on what we can know about the physical bases of mental processes. One might well entertain the first while harboring skepticism about the second. Papineau addresses the question as to whether (macroscopic theories in) the human sciences are threatened by the authority of physical laws, and discusses how reductionist theories might avoid this loss of autonomy. His concern is shown by his discussion of the concept of heat:

> Compare the way that heat was reduced by the kinetic theory of gases, rather than eliminated. The kinetic theory showed that all the *supposed* effects of heat can be explained by the motion of molecules. But science didn't conclude that therefore there is no heat. Rather it said that heat is *nothing more* than molecular motion. (Papineau, this book, p. 104, italics added)

Papineau's disdain for "the supposed effects of heat" which "is nothing more than" etc. is countered by the realization that the laws about heat and energy have continued to be more and more useful, as, for example, the discoveries of electromagnetism and radioactivity extended their applicability. Thermodynamics measures and describes the macroscopic state of material systems. It was perfectly clear in the nineteenth century that there were microscopic states which might describe systems in terms of the properties of the molecular or atomic components. Yet as Gibbs said about thermodynamics "we do not mean a state in which each particle shall occupy more or less exactly the same position as at some previous epoch, but only a state that shall be indistinguishable from the previous one in its sensible properties" (Gibbs 1876: 228).

On what basis, other than admiration for the reducing powers of statistical mechanics, does Papineau patronize the macroscopic concept of heat? Even if we didn't have a statistical explanation of heat, with its many implications, the Carnot cycle would still describe the efficiency of steam engines, the free energy of a chemical reaction would still be calculable by thermodynamics, and the heat from a hot pot would still flow to a cooler hand with "nothing more" perhaps than a bad burn.

Empirical social science seeks properties of the macroscopic state. If we had theories about the properties of the microscopic states of the social sciences, whatever they might be – individual people, the brain, rationality, or economic self-interest, theories which had the physical reliability of the unifying theories now available for inorganic materials – they might support physicalism epistemologically as well as metaphysically. Of course we don't have those unifying theories but to the extent that the present understanding was empirically valid, it would be supported by them and we would see that understanding macroscopic properties is pretty damn useful.

At this point the philosophical argument loses relevance for us, a political scientist and a neuroscientist. The relevant reducing theories have not been found and we have no idea as to what they might be or how they might affect our empirical results. Instead we are more inclined to follow Niels Bohr's response to the loss of physical causality by the reducing theory of quantum uncertainty, when he said: "It is wrong to think that the task of physics is to find out how nature is. Physics concerns what we can say about nature" (Petersen 1963). Until a unifying physical theory tells us what more can be said about our sciences, we propose to continue studying phenomena at the macroscopic level. In the present absence of unifying theories our sciences are, like thermodynamics in the nineteenth century, working at the macroscopic level. Heat is really heat, not merely heat, and Gresham's law has value. Our sciences will continue to seek connections with all information, including that offered by physical science.

Papineau's formulation of the powers of the yet to be found reducing theories are not merely irrelevant to our fields. Insofar as they form a model for proposed reducing theories, they are inimical to contemporary research. Neuroscience is just beginning to recover from the futile use of its powers in the service of a unifying theory of the mind developed by Fodor and others, which assumed a computer-like brain that works by processing innate mental concepts (Shulman 2001).

Some social scientists have defined their enterprise by looking for the sorts of reductive feature that Papineau regards as defining the scientific enterprise. Utilitarians, Marxists, Freudians, structural-functionalists, rational choice theorists, and, most recently, those who are besotted by the apparent social-scientific possibilities of brain-imaging, have all sought to explain social and political phenomena by reducing them to one type of explanatory variable – though only the last of these have thought seriously about locating it in physiological processes. The results have been dismal. When advances in the social sciences have occurred, this has been via the systematic empirical testing of hypotheses against the most plausible alternatives, or by coming up with new or improved descriptions of dependent variables that facilitate better empirical evaluation (Green and Shapiro 1994: chs. 4–6 and Shapiro 2002).

Most practicing scientists are instrumentalists about theory – whether they are Popperians, Lakatosians, or Friedmanites.[3] They are less interested in where hypotheses come from than what they can explain. For the most part, researchers proceed inductively: looking for middle-level generalizations that account for the data better than existing theory.[4]

And with good reason. The idea that the certainty that accompanies theorems is the only hallmark of science is an obsolete hangover of the early Enlightenment. Descartes wanted it, and so did Kant, but few other philosophers have taken it seriously, and almost no empirical scientists do. Verification was displaced by falsification and pragmatism as the more tentative hallmarks of scientific advance as people came to recognize that all knowledge claims are corrigible, so that even the most accomplished scientists expect their work to be superseded by future discoveries. Papineau seems as oblivious of this as he is of what practicing scientists actually do. It's hard to see how anything he says will have any bearing on *that* – whether or not he is right about how much of the social sciences turn out to be reducible in his sense.

REFERENCES

Fodor, J. 2001. *The Mind Doesn't Work That Way: The Scope and Limits of Computational Psychology.* Cambridge, MA: MIT Press.
Gibbs, J. W. 1876. "Equilibrium of heterogeneous substances," *Trans. Connecticut Academy Sci. 3.*

[3] See Green and Shapiro (1994, ch. 2) for discussion of how their different philosophies of science nonetheless leave them in the same position of evaluating hypotheses by their empirical payoff.
[4] There are a few exceptions. John Nash famously came up with his equilibrium concept without having a single empirical application in mind, yet it has since become widely deployed in economics and political science. But this is surely an exception that proves the rule. One would be hard-pressed to find other examples.

Green, D. and I. Shapiro 1994. *Pathologies of Rational Choice Theory: A Critique of Applications in Political Science*. London: Yale University Press.

Kaplan, J. T., J. Freedman, and M. Iacoboni 2007. "Us versus them: Political attitudes and party affiliation influence neural response to faces of presidential candidates," *Neuropsychologia* **45**: 55–64.

Petersen, A. 1963. "The Philosophy of Niels Bohr," *Bulletin of the Atomic Scientists*, pp. 8–14.

Shapiro, I. 2002. "Problems, methods, and theories in the study of politics, or: What's wrong with political science and what to do about it," *Political Theory* **30**(4): 588–611.

Shulman, R. G. 1996. Interview in *Journal of Cognitive Neuroscience* **8**: 474–480.

   2001. "Functional imaging studies: Linking mind and basic neuroscience," *American Journal of Psychiatry* **158**: 11–20.

Van Eijsden, P., F. Hyder, D.L. Rothman, and R.G. Shulman 2009. "Neurophysiology of functional imaging," *Neuroimage* **4**: 1047–1054.

5   Complexity and Explanation in
the Social Sciences

Sandra Mitchell

> *The only foundation for the knowledge of the natural sciences is the idea that the general laws, known or unknown, which regulate the phenomena of the Universe, are necessary and constant; and why should that principle be less true for the intellectual and moral faculties of man than for the other actions of nature?*[1]

1   Introduction

There is an ongoing debate amongst philosophers of social science about the possibility of laws of the kind Condorcet recommends. The stakes are high, as laws are what science is said to search for and are at the core of traditional accounts of explanation. On the deductive nomological account of explanation (Hempel 1965) laws are required for the logical deduction of the explanandum, the statement of the event to be explained, from statements describing the antecedent conditions. To explain why a particular individual makes certain market choices, a law of the rationality of maximization is required. Alan Nelson, in discussing economics, points out further that "when we work with defective laws, we often wind up with defective accounts of facts" (Nelson 1986: 163). The logical representation of what counts as an explanation has come under strong criticism. It is characteristically replaced by claiming that explanations may not be derivations, rather they appeal to causes. To answer why something occurred, what is needed is to identify the cause of the event. Even with explanation by causes, laws enter the scene, as it is laws that describe the causal relations that hold between events in the world. Laws are, on the standard account universal, exceptionless, naturally necessary truths about our world, and it is to this type of law that Condorcet's question refers.

To answer Condorcet, in this chapter I will investigate what it is about the social world that makes the universal, exceptionless

[1] Condorcet, *Rapport et projet de décret sur l'organisation générale de l'instruction publique* (1792) G. Compayre (ed.), Paris 1883, p. 120 quoted in F. A. von Hayek (1941).

130

generalizations that are heralded as the foundation of knowledge of the physical world so elusive. I am not going to rehearse all the arguments for and against the possibility of laws in the social realm. What I aim to do is not to take either side of the debate, that is, not to say – "YES! Social science does have laws just like physics (or close enough anyway)" or "NO! Social science can never have laws like those of physics; knowledge of the social has a wholly different character." Rather I will suggest replacing the standard conception of laws that structure the debate with a more spacious conceptual framework that not only illuminates what it is about knowledge of the social that is similar to knowledge of the physical, but also explains what is so different in the two scientific endeavors.

The route I will take involves a detour through understanding the character of knowledge claims in biology, as my research has focused on related arguments about the absence of laws in biology. Having laid out the position in that domain, I will then suggest how it would naturally be extended to considerations of causal explanation for social science. I will argue against a common way of modifying the standard account of strict laws by means of the addition of a *ceteris paribus* clause in order to squeeze the non-universal, exception rich causal structures that populate both the biological and social domains under the cloak of scientific lawfulness. The types of knowledge gained of the social world are much like the types of knowledge we can claim of the biological world. The language of "strict law" and "*ceteris paribus* law" that have been imported from their home in the analysis of fundamental physics fail to adequately represent the relationship between the types of knowledge, and the methodological consequences for acquiring and using knowledge that distinguish fundamental physics from evolutionary biology or from economics or political science.

Like knowledge of the social world, biological knowledge does not appear to fit the image of scientific law advocated by many philosophers. As a consequence, it has long been argued that biology has no laws (Smart 1968 and Beatty 1995). Yet, biologists speak of "laws" in their writings. One of Mendel's "laws" claims that with respect to each pair of alleles at a locus on the chromosome of a sexually reproducing organism, 50% of the organism's gametes will carry one representative of that pair, and 50% will carry the other representative of the pair.

Why do some philosophers choose not to count these results of biological investigation as laws? How are they different from Proust's law of definite proportion, or Galileo's law of free fall or the conservation of mass-energy law? Those who argue that there are no laws in biology point to the *historical contingency* of biological structures and the *complexity* of biological causation as grounds for excluding the law designation. These are also identified as two of the culprits in the parallel argument against the possibility of social scientific laws (McIntyre 1996 and Kincaid 1996).

I believe that we need to think about scientific laws in a very different way – to recognize a multidimensional framework in which knowledge claims may be located and to use this expanded framework to explore the variety of epistemic practices that constitute science. Dichotomous oppositions like "law vs. accident" and "necessity vs. contingency" produce an impoverished conceptual framework that obscures much interesting variation in both the types of causal structures studied by the sciences and the types of representations used by scientists.

In defending a multidimensional account of scientific knowledge I will expose limitations of traditional philosophical analyses and representations of knowledge of causal structures in nature in the hopes of showing how a different sort of enterprise promises a better understanding of the diversity of scientific beliefs and practices.

2   Laws: the Traditional Account[2]

There is general agreement that laws allow us to explain, predict, and successfully intervene in the world. The features which are alleged to permit them to accomplish these functions are:

(a) logical contingency (have empirical content);
(b) universality (cover all space and time);
(c) truth (exceptionless); and
(d) natural necessity (not accidental).

Traditionally, philosophers represented scientific claims originally formulated in either natural language or mathematical formula in some formal logic. Facts are translated into propositional claims and laws are rendered as universally quantified conditionals (or some properly modalized version of such). $(x)(Px \rightarrow Qx)$ (for all x if x has the property P then x has the property Q) is the familiar reflection of a scientific law in this schema. The functions of laws (i.e., explanation and prediction) are then rendered as deductive (or sometimes inductive) patterns of inference from the suitably formalized law statements to suitably rendered fact statements simply by binding the variable "x" to some object in the world.

The defining characteristics of laws as they are traditionally understood and the standard ways of representing them have blinded us to

---

[2] These arguments are developed in my paper (2000).

important features of scientific knowledge. Part of the problem is the dichotomous character of the representational framework of standard logic. Statements are either true or false, and the truth of a statement either follows necessarily from the truth of some other statements (in virtue of its form) or it does not (i.e., it is only contingently true). As representations of knowledge of our world, when we are given only two options – law or non-law – the knowledge that scientists acquire about the causal structure of the biological and social worlds invariability ends up in the non-law box.

The first and third criteria – that laws are empirical and not logical truths about our world – are warranted features of the knowledge scientists seek to discover. However, some have challenged the first require ment, arguing that mathematical truths can be laws, for example the Hardy-Weinberg law of equilibrium in population genetics (Sober 1997 and Elgin 2003) and the use of idealized models in prediction and explanation also raise questions for the truth of explanatory claims. However, these are not the features of lawfulness for which I am concerned in this chapter. Rather, what is a particular concern for both explanatory claims in biology and the social sciences are the requirements of universality, exceptionlessness, and necessity.

Universality is standardly represented by the universal quantifier (x) in (x) (Px –> Qx). The scope of the quantifier is taken to be all space and all time. Once the scope of the law is understood as universal in this broadest sense, then it is clear that a true law will permit no exceptions. That is, any point in space/time that is described as Pa and ~ Qa excludes the general claim from qualifying as a universal, true law.

So-called accidental generalizations are a problem for the standard account of laws. "All gold spheres are less than a mile in diameter" and "All uranium spheres are less than a mile in diameter" both satisfy the formal requirements of lawfulness. However, only the second is deemed a genuine law, the first being merely accidentally true. Many have appealed to some form of natural necessity to distinguish them. Indeed natural necessity is supposed to account for the explanatory power of laws, allow confirmation by instances and permit laws to support counterfactuals. The presumption that laws are naturally necessary begs the epistemic question of what counts as evidence and warrants the detachment of the claim from the instances of evidence that typically confirm a generalization. If a relationship between a cause and effect is thought to be lawful, then having garnered sufficient instances in support allows one, on the assumption of necessity, to expect that relationship to be true of all naturally possible instances – past, future or imagined – independent of the features of the context of the confirming instances.[3]

A problem derives from thinking about natural necessity as isomorphic to logical necessity. Logical necessity carries the strongest possible warrant from truth of premises to truth of conclusion – the conclusion could not be false. A similar, though not identical, kind of warrant is desired to carry one from occurrence of cause to occurrence of effect in the expression of laws about the natural world – the effect could not be otherwise. Thus if the predicted effect does not occur, the generalization is not truly a law. It lacks natural necessity.

Modeling natural necessity on logical necessity carries with it the presumption that the latter, like the former, is an all-or-nothing property. Logically, a statement is either necessary or contingent. So nomologically a relation between two events in the world is taken to be either necessary or accidental. The dichotomous character of logical truth/falsity and necessity/contingency is mirrored in the empirical truth/falsity and nomological necessity/accidental contingency.

This leaves no place, except the vast category of non-laws, in which to locate a generalization that describes a strong causal relation between events yet fails to exhibit the strongest conditions of nomological connection. Mendel's law of 50:50 segregation pertains to contingently evolved organisms and, even so, has exceptions among those. Thus, on the traditional account, it fails to satisfy the strong warrant attached to strict, necessary laws. It follows from judging biological generalizations by this conception of what it is to be a law, that biology has no laws. The same inference holds for true generalizations that capture the contingent causal relations studied by the social sciences.

A question may be raised at this point as to what the philosophical enterprise of providing an account of laws of nature aims to accomplish. I believe we should begin with what science has discovered about our world that allows us to explain, predict, and successfully intervene. It is clear that scientists, at least sometimes, use the language of laws to capture the causal patterns detected in or justified by the results of individual observations and experimental set ups they investigate. In general, what is required for usable knowledge is some claim that one can detach from the particulars of a given observational or experimental situation and export to other contexts. What kinds of information may be used for this purpose? The best case, the ideal kind of information, would be applicable to all contexts outside of the evidentiary ones. It seems to me that this is the type of law described in the standard account. A law

---

[3] See also Goodman (1966) for the relationship of laws, accidents, and induction.

is universal, exceptionless, and necessary, and hence is guaranteed to apply everywhere and for all time. This type of claim can function to predict, explain and allow us to successfully intervene. Yet when one looks to the actual products of scientific practice, one is hard pressed to find examples that fit that ideal image.

That is not to say that strict laws cannot be found. Earman *et al.* (2002) have argued that:

... typical theories from fundamental physics are such that *if* they were true, there would be precise proviso free laws. For example, Einstein's gravitational field law asserts – without equivocation, qualification, proviso, *ceteris paribus* clause – that the Ricci curvature tensor of spacetime is proportional to the total stress-energy tensor for matter-energy; the relativistic version of Maxwell's laws of electromagnetism for charge-free flat spacetime asserts – without qualification or proviso – that the curl of the E field is proportional to the partial time derivative, etc (1999: 446).

But claims in biology and the social sciences, Mendel's law of gamete segregation being 50:50, or the "Law of Supply" that quantity supplied is directly proportional to price do not fit the ideal image of a law. In the biology example there are cases of meiotic drive, or segregation distortion in which some genes get greater representation in the gametes than others. In economics, there are externalities, or issues of public goods, that constitute exceptions to the law of supply. Instead of universal, exceptionless, necessary laws, what one finds in scientific practice is a range and variety of models, explanations, and theories that provide us with the tools for explaining, predicting, and intervening in our world.

Two questions arise given the mismatch of the standard view of law and the facts about the scientific practices of biology and the social sciences (and much of physics). First, why doesn't the knowledge discovered of biological and social systems fit the ideal model? Second, what conclusions should we draw? In the end, I will endorse a pragmatic strategy that directs one to look at what can do the job or perform the function that laws under the standard description are supposed to do – and then ask whether there are other types of knowledge claims that can perform that same function. If there are, then we have two options. We can define a law functionally as a claim that permits explanation, prediction, and intervention. In that case, I would argue, claims that are less than universal, exceptionless generalizations can still qualify as laws. Or we can demote the significance of laws as the only knowledge claims that do the scientific work we are interested in. Indeed, given that strict laws are as rare as hen's teeth, interest in finding them becomes an esoteric task, rather than the defining goal of scientific practice. The important question, in either case, is how less than universal, exceptionless generalizations can do the jobs of science, since it is clearly not simply by means of generalization from one or a few instances and instantiation to all regions of space and time.

3   Laws and Contingencies in Biology

Beatty (1995, 1997) has argued that distinctively biological generalizations, while true, cannot be laws because they are contingent on a particular historical pathway traversed as a result of evolutionary dynamics.

To say that biological generalizations are evolutionarily contingent is to say that they are not laws of nature – they do not express any natural necessity; they may be true, but nothing in nature necessitates their truth (Beatty 1995: 52).

The idea of evolutionary contingency is meant to capture the meaning of Steven J. Gould's metaphoric appeal to Frank Capra's "It's A Wonderful Life" (Gould 1990). That is, if we rewound the history of life and "played the tape" again, the species, body plans, and phenotypes that would evolve could be entirely different. The intuition is that small changes in initial "chance" conditions can have dramatic consequences downstream. Sexual reproduction itself is thought to be a historically contingent development and hence the causal rules that govern gamete formation, for example, are themselves dependent on the contingent fact that the structures that obey those rules evolved in the first place. Biological contingency denotes the historical chanciness of evolved systems, the "frozen accidents" that populate our planet, the lack of necessity about it all.

Mendel's law of the 50:50 ratio of gamete segregation is true (when it is) only because the genes determining that ratio had been selected for in a particular episode in the evolutionary history of life on this planet. It could have been otherwise, hence the generalization is contingent on that particular evolutionary history and, for Beatty, therefore not a law. Indeed those historical conditions on which the truth of the generalization is contingent (e.g., those determining the selective advantage of the 50:50 segregation gene) may vanish in the future, rendering the generalization no longer capable of truly describing the state of nature. This feature of biological rules Beatty (1995) calls "weak contingency."

In addition, by "strong contingency" Beatty denotes the situation in which from the same set of conditions with the same selection pressures operating variant functionally equivalent outcomes may be generated. Thus a particular one of the possibly multiple rules describing these variant outcomes appears not to be necessitated by the prior conditions which gave rise to it.

The evolutionary contingency that Beatty attributes to biological generalizations does not separate out biological generalizations from those of the other sciences. All scientific laws or laws of nature are contingent in two senses. First they are clearly logically contingent. Second, they are all "evolved" in that the relations described in the law depend upon certain other conditions obtaining. That Galileo's law of free fall truly describes relations of bodies in our world requires that the mass of the earth be what it is. If, for example, the core of the earth were lead instead of iron, the quantitative acceleration would be four times what it is (though it would still be an inverse square relation). That the earth is configured the way it is, is the result of the origin of the universe and the subsequent creation of the stars and planets. Generally stated, there are conditions in our world upon which the truth of laws, like Galileo's law of free fall, depend. They all could have been otherwise. This is the case whether or not those conditions are the result of particular episodes of biological evolution and are subject to further modification, or whether they are conditions that were fixed in the first three minutes after the birth of the universe (Weinberg 1993). Whatever else one believes, scientific laws describe our world, not a logically necessary world. All laws are logically contingent, and yet there is still a difference between Mendel's law of 50:50 segregation and Galileo's law of free fall. How can we represent that difference? That there is a difference between Mendel's laws and Galileo's law should be explained, but it is not the difference between a claim that could not have been otherwise (a "law") and a contingent claim (a "non-law"). What is required to represent the difference between these two laws is a framework in which to locate different degrees of stability of the conditions upon which the relation described is contingent. The conditions upon which the different laws rest may vary with respect to stability in either time or space or both.

The dichotomous opposition between natural contingency and natural necessity is a product of framing natural relations in logical terms. The difference between generalizations in physics and those in biology or the social sciences is inadequately captured by the dichotomy between necessity and contingency. Any empirical truth describes events that could have been otherwise. What it would take to make them otherwise, however, varies. Changing the conditions upon which Galileo's law depends has more downstream consequences than changing the historical conditions upon which Mendel's law depends. The lawful relationship between free falling bodies on the earth and parent and gamete frequency, therefore, have different degrees of stability. The actual acceleration of falling bodies, given those conditions is deterministic, while Mendel's law is probabilistic. Thus, they differ both in the degree of contingency they display, or the stability of the conditions upon which they depend, and the strength of the relationship described. They do not differ in kind. Both causal relationships described in the two general claims are contingent and both are true of our world.

The stability of the conditions upon which a causal relationship depends constitutes a continuum, rather than a singly partitioned space of the necessary and the contingent. There is no clear metric to use to measure degrees of stability, but there is, I believe an ordering amongst the domains of the sciences that sheds light on the differences in the explanatory problems that face biologists and social scientists compared to physicists.

Consider an ordering in terms of constraints on what is possible and the relationship of this set to what is actualized. In evolutionary biology and in social organization in particular, what is actualized depends on the history of how things developed on our planet, and the composition relations describing how complex systems are built out of their components.

Without giving an account, let's ignore the contingencies in the development of the "physical stuff" in our universe and accept that it realizes some subset of the logically possible relations and structures that includes those that might have come into existence but in fact did not (see Figure 5.1).

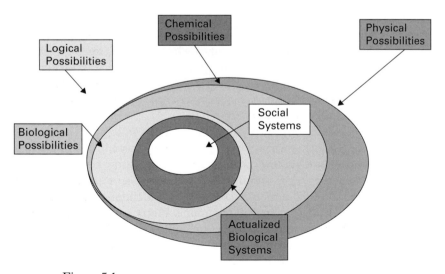

Figure 5.1

The domain of the physically possible then contains in it the subset of the physically actual – which I presume does not in fact occupy all of physical possibility. For example there might have been different quantities of matter in the primordial atom at the time of the big bang and this would have downstream consequences on the types of relationships and causal laws describing them that would have obtained. Perhaps what is meant by physically or naturally necessary is that once some fundamental features are fixed then the further relationships that will hold among physical properties and events become necessary contingent on those most fundamental features.

One can then ask, presuming all of this, what kinds of biological forms would be possible? What kinds have been actualized? At any point in this history – if it went a different way, path-dependence would force the actual biological domain into different regions of the possibility space. A parallel argument could be applied to the realized social configurations, a subset of the constrained possibilities permitted by actual biology and further down the hierarchy. What biologists primarily investigate is not what is biologically necessary given the constraints upon biological form and behavior from the actual physical and chemical relationships available. They do not explore the entire domain of the biologically possible and explain why features that we observe are found within that domain. Rather, biologists are concerned to explain the subset of that space, populated by the actual biological forms and rules that occupy our planet historically and currently. The causes that drive biological form into the region it occupies are evolution by natural selection, conferring evolutionary contingency on the forms, developmental constraints that confer a different sort of contingency, and chance. Evolutionary contingency points to the fact that none of the forms and rules that govern actual biological systems are necessary even given the constraints from below; they could have been different even granting the same physical components they are made from, and they may well be different in the future.

One could have a change in the behavior of biological structures without necessarily a change in the underlying physical stuff if the change was one that was built solely on structure and organization. The very same underlying biochemistry of a bee, for example, remains the same for solitary and social insects. But the behavior of the solitary and social are different, obey different rules, and that is a result of the social organization in which they live, not their basic biology. Reductive approaches to explaining the behavior of complex systems get you something, but not everything we want to explain.

4   The *Ceteris Paribus* Alternative

Perhaps there is a way in the traditional understanding of laws to save the phenomena of knowledge claims in biology and the social sciences. The contingency might be managed without losing the universality and necessity of lawfulness. Perhaps we should not be too quick to abandon the standard. There is, after all, a well-worn strategy for converting domain restricted, exception ridden claims into universal truths and that is by means of the addition of a *ceteris paribus* clause. Take the causal dependency described by Mendel's law of segregation. That law says: in all sexually reproducing organisms, during gamete formation each member of an allelic pair separates from the other member to form the genetic constitution of an individual gamete. So, there is a 50:50 ratio of alleles in the mass of the gametes. In fact, Mendel's law does not hold universally. We know two unruly facts about this causal structure. First, this rule applied only after the evolution of sexually reproducing organisms, an evolutionary event that, in some sense, need not have occurred. Second, some sexually reproducing organisms don't follow the rule because they experience meiotic drive, whereby gamete production is skewed to generate more of one allele of the pair during meiotic division. Does this mean than Mendel's law of segregation is not a "law"? We can say that, *ceteris paribus*, Mendel's law holds. We can begin to spell out the *ceteris paribus* clause: provided that a system of sexual reproduction obtains, and meiotic drive does not occur, and other factors don't disrupt the mechanisms whereby gametes are produced, then gamete production will be 50:50. Finer specifications about possible interference, *especially when they are not yet* identified, get lumped into a single phrase – "*ceteris paribus*" – when all else is equal, or provided nothing interferes. This logical maneuver can transform the strictly false universal claim of Mendel's law into a universally true, *ceteris paribus* law. With the *ceteris paribus* clause tacked on, even biological generalizations have the logical appearance of laws.

Harold Kincaid adopts this strategy explicitly with respect to social regularities. He compares knowledge of the social with knowledge of the biological and concludes that if the social is like the biological then generalizations in social science are just as lawful. But this is a house of cards, for if the biological is not warranted, neither is the social. What makes the causal structures found in the social world explanatory cannot be determined merely by sccing if they are similar to the causal structures in the biological world, without a defense of why and how the biological are genuinely explanatory. The more basic question, what makes a claim lawful or explanatory is left unexplored in Kincaid's approach.

In his 1996 book *Philosophical Foundations of the Social Sciences* Kincaid defends the claim that the social sciences can produce well-confirmed causal explanations and laws. He engages a number of objections to the possibility of social science laws, but my concern here is with his use of the *ceteris paribus* clause in defending laws in social science. *Ceteris paribus* laws are problematic, since "other things being equal" may not in fact occur in situations for empirical test and explanation. Kincaid's defense is to say that *ceteris paribus* laws occur also in physics and particularly in biology and that alone is sufficient to claim "there no *inherent* obstacle" for their use in social science. The paradigmatic character of Kincaid's arguments by itself, I believe, fails to give a principled defense.

However, Kincaid goes on to argue that *cp* laws do explain and are confirmed when the *cp* clause is met. However, unspecified *cp* clauses cannot be met, or at least we would never be in a position to know if they were or were not for purposes of confirmation of explanation. Kincaid suggests that the way in which a law like "A causes B, *ceteris paribus*" explains and can be confirmed is if we consider it to describe a tendency or a partial cause. However, Kincaid marries this claim with the view that tendencies do not entail counterfactuals. If that is so then how can A having the tendency to produce B, *ceteris paribus*, explain what happens when the *cp* clause is not met, which, everyone agrees, is always (or nearly always) the case?

Kincaid's solution to the problem of laws in the social sciences is to suggest that laws are about tendencies and are *ceteris paribus* laws. However, the confirmation and explanation requirements for lawfulness are only met when nothing else interferes, or when the confounders are specified and managed (i.e., when there is no open-ended *cp* clause). So Kincaid's defense of laws in social science turns out to be rather limited : the laws – universal, exceptionless generalizations – are universal and exceptionless only when a *ceteris paribus* clause is attached, and these *cp* laws are confirmable and explanatory only when the *cp* clause is either met or specified. Since they are never met, then the appeal to laws takes a "derivative role" in explanations in the social sciences and only when they are specified, where the primary role is played by the constellation of possibly unique singular causes interacting to bring about the condition we wish to explain. Kincaid's defense of laws in the social sciences turns out to be a rejection of the usefulness of the notion of law (made applicable only with a *cp* clause) to the social sciences. There is good reason to reject *cp* laws as the solution to the problem of laws in the social sciences.

The cost of the *ceteris paribus* clause is high.[4] First, although making a generalization universally true in this way can always be done, it is

---

[4] This argument is developed in Mitchell 2002a.

at the risk of vacuity. Some, like Kincaid or Pietroski and Rey (1995) have suggested that there are ways to fill out the *ceteris paribus* clause to make it contentful. However, the ability to fully fill in the conditions that could possibly interfere may well be an impossible task. Indeed, in evolutionary systems new structures accompanied by new rules may appear in the future, and hence we could never fully specify the content of potential interfering factors. Indeed, filling in the *cp* clause may well be beside the point, since what it needs to do is specify all the ways in which the causal relations described by the law can fail to be operative. This requires the exploration of all of the biologically or socially possible space, not just the actualized structures the biologists and social scientists aim to understand. Earman *et al.* (2002) maintain that there are strict laws to be found, at least for fundamental physics, so there is no need for *ceteris paribus* laws there. Furthermore, they argue, that although the special sciences cannot discover strict laws, there are no such things as *ceteris paribus* laws. Their challenge leaves us with the problem of how to account for the explanatory and predictive power of biological generalizations if, as their account would entail, there are no laws of any kind in these domains.

Critics of the *ceteris paribus* clause correctly identify the fact that the clause violates the logical spirit of the concept of "law." I argue that, more importantly, it violates a pragmatic aspect of "laws" in that it collapses together interacting conditions of very different kinds. The *logical* cloak of *ceteris paribus* hides important differences in the *ontology* of different systems and the subsequent differences in epistemological practices. Where as *ceteris paribus* is a component of the statement of a causal regularity, what it marks in the world is the *contingency* of the causal regularity on the presence and/or absence of features upon which the operation of the regularity depends. Those contingencies are as important to good science as are the regularities that can be abstracted from distributions of their contextualized applications.

We need to rethink the idea of a scientific law pragmatically or functionally, that is, in terms of what scientific laws let us do rather than in terms of some ideal of a law by which to judge the inadequacies of the more common (and very useful) truths. Woodward (2002) also adopts a strategy to reconsider the nature of laws in the special sciences, rather than forcing those claims uncomfortably into the standard view, wedged in with the help of *ceteris paribus* clauses. He has developed an account of explanation that requires generalizations less than universal in scope, but which can, nevertheless, support counterfactuals.

I see my approach as complementary to Woodward's. I suggest that we need to reconfigure our framework for understanding knowledge claims to accept that the world is occupied by causal relations that form a continuum of degrees of contingency and stability. As I have argued, this anti-dichotomous framework permits a clearer understanding of the character of the contingencies of the causal dependencies in biological and social systems that are often lumped into a single abstract concept of "contingency" and singled out as *the* culprit preventing lawful generalizations. The general truths we discover about the world vary with respect to their degree of contingency on the conditions upon which the relationships described depend. Indeed, it is true that most of the laws of fundamental physics are more generally applicable (i.e., are more stable over changes in context) in space and time, than are the causal relations we discover hold in the biological world. They are closer to the ideal causal connections that we choose to call "laws". Yet, few of even these can escape the need for the *ceteris paribus* clause to render them logically true.

The difference between fundamental physics and the special sciences is *not* between a domain of laws and a domain of no laws. Yet, I would agree, there are differences and those differences can inform our understanding of not only the special sciences but of the very notion of a "law" and its function. By broadening the conceptual space in which we can locate the truths discovered in the various scientific pursuits we can better represent the nature of the actual differences. The interesting issue for biological knowledge is not so much whether it is or isn't just like knowledge of fundamental physics, but how to characterize the types of contingent, complex causal dependencies found in that domain.

If explanatory generalizations were universal, then when we detect a system in which A is correlated with B, and we determine that this is a causal relationship, we could infer that A would cause B in every system. But we often cannot. The difficulty goes beyond the correlation–causation relationship. Even when we gave good evidence that A causes B in a system, say through controlled intervention, we still can't say that A would cause B in every system. When we look at the tidy behavior of Mendel's pea plants, where internal genetic factors assort independently and segregate fairly, we might wish to infer that would always happen, in any sexually reproducing population. But it doesn't. And since it doesn't, we need to understand more about the system of Mendel's peas and their relationship to other systems to know what about the original test case is exportable to the new domains. If we were so lucky as to have detected a universal exceptionless relationship, constitutive of the strict interpretation of law, we would know it would automatically apply to all times and all places. But that is not the world of the special sciences.

In systems that depend on specific configurations of events and properties which may not obtain elsewhere, and which include the interaction of multiple, weak causes rather than the domination of a single, determining force, what laws we can garner will have to have accompanying them much more information if we are to use that knowledge in new contexts. These are precisely the domains that the special sciences take as their objects of study. Thus the central problem of laws in the special sciences, and perhaps for all sciences, is shifted from what is a strict law, *ceteris paribus,* or no law at all, to how do we detect and describe the causal structure of complex, highly contingent, interactive systems and how do we export that knowledge to other similar systems (Mitchell 1997, 2002a, 2002b, 2002c).

I have tried to provide an answer to Condorcet's original question, why should explanation by universal laws be less true for the intellectual and moral faculties of man than for the other actions of nature in terms of the kinds of contingency that are found in complex, historically and context dependent parts of nature. This approach requires a conceptual shift, from Newtonian exceptionless universality to considering causal structures that permit of multiple degrees of stability and strength. Thus our conceptual resources are more expansive, replacing the dichotomy of law vs. accident with a continuous domain of degrees of stability. Thus we are in a position to explain the difference in degree, rather than in kind, between the physical law of conservation of mass/energy and Mendel's law of segregation and the social science law of supply.

This conceptual shift brings with it methodological changes. Without the presumption of universal, exceptionless laws, we are no longer able to infer from one or a few observed and studied instances that all similar situations will behave in exactly the same way. Instead, we will have to carry the evidence and information about the context from one location of confirmation to see if and how the causal relation discerned applies in other circumstances.

REFERENCES

Beatty, J. 1995. "The Evolutionary Contingency Thesis." In G. Wolters and J.G. Lennox (eds.) *Concepts, Theories, and Rationality in the Biological Sciences.* Pittsburgh: University of Pittsburgh Press, pp. 45–81.
    1997. "Why do biologists argue like they do?," *Philosophy of Science* **64**: 432–443.
Condorcet, N. de 1792. *Rapport et projet de décret sur l'organisation générale de l'instruction publique.* Compayre, G. (ed.) Paris 1883, p. 120 quoted in F.A. von Hayek "The counter-revolution of science" *Economica* 1941, pp. 9–36.

Earman, J. and J. Roberts 1999. "Ceteris paribus, There is no problem of provisos," *Synthese* 439–478.

Earman, J., J. Roberts, and S. Smith 2002. "Ceteris paribus post," *Erkenntnis* 281–301.

Elgin, M. 2003. "Biology and a priori laws," *Philosophy of Science* **70**: 1380–1389.

Goodman, N. 1966. "The new riddle of induction," *Journal of Philosophy* **63**: 281–331.

Gould, S. J. 1990. *Wonderful Life: The Burgess Shale and the Nature of History.* Harmondsworth: Penguin.

Hayek, F. A. von 1941. "The counter-revolution of science", *Economica*, pp. 9–36.

Hempel, C. G. 1965. *Aspects of Scientific Explanation and other Essays in the Philosophy of Science.* New York, NY: Free Press.

Kincaid, H. 1996. *Philosophical Foundations of the Social Sciences.* Cambridge: Cambridge University Press.

McIntyre, L. C. 1996. *Laws and Explanation in the Social Sciences: Defending a Science of Human Behavior.* Boulder, CO: Westview Press.

Mitchell, S. D. 1997. "Pragmatic Laws." In L. Darden (ed.) *PSA 1996: Part II, Symposia Papers, Philosophy of Science (special issue)*, pp. S468–S479.

2000. "Dimensions of scientific law," *Philosophy of Science* **67**: 242–265.

2002a. "Ceteris Paribus – An Inadequate Representation for Biological Contingency." In Earman, J., C. Glymour, and S. D. Mitchell (eds.) *Erkenntnis* **57**: 329–350.

2002b. "Contingent Generalizations: Lessons from Biology." In R. Mayntz (ed.) *Akteure, Mechanismen, Modelle. Zur Theoriefähigkeit makro-sozialer Analysen.* Frankfurt: Campus Verlag, pp. 179–195.

2002c. "Integrative pluralism," *Biology and Philosophy* 17(1): 55–70.

Nelson, A. 1986. "Explanation and justification in political philosophy," *Ethics* **97**(1): 154–176.

Pietroski, P. and G. Rey 1995. "When other things aren't equal: Saving ceteris paribus laws from vacuity," *British Journal for the Philosophy of Science* **46**: 81–110.

Smart, J. J. C. 1968. *Between Science and Philosophy: An Introduction to the Philosophy of Science.* New York, NY: Random House.

Sober, E. 1997. "Two outbreaks of lawlessness in recent philosophy of biology," *Philosophy of Science* **64**: 458–467.

Weinberg, S. 1993. *The First Three Minutes: A Modern View of the Origin of the Universe.* New York, NY: Basic Books.

Woodward, J. 2002. "There is no such thing as a ceteris paribus law," *Erkenntnis* **57**: 303–328.

5 – Comment
Conditional Knowledge: An Oxymoron?

James Alt

Sandra Mitchell's program to keep the "science" in "social science" emphasizes that explanation – answering the question "Why?" – is the job of science. For her, the main question facing social science is "How do we detect and describe the causal structure of complex, highly contingent, interactive systems and how do we export that knowledge to other similar systems?" in order to "explain, predict, and successfully intervene in the world" when it is inevitably the case that "The general truths we discover about the world vary with respect to their degree of contingency on the conditions upon which the relationships described depend." I think it matters whether they are more or less general or more or less true, but in any case she argues that "The important question is how less than universal, exceptionless generalizations can do the jobs of science, since it is clearly not simply by means of generalization from one or a few instances and instantiation to all regions of space and time." That gives us an enormous amount to think about. I will de-emphasize interactions, predictions, and interventions, and focus mostly on how clearly Mitchell's arguments bring out some important points about the role in political science explanations of "causal" effects, complexity, and models.

Partly because political scientists train in a range of traditions like economics, journalism, philosophy, psychology, or sociology, the field of political science is not defined by any agreement on which is the best method of inference and so methodological diversity runs deep. If this were a comment on Searle's chapter, for instance, I could devote my space to discussing the roles of agency and identity in political science explanations; topics on which I am not even going to touch. However, like Mitchell, I personally believe that the model of the natural scientist who provides causal explanations of phenomena is an appropriate one for the budding social scientist, provided we understand what it means to provide a causal explanation, and that reliance on statistical techniques, probability statements, and the observation of "imperfect correlations" does not make social science somehow incapable of offering

146

causal explanations. So when Mitchell asks "What to call a generalization that describes a strong causal relation between events yet fails to exhibit the strongest conditions of nomological connection?" then calling it a "strong causal relation" sounds good to me, unless you prefer "an explanation" which adds to the causal relation some further degree of insight about "what else" obtains.

Mitchell recommends that we reconfigure our framework for understanding knowledge claims to accept that the world is occupied by causal relations that form a continuum of degrees of contingency and stability. Contingency seems to be a matter of scope, or limited domain of applicability. There is not much detail on what metric to use to measure degrees of stability. Two important considerations are that it will require a "multidimensional" framework and that the behavior of the solitary and social are different. The latter is because of the social organization in which individual humans live, not basic biology: they obey different rules, so this means we have to deal with complications of aggregation problems and the problem of institutional endogeneity, at least. However it can be measured, though, stability reiterates two vital questions about any explanation: "How many anomalies are there?" and "How puzzling are they?"

Social research is inordinately complicated. With a trillion neurons in the human brain even an FMRI makes the biological part alone complex, and the influence of environment, genetics, and interaction between one human being and another makes one wonder how we can ever model crime or foresee the next genocide. However, Mitchell reminds us that we want theories to be correct (and also maybe beautiful) and the only way to evaluate "correct" is constantly to compare them to data, always looking to see if there is any way that what we observe can prove them wrong. Both a deep seated skepticism and openness to new data are therefore central to science. I do however also believe that while it is important to know when something you see says "you're wrong," if we only have probabilistic statements we can't say this easily. But that does not mean we have to give up, either.

1   Laws, Contingency, and Modeling in the Social Sciences

Mitchell forces us to think carefully about the role of laws in political science. What are they like? Some are regularities amenable to being captured by mathematical formulae similar to those used by physicists or by economists. Such purely formal results (like the theorems of McKelvey–Schofield, Arrow, or Gibbard–Satterthwaite) are logically true rather than empirically true. They are logical results. What is the place of such theorems that are universally true in a positive theory of politics? Positive theory like this, formal results within models, is central to empirical political science. Positive theory establishes things like puzzling anomalies, observational equivalence, and impossibility results, all of which are logical implications following from *a priori* assumptions. McKelvey–Schofield (how group preferences aggregated by simple majority rule yield little outcome-predictive leverage at the aggregate level) or Gibbard-Satterthwaite (how the incentive to vote strategically necessarily accompanies democratic decision making) both hold in any environment in which there are three or more alternatives from which a group must choose. Note though that neither proves that people vote strategically, let alone whether such behavior is good or bad.

One might like to believe that historical episodes are unique contexts that reveal the applicability of general principles. Nevertheless, as Richard McKelvey (Aldrich *et al.* 2007a), pointed out very clearly, theorists frequently doubt whether empirical studies have "adequately" tested the theoretical models that they are based on. Why? Paraphrasing his argument, for one thing, the world that theorists look at is necessarily abstracted and simplified, and focuses on the effects of some variables to the exclusion of others. In the real world, one seldom can find empirical situations where *only* the variables of theoretical interest are active. Moreover, there are problems with measuring variables of theoretical interest like utilities or beliefs in natural settings. Thus, it is frequently hard to get the data needed to test theoretical models. This might be in part because the theories are too primitive to address real-world situations in which empirical researchers are interested. However, empirical studies also often do not provide natural experiments. As a consequence, modeling is ubiquitous, but so is uncertainty, and this is the rationale for a skeptical posture. There is also a lot of room at this point for imagination: the phrase "suppose it were true that ... " is an important source of the counterfactuals needed to support explanation.

However, other things in political science are sometimes called laws, like those of Duverger or Gamson. What are these? They are what Mitchell talks about: models, clearly logically contingent, and all "evolved." They depend upon certain other conditions obtaining, and describe our world, not a logically necessary world. How do we use empirical evidence with laws like this to build up an explanatory edifice? I want to look at another field (geology rather than Mitchell's biology) to sketch the contours of this process, and then I will give some examples of "good practice" in political science.

In our lifetime the field of plate tectonics tells a good tale about this sort of development. It started with an intuition about 150 years ago: the shape of the Atlantic coasts, continuities in strata, but all widely discredited until the 1960s when scientists discovered younger rocks toward the middle of the Atlantic. This they called "continental drift," subsequently generalized to plates, not continents: this was a uniform process, broadened to explain mountains whose upthrust was inexplicable (or only *ad hoc* explicable). The "Pacific rim of volcanic activities" is now "one plate's boundaries" and the Hawaiian islands are not a chain of volcanoes but a thin spot in the earth's mantle with a plate drifting over it. In fact, lasers can now measure the distance (perhaps 1 cm/year) that plates move.

John Stewart (1990) analyzes the transition from the static crust, shrinking earth view to plate tectonics. He is a sociologist, so persuasion and interaction are his main interests. The process was complete in at most ten years. He mentions half a dozen features worth noting:

- (abstraction) The features we observe like seafloors and continents were the initial units of observation, but turn out not to be the ultimate theoretical units. (This is a rebuke to micro-oriented political scientists who believe that observing units in the world like revolutions or countries *must* be an obstacle to theoretical progress.)
- (resistance) It was useful to have skeptics at Lamont, the main Data Center; science was not harmed by having the main data owners take a conservative position.
- (parsimony) Lamont scientists attached enormous importance to being able to fit magnetic anomalies to *constant* rates of spread, generalizing beyond the *ad hoc*.
- (scope) The theory starts to account for magnetic reversal further back in time and then they added sedimentary cores (more recent phenomena).
- (accuracy/consistency) Chronologies from land, cores, and seafloor studies were reasonably close (these were correlations!).
- (abstraction/parsimony) Theory then shifts to plates, not linked to seas or continents specifically.

For us, the point is to observe how the dynamics of change from ongoing research re-shaped the very view of the evidence. At each point indeed a "discordant" fact was only discordant in the context of some alternative explanation, and the alternative's extensions had to be supplied. But nevertheless there were "rules." It was not an irrational process, as long as there were alternative conjectures.

What rules governed the process? Trade more scope for the same degree of parsimony or accuracy (i.e., for the same number of propositions, with the same facts covered, drop an antecedent condition); or gain more parsimony with the same scope and accuracy (with the same antecedent conditions and same facts, drop a proposition); or more accuracy with the same scope and parsimony (with the same antecedent conditions and number of propositions, add more covered facts). This is an incremental form of what King *et al.* (1994) meant by "maximizing leverage," in the language of their influential text.

I believe this is how to think about stability. However, the theory of multi-criteria decision making reminds us that with three criteria – scope, accuracy, parsimony – there will be competing claims, cycles, and no deterministic answer about what is "best" at any time. In order to deal with anomalies, we have to be self-conscious about our models. An appropriate way to evaluate a model's empirical implications is to ask a series of questions like de Marchi's (2005): What are the assumptions/parameters of the model? How sure can we be that the main results of the model are immune to small perturbations of the parameters? How central are these particularly brittle parameters to the theoretical questions at hand? Do the results of the model map directly to a dependent variable, or by analogy from the model to an empirical referent? Are the results verified by out-of-sample tests? Is the parameter space of the model too large to span with available data (the "curse of dimensionality")?

How far can experiments and model-building mitigate instability problems? At one extreme, some econometricians argue that we can differentiate among the multiple causal claims consistent with observed correlations by proper specification and estimation (Heckman 2008). At the other, some believe that experimental design and control are essential to "causation," avoiding probabilistic statements. Nevertheless, there is still the fundamental problem that one cannot observe treatment and control on the same unit, and this is true of experiments as well as observations procured in other ways. Consider Holland's (1986) (statistical) account of causation: that even when we stir a solution, split it, and treat half, we are making assumptions about the (very low) probability that the two halves differ in some way, which results in measurements erroneously attributed to the treatment. However, the two halves are still only conditionally independent (conditional on mixing, splitting, randomization), just as would be averages over separate samples of observations where replication and treatment were not possible. In general one must assume either *unit homogeneity* (two units are the same in all respects save one, often weakened to assuming constant causal effects across units instead) or *conditional independence*, which means that values are assigned to explanatory variables independent of values of dependent variables. This supports the methods that King *et al.* (1994)

use to define a quantity of interest as the difference in one characteristic of a unit which is "attributable" to another. From here it is a short step to defining a "causal effect" as the difference between the actual (systematic part of the observed unit) and the counterfactual case in which we hold everything constant except the treatment variable and everything that might be (partly) a consequence of it.

This touches on the general problem raised by Mitchell of incorporating specific details of the situation into more general models. A causal effect defined this way has interesting affinities to Mackie's (1965) INUS condition, in which each element a_i is necessary for the antecedent condition A to be sufficient for all known posteriors P but not all a_i are necessary for any P_j and no individual a_i is sufficient for all P. This less quantitative, more historical approach might appropriately identify among the individual causes of plane crashes such a factor as metal fatigue, if at least one plane crash would not have happened without metal fatigue but metal fatigue is absent also in at least one case. What Mackie brings out is the importance of completeness in causality: adding "pilot error" sacrifices parsimony for "accuracy" even if it initially looks like adding special cases. To reintegrate this approach with one more statistical, we can think of causal explanations as containing a systematic component and a residual, and the idea is of course to minimize the residual without adding too many special cases. A commonly cited piece of good practice that follows is reporting uncertainty. Note also that on this view even with a "perfect" causal model of the effect of x on y there can be imperfect correlation between y and each x.

2   Good Practices in Political Science: Bringing Together Theory and Data

We never avoid contingency, Mitchell's version of the problem of appropriate domain, since " ... often the causal patterns are dramatically different across the cases. In those instances, subsetting the sample and doing the statistical analysis separately for each distinct causal pattern ... will need far fewer control variables ... ." (Achen 2002: 447). One might recommend subsetting observations into two or more theoretically-generated sub-samples, with modest controls in each while further controls are what separate the sub-samples. In a model one can add detail on conditionalities (or Achen's interactions) on the formal theory side or take the statistical model and add some detail on what you don't know. Either way one hopes to get enough structure so that the theoretical model simplifies but still generates conjectures, while the statistical model has enough specification detail to identify instruments and address endeneity problems. Or at least that should be so for enough important cases to make this approach worthwhile.

How does it look in practice? How big is the gain in understanding what is theoretically at stake when an empirical regularity is accompanied by a carefully worked-out micro-model? Consider how Carroll and Cox (2007) outline the evolution of recent work on *Gamson's Law*, the proposition that parties forming a coalition government each get a share of portfolios proportional to the seats that each contributed to the coalition (Gamson 1961). Empirically corroborated by many scholars, Gamson's Law conflicts with standard bargaining theories, in which a party's ability to pivot between alternative minimal winning coalitions and its ability to propose governments determine its portfolio payoff.

Laboratory experiments that confront Gamson's Law broadly support the bargaining models. Carroll and Cox specify a theoretical reason why portfolios should be handed out in proportion to seat contributions (by assuming some parties can make binding commitments to one another). This provides an empirical model that encompasses Gamson as well as bargaining models. It also yields a refinement on earlier results and even produces an entirely new prediction. This is an extraordinary increase in being able to examine "what obtains if . . . " in empirical work, compared to the earlier recording of underspecified correlations.

Schofield's (2007) study of small parties is another, but different, example. Here one sees theoretical context and "stability" carefully specified; he writes (2007: 223):

Perhaps the most important application of the McKelvey-Schofield symmetry conditions for existence of a core is . . . to existence of an equilibrium in coalition bargaining, where a small number of parties have policy preferences, and differing political weights.

He uses qualifiers to describe Mitchell's stability: examples include "It seemed *likely* that democratic polities . . . " and "I shall argue that a *useful* model of elections is *essentially* stochastic." Nevertheless there is a clear statement of expectations: "The framework is thus compatible with the observation, made particularly with respect to European multiparty polities, that there may be many small 'radical' parties positioned far from the electoral center" and further that the "theory indicates that these parties are small because their valence is low, and their positions on the electoral periphery are chosen to maximize their electoral support." There is also a possible extension to other theoretically derived expectations: "This 'feedback effect' may be the underlying mechanism that can be used to explain the Riker-Duverger thesis about the relationship between the plurality electoral mechanism and the stability of the two party system." In general there is real clarity about multiple possibilities, mechanism, and observational equivalence.

To return to stability and contingency, it is gratifying that the political science interaction of theoretical models and empirical implications described above is philosophically embraced by Mitchell's arguments. Anomalous results, puzzles, and carefully-limited conditional domain statements figure in empirical research. But some propositions are more important, or have greater extent of applicability, while others at most seem likely or useful. Keeping track of that clearly, and exchanging ideas about why, is the essence of the social science enterprise.

REFERENCES

Achen, C. 2002. "Towards a new political methodology: Microfoundations and ART," *Annual Review of Political Science* 5: 423–450.

Aldrich, J. H., J. E. Alt, and A. Lupia 2007a. "Introduction." In Aldrich, Alt, and Lupia (2007b), pp. 1–8.

(eds.) 2007b. *Positive Changes in Political Science: The Legacy of Richard D. McKelvey's Most Influential Writings*. Ann Arbor, MN: University of Michigan Press.

Carroll, R. and G. Cox 2007. "The logic of Gamson's Law: Pre-election coalitions and portfolio allocations," *American Journal of Political Science* 51(2): 300–313.

de Marchi, S. 2005. *Computational and Mathematical Modeling in the Social Sciences*. Cambridge, New York, NY: Cambridge University Press.

Gamson, W. A. 1961. "A theory of coalition formation," *American Sociological Review* 26(3): 373–382.

Heckman, J. 2008. "Econometric causality," *International Statistical Review* 76(1): 1–27.

Holland, P. W. 1986. "Statistics and causal inference," *Journal of the American Statistical Association* 81: 945–960.

King, G., R. Keohane, and S. Verba 1994. *Designing Social Inquiry: Scientific Inference in Qualitative Research*. Princeton, NJ: Princeton University Press.

Mackie, J. L. 1965. "Causes and conditions," *American Philosophical Quarterly* 2(4): 245–264.

Schofield, N. 2007. "Social Choice and Elections." In Aldrich et al. (2007b), pp. 221–241.

Stewart, J. A. 1990. *Drifting Continents and Colliding Paradigms: Perspectives on the Geoscience Revolution*. Bloomington, IN: Indiana University Press.

6 The Heterogeneous Social: New Thinking About the Foundations of the Social Sciences

Daniel Little

1 Introduction: Reconsideration of the Foundations of Social Science Research

This chapter is an effort to contribute to ongoing discussions about the future of the social sciences. It has several purposes: to bring the perspective of the philosophy of social science into closer engagement with social scientists; to review some very interesting current areas of innovation in American social science research communities; and to identify some of the threads that might guide a "post-positivist" social science paradigm.[1]

This is a particularly important time for us to develop effective and innovative programs of social science research, if scholars and policy-makers are to have a reliable basis for understanding and managing the changes the world is currently undergoing. Many societies today are undergoing processes of social change that are both momentous and globally unprecedented. The processes themselves are complex and large. The causal connections between one set of changes and another set of consequences are obscure. The social, political, and human effects of these changes are large and unpredictable. And these processes are not adequately understood. We do not have comprehensive social theories for which a particular experience is the special case – whether "modernization theory," "world systems theory," "social functionalism," "rational choice theory," or "theory of exploitation." In fact, it is a radical misunderstanding of the nature of the social, to imagine that there might be such comprehensive theories. Rather, social development is a

[1] An earlier version of this chapter was presented to faculty and graduate students at Tsinghua University (Beijing) and Northwest University (Xi'an). These were delightful opportunities for intellectual exchange, and the discussions displayed a surprising level of agreement by Chinese graduate students concerning the need for a rethinking of the foundations of sociology for China. I am grateful to my hosts, Professors Li Bozhong and Liu Beicheng in the Department of History at Tsinghua University, and Professor Chen Feng in the Department of History and Cultural Heritage Protection at Northwest University.

contingent, multi-threaded social fabric. We should expect unpredictable twists and turns, the forming and dissolving of institutions and social compromises; regional and sectoral differences in social processes; changing interactions between the state and civil society; the strategic behavior of various social groups; and myriad other contingent and variegated forms of social processes.

Plainly, there is an urgent need for a new surge of effective social-science research. But equally, it is clear that the many areas of change currently underway represent a mix of different kinds of social processes and mechanisms, operating according to a variety of temporal frameworks, with different manifestations in different regions and sectors of society. So we should not expect that a single sociological framework, a unified sociological theory, or a unique sociological research methodology will suffice. Instead, sociological research needs to embrace a plurality of methods and theories in order to arrive at results that shed genuine light on social development. A social science research community will be most successful when it has a wide variety of methods and theories at its disposal, and is thereby able to match its inquiry and explanatory strategies to the particular features of the domain of phenomena to be understood.

2   A Role for the Philosophy of Social Science?

The philosophy of social science is especially important in these circumstances. Most broadly, the goal of this discipline is to provide a careful, analytical treatment of the most basic problems that arise in the scientific study of society and social behavior: for example, ontology, methodology, theory, and explanation. I believe that the intellectual challenges posed by the social sciences are, if anything, more difficult and obscure than those in other areas of the sciences. What do we mean by a "social" "science"? What is the nature of the social? How can we investigate its properties and structures? What is involved in a scientific study of the social? What is the role and scope of social theory in investigating and explaining the social world? How are social science hypotheses justified, confirmed, or tested? What, after all, is a good social explanation?

Some philosophers have approached this family of questions on the basis of abstract philosophical theory, much as logical positivism derived from the tenets of British empiricism. I take a different view. I believe that philosophy of social science will only make genuine progress when it arrives at its formulation of the issues and approaches in deep engagement with working social scientists. I believe that we cannot pose these questions in the abstract; it is necessary to develop our questions and answers in close partnership between practitioners and philosophers. In other words, we need to probe in depth, the ontologies, research methodologies, puzzles, debates, and theories of sociology, political science, economics, anthropology, the area studies – and then arrive at some more informed ideas about what is involved in studying and explaining aspects of the social world. What are the real and practical issues of methodology that social scientists are facing? How can the tools of philosophical analysis shed new light on these issues? And to what extent can we determine the underlying and often erroneous assumptions about "science" that motivate some social science doctrines (e.g., logical positivism or behaviorism)? There is no master theory of science that can simply be "applied" to the social sciences. Instead, it is necessary for the philosopher to bring some general ideas about reasoning, rigor, and rationality into engagement with concrete problems of social science problem formation and research, and to construct new representations of the forms of knowledge that can emerge from social science research and theory.

The aspiration represented by the philosophy of social science is an important one. Good work in this area can contribute to better social science frameworks and theories. It can contribute to more effective research methodologies. And it can help formulate a more adequate background "metaphysics" on the basis of which social scientists can approach their research.

3   Current Discussions of the Social Sciences in the United States

So what can we learn from some innovative social scientists at work today, that might be helpful in addressing the problems of social study currently encountered in today's rapidly changing societies? The past several years have witnessed a surge of path-breaking discussions of the nature and direction of the social sciences among a wide range of innovative social scientists and historians in the United States. A range of authoritative voices have called for fundamental rethinking of the foundations of the social sciences – beginning with Alvin Gouldner's work, *The Coming Crisis of Western Sociology* (Gouldner 1970), but gaining new impetus in the 1990s with the Gulbenkian Commission on the Restructuring of the Social Sciences (Gulbenkian Commission on the Restructuring of the Social Sciences 1996; Wallerstein 1999), handbooks on social science methodology (Turner and Risjord 2006), and a variety of reports to the US Social Science Research Council. Debates within the social sciences in the United States are taking up this challenge. Particularly important are contributions by Adams, Clemens, and Orloff (2005), Steinmetz (2005), Sewell (2005), McDonald (1996), Mahoney and Rueschemeyer (2003), Ortner (1999), Lieberson (1987), and Abbott (1999, 2001). Each of these works (including especially the extensive introductions provided in Adams *et al.* and Steinmetz) provides an analysis of the new currents that are emerging in social science research.

A series of important methodological and substantive insights have emerged from the methodological ferment of the past two decades. Several topics have particular value as we consider what the sociology of the next twenty years might look like. First, there has been lively discussion of the importance and promise of the intellectual framework of *comparative historical sociology*.[2] The core of the approach is a concern for large historical structures embodied in different social settings, and a conviction that we can discover historical causes by comparing in detail the unfolding of some important historical processes in different social settings. The approach insists that "history matters" – that we need to consider the historical context and contingent pathways from which a given complex of social structures emerged. And the approach postulates that we can discover the concrete social causal mechanisms that combine to produce macro-outcomes through careful study of a small set of detailed historical cases. The most studied topic within comparative historical sociology is that of revolution; what historical circumstances combine to make successful revolution more likely (Skocpol 1979; Goldstone 1991)? But the approach has been also used fruitfully to study such mid-level topics as official corruption, union organization, and public welfare systems.

A second important development in contemporary sociology is a focus on concrete *social causal mechanisms*. Charles Tilly's recent work illustrates this approach. His current work takes up once again the topic of social contention (McAdam, Tarrow, and Tilly 2001, Tilly 2003). Now, rather than turning to large sociological theories to explain social outcomes, Tilly has come to emphasize social variability, path-dependence, and a methodology emphasizing the discovery of specific social mechanisms that combine in novel ways in different historical circumstances. In *The Politics of Collective Violence* he puts his approach this way: "We are looking for explanations of variability: not general laws or total explanations of violent events, but accounts of what causes major variations among times, places, and social circumstances in the

---

[2] Extensive, insightful discussion of the methods and problems of CHS are found in McAdam, McCarthy, and Zald (1996); Mahoney and Rueschemeyer (2003); Adams, Clemens, and Orloff (2005).

character of collective violence. We search for robust mechanisms and processes that cause change and variation. Mechanisms are causes on the small scale: similar events that produce essentially the same immediate effects across a wide range of circumstances" (Tilly 2003: 20). What is most important about this approach is Tilly's abandonment of the ideal of discovering regular historical patterns – "food crisis in the presence of local social organization leads to social contention" – in favor of specific historical trajectories in specific settings.

Related to both these threads of discussion is an approach that attempts to discover causal relationships through the study of a single important historical case (*case study methodology*). If we believe that there are effective social causal mechanisms that occur in multiple settings (causal realism), then it is credible that we can examine single instances, and the events and conditions that led up to them, in order to discern some of the causal mechanisms that were instrumental in bringing the event about. An important current body of methodology and theory in the social sciences provides support for the case-study method, including especially Goertz and Starr (2003) and George and Bennett (2005). Analysis of the causes involved in a single outcome is referred to as "process-tracing." George and Bennett describe process-tracing as a method "which attempts to trace the links between possible causes and observed outcomes. In process-tracing, the researcher examines histories, archival documents, interview transcripts, and other sources to see whether the causal process a theory hypothesizes or implies in a case is in fact evident in the sequence" (George and Bennett 2005: 6). A variety of scholars have argued, then, that it is possible to provide evidence in favor of, or contrary to, singular causal claims, and that this approach can provide substantial and systematic insight into important social transitions.

The historical turn in sociology has given a major impetus to detailed study of specific institutions as a primary causal mechanism of social processes (North 1990; Knight 1992; and Brinton and Nee 1998). The *new institutionalism* has made the point very convincingly that the specific rules that constitute a given institution make a substantial difference to the behavior of persons subject to the institution – and this has major social consequences. So detailed sociological study of the specific institutions surrounding a given kind of social behavior can provide the basis for a compelling causal explanation of a large resulting social outcome. And differences among roughly similar institutions can likewise explain substantial differences in the social outcomes observed in the different settings. For example, different systems for collecting and verifying taxes have very different collection rates and very different patterns of evasion. Historical sociologists have pushed this insight deeper, by examining specific institutions in detail, corruption (Klitgaard 1988), training regimes (Thelen 2004), technology regimes (Dobbin 1994), and conscription regimes (Levi 1988), examining the social factors that sustain them over time; and examining the factors that lead to the creation and development of new institutions. Once we abandon the reifying assumption that institutions are stable, coherent entities, we are obliged to consider the ways in which institutions maintain some degree of stability over time, and the ways in which specific institutions change over time. We should look at institutions as a dynamic fabric of overlapping behaviors, cognitive frameworks, and networks of actors, rather than as an abiding and unchanging structure (Abbott 1999).

There has been valuable discussion of the nature of *social entities* in the recent sociology literature. Andrew Abbott is one of the most innovative and fruitful sociologists working in the United States today. His primary field of research is on the sociology of the professions: medicine, psychiatry, university professoriate, law, etc. (Abbott 1988, 2001). Beyond his empirical studies, he has made a very important contribution through his formulation of the ontological questions that sociologists need to raise – but seldom do: What is a social structure or entity? How do actors create, embody, and change the social world around them? His ontological insights are most clearly expressed in *Department and Discipline*, a micro-study of the "Chicago school of sociology." He analyzes the historical twists and turns of this "school" – the founders, the theories, the department politics, the *American Journal of Sociology*, the American Sociological Association. But more profoundly, he draws sharp attention to the fluidity and permutability of this social "entity" (Abbott 1999: 223). A social entity is not a fixed thing with stable properties. It is rather a continuing swirl of linked social activities and practices, themselves linked to other "separate" social traditions. And particularly important, he endorses a social localism – that all social facts are carried by socially constructed individuals in action. For these reasons, Abbott is deeply critical of the prevailing positivist paradigm for sociology – which he refers to as the "variables paradigm" – according to which the task is to discover social variables and assess relationships among them (Abbott 1999: 222).

Another important theme in recent debates is the importance of the *cultural turn* in social research. Social scientists and historians have come to recognize the irreducibly "cultural" nature of social behavior (Darnton 1984; McDonald 1996; and Sewell 2005). Human beings construct their worlds and their actions around a set of understandings, values, and identities that are variable from place to place, and that make a difference in large social outcomes. It is always important for social researchers to be aware of the constitutive role that cultural formations play within social processes, and sometimes exploration of this role is crucial to an adequate understanding and explanation of a social pattern. The "cultural turn" in the social sciences has come to highlight that culture is a feature of all social life, and that every area of social science research needs to have some theoretical and methodological ability to analyze the role of culture in a given domain. This entails that there is an important place for qualitative research in all areas of social investigation, complementary to other methods of research and analysis.

I turn, finally, to some new critical thinking about traditional *quantitative sociological research*. Stanley Lieberson is a gifted and prolific sociologist, skilled in standard methods of quantitative sociology. He has written extensively on the sociology of race and urban life in the United States (Lieberson 1963, 1980; Lieberson and Waters 1988). But he takes on distinctly non-standard sociological questions as well: can we give causal explanations of the success rates of professional sports teams (Lieberson 1997)? Are there regularities in the "tastes" that a population exhibits (manifested, perhaps, in changing frequencies of first names over time) (Lieberson 2000)? Can we offer causal hypotheses about these sociological changes, and can we rigorously evaluate these hypotheses? Lieberson's work is important because he asks novel questions, and because he casts doubt on the methodological common sense of sociology as a discipline. His book, *Making it Count* and numerous methodological articles (Lieberson and Lynn 2002), provide sharp and sustained critique of standard assumptions involved in quantitative sociology. He writes that it is necessary to come to "a different way of thinking about the rigorous study of society implied by the phrase "science of society" (Lieberson 1987: 3–4). His central criticisms of standard assumptions about quantitative methods in sociology are, first, that standard quantitative models assume that sociology is an experimental science, whereas it is almost always impossible to perform experiments on sociological topics; that sociology has not clearly specified the types of causal questions that are accessible to sociological research; that quantitative sociologists usually misunderstand the logic of selectivity and control variables; and that sociologists have not given enough attention to the assumptions they make about sociological causation. Lieberson does not reject quantitative reasoning. Rather, he insists upon clearer thinking about causes and mechanisms, and he rigorously criticizes sloppy thinking about quantitative methods. His approach demands that the researcher give careful attention to the possible social mechanisms that may be involved in the sociological outcomes of interest, and then to design a study that permits elucidation and evaluation of those causal hypotheses.

3.1   Assessment

This brief review of some important recent work within the social sciences on foundational issues is relevant in several ways. First, these authors provide substantive and innovative insights into some very important foundational issues, about the nature of the social and some good ways of investigating and explaining the social. Second, this review demonstrates very clearly the value of interaction between philosophers and social scientists. All these thinkers are engaging in clear, rigorous, and intelligent interrogation of foundational issues of ontology and epistemology of social science. They are doing the philosophy of social science, whether they call it that or not. And therefore it is highly productive, on both sides of the equation, for working sociologists and philosophers to interact in a sustained way on these important foundational issues. Better social science and better philosophy will be the result.

Several specific insights about the social sciences emerge from these debates. There is the "historic turn" in the social sciences: the recognition that the history of social processes and institutions is deeply significant for understanding their current functioning. What are the contingent pathways that have brought a given phenomenon to its current configuration? How does this history inform and condition the current workings? There is the "cultural turn" in history and several areas of the social sciences: the idea that social phenomena are always invested with meaning, and it is important for social scientists to consider some of the methods through which it is possible to investigate and explore these cultural meanings. Individuals act in the context of their own understandings of the social world and their position within it; this requires an analysis of culture. There is a surging interest in identifying social causal mechanisms as a basis for explanation of social outcomes and processes: the idea that social outcomes have come to be as the result of specific social mechanisms that can be uncovered through case studies, comparative analysis, and the resort to various areas of social theory (rational choice theory, new institutionalism, learning theory). There is strengthening focus on institutions – how they constrain social processes, and how they emerge and change over time. And there is renewed attention to the role of comparison within the social sciences – comparative historical analysis.

At a more abstract level, we can identify a set of ontological insights that are emerging within these debates: doubt about the availability of strong universal laws among social phenomena; attention to the multiple pathways and structural alternatives that exist in large-scale historical development; awareness of the deep heterogeneity of social processes and influences – in time and place; attention to the substantial degree of contingency that exists in historical change; attention to the plasticity of social organizations and institutions; and attempts at bridging between micro- and macro-level social processes. Almost all of these points amount to a fundamental challenge to positivist social science. The skepticism that is now well established about social regularities is a deep blow to the positivist program. The emphasis on causal mechanisms suggests a post-positivist, realist approach to scientific explanation. The point about heterogeneity of social processes suggests the importance of theoretical pluralism rather than unified grand theories of social phenomena.

4   A Philosophy for the Social Sciences

In the remainder of this chapter I offer a particular approach to the philosophy of social science, because I believe these ideas can help to organize our thinking about the social world and the strategies we use to investigate social phenomena. It is a philosophy of social science that is highly consonant with the themes just described from contemporary research in sociology. These observations are organized within a perspective I label "methodological localism," according to which all social facts, social structures, and social causes must be understood to possess their characteristics by virtue of the "socially situated individuals" who make them up. I argue that social conjectures need to be supported by analysis of the "microfoundations" of the social structures and processes that are postulated. I argue that social explanation depends on identifying the concrete social causal mechanisms that bring about social outcomes and types of outcomes. I argue that we should think of the social world as a fabric built up out of a myriad of overlapping "local social environments." And I argue against the idea of there being strong "social laws" that could be discovered, in analogy with laws of nature. These views support a perspective on the social world that emphasizes the contingency of social outcomes, plasticity of social structures, and variability of social behavior.

4.1   *Methodological Localism*

I offer a social ontology that I refer to as *methodological localism* (ML) (Little 2006). This theory of social entities affirms that there are large social structures and facts that influence social outcomes. But it insists that these structures are only possible insofar as they are embodied in the actions and states of socially constructed individuals. The "molecule" of all social life is the socially constructed and socially situated individual, who lives, acts, and develops within a set of local social relationships, institutions, norms, and rules. With methodological individualism, this position embraces the point that individuals are the bearers of social structures and causes. There is no such thing as an autonomous social force; rather, all social properties and effects are conveyed through the individuals who constitute a population at a time. Against individualism, however, methodological localism affirms the "socialness" of social actors. Methodological localism denies the possibility or desirability of characterizing the individual extra-socially. Instead, the individual is understood as a socially constituted actor, affected by large current social facts such as value systems, social structures, extended social networks, and the like. In other words, ML denies the possibility of reductionism from the level of the social to the level of a population of non-social individuals. Rather, the individual is formed by locally embodied social facts, and the social facts are in turn constituted by the current characteristics of the persons who make them up.

This account begins with the socially constituted person. Human beings are subjective, intentional, and relational agents. They interact with other persons in ways that involve competition and cooperation. They form relationships, enmities, alliances, and networks; they compose institutions and organizations. They create material embodiments that reflect and affect human intentionality. They acquire beliefs, norms, practices, and world views, and they socialize their children, their friends, and others with whom they interact. Some of the products of human social interaction are short-lived and local (indigenous fishing practices); others are long-duration but local (oral traditions, stories, and jokes); and yet others are built up into social organizations of great geographical scope and extended duration (states, trade routes, knowledge systems). But always we have individual agents interacting with other agents, making use of resources (material and social), and pursuing their goals, desires, and impulses.

At the level of the socially constituted individual we need to ask two sorts of questions: First, what makes individual agents behave as they do? Here we need accounts of the mechanisms of deliberation and action at the level of the individual. What are the main features of individual choice, motivation, reasoning, and preference? How do these differ across social groups? How do emotions, rational deliberation, practical commitments, and other forms of agency influence the individual's deliberations and actions? This area of research is purposively eclectic, including performative action, rational action, impulse, theories of the emotions, theories of the self, or theories of identity.

Second, how are individuals formed and constituted? Methodological localism gives great importance to learning more about how individuals are formed and constituted – the concrete study of the social process of the development of the self. Here we need better accounts of social development, the acquisition of world views (cf. p. 163), preferences, and moral frameworks, among the many other determinants of individual agency and action. What are the social institutions and influences through which individuals acquire norms, preferences, and ways of thinking? How do individuals develop cognitively, affectively, and socially? So methodological localism points up the importance of discovering the microfoundations and local variations of identity formation and the construction of the historically situated self.

So far we have emphasized the socially situated individual. But social action takes place within spaces that are themselves socially structured by the actions and purposes of others – by property, by prejudice, by law and custom, and by systems of knowledge. So our account needs to identify the local social environments through which action is structured and projected: the interpersonal networks, the systems of rules, the social institutions. The social thus has to do with the behaviorally, cognitively, and materially embodied reality of *social institutions*. An institution, we might say, is an embodied set of rules, incentives, and opportunities that have the potential to influence agents' choices and behavior.[3] An institution is a complex of socially embodied powers, limitations, and opportunities within which individuals pursue their lives and goals. A property system, a legal system, and a professional baseball league all represent examples of institutions.[4] Institutions have effects that are in varying degrees independent from the individual or "larger" than the individual. Each of these social entities is embodied in the social states of a number of actors – their beliefs, intentions, reasoning, dispositions, and histories. Actors perform their actions within the context of social frameworks represented as rules, institutions, and organizations, and their actions and dispositions embody the causal effectiveness of those

[3] "Institutions are the humanly devised constraints that structure human interaction. They are made up of formal constraints (e.g., rules, laws, constitutions), informal constraints (e.g., norms of behavior, conventions, self-imposed codes of conduct), and their enforcement characteristics. Together they define the incentive structure of societies and, specifically, economies" (North 1998: 247).

[4] See Brinton and Nee (1998), North (1990), Ensminger (1992), and Knight (1992) for recent expositions of the new institutionalism in the social sciences.

frameworks. And institutions influence individuals by offering incentives and constraints on their actions, by framing the knowledge and information on the basis of which they choose, and by conveying sets of normative commitments (ethical, religious, interpersonal) that influence individual action.

It is important to emphasize that ML affirms the existence of social constructs beyond the purview of the individual actor or group. Political institutions exist – and they are embodied in the actions and states of officials, citizens, criminals, and opportunistic others. These institutions have real effects on individual behavior and on social processes and outcomes – but always mediated through the structured circumstances of agency of the myriad participants in these institutions and the affected society. This perspective emphasizes the contingency of social processes, the mutability of social structures over space and time, and the variability of human social systems (norms, urban arrangements, social practices, and so on).

This approach highlights the important point that all social facts, social structures, and social causal properties depend ultimately on facts about individuals within socially defined circumstances. Social ascriptions require *microfoundations* at the level of individuals in concrete social relationships. According to this way of understanding the nature of social ontology, an assertion of a structure or process at the macro-social level (causal, functional, structural) must be supplemented by two things: knowledge about the local circumstances of the typical individual that leads him or her to act in such a way as to bring about this relationship; and knowledge of the aggregative processes that lead from individual actions of that sort to an explanatory social relationship of this sort.[5] So if we are interested in analysis of the causal properties of states and governments, we need to arrive at an analysis of the institutions and constrained patterns of individual behavior through which the state's characteristics are effected. We need to raise questions such as: How do states exercise influence throughout society? What are the institutional embodiments at lower levels that secure the impact of law, taxation, conscription, contract enforcement, and other central elements of state behavior?[6] If we are concerned about the workings of social

---

[5] We may refer to explanations of this type as "aggregative explanations." An aggregative explanation provides an account of a social mechanism that conveys multiple individual patterns of activity and demonstrates the collective or macro-level consequence of these actions. Thomas Schelling's *Micromotives and Macrobehavior* (Schelling 1978) provides a developed treatment and numerous examples of this model of social explanation.

[6] An excellent recent example of historical analysis of Chinese local politics illustrates the value of this microfoundational approach: "But the villages were not totally out of the government's reach; nor was the subcounty administration necessarily chaotic,

identities, then we need to inquire into the concrete social mechanisms through which social identities are reproduced within a local population – and the ways in which these mechanisms and identities may vary over time and place. And if we are interested in analyzing the causal role that systems of norms play in social behavior, we need to discover some of the specific institutional practices through which individuals come to embrace a given set of norms.[7]

The microfoundations perspective requires that we attempt to discover the pathways by which socially constituted individuals are influenced by distant social circumstances, and how their actions in turn affect distant social outcomes. There is no action at a distance in social life; instead, individuals have the values that they have, the styles of reasoning, the funds of factual and causal beliefs, etc., as a result of the structured experiences of development that they have undergone as children and adults. On this perspective, large social facts and structures do indeed exist; but their causal properties are entirely defined by the current states of psychology, norm, and action of the individuals who currently exist. Systems of norms and bodies of knowledge exist – but only insofar as individuals (and material traces) embody and transmit them. So when we assert that a given social structure causes a given outcome, we need to be able to specify the local pathways through which individual actors embody this causal process. That is, we need to be able to provide an account of the causal mechanisms that convey social effects.

It is evident that methodological localism implies a fairly limited social ontology. What exists is the socially constructed individual, within a congeries of concrete social relations and institutions. The socially constructed individual possesses beliefs, norms, opportunities, powers, and capacities. These features are socially constructed in a perfectly ordinary sense: the individual has acquired his or her beliefs, norms, powers, and desires through social contact with other individuals and institutions, and the powers and constraints that define the domain of choice for the individual are largely constituted by social institutions (property

---

inefficient, and open to malfeasance. In fact, during most of the imperial times, the state was able to extract enough taxes to meet its normal needs and maintain social order in most of the country. What made this possible was a wide variety of informal institutions in local communities that grew out of the interaction between government demands and local initiatives to carry out day-to-day governmental functions" (Li 2005: 1).

[7] "Explanations of social norms must do more than merely acknowledge the constraining effects of normative rules on social action. Such explanations must address the process that culminates in the establishment of one of these rules as the common norm in a community. One of the keys to the establishment of a new norm is the ability of those who seek to change norms to enforce compliance with the new norm" (Knight and Ensminger 1998: 105).

systems, legal systems, educational systems, organizations, and the like).[8] Inevitably, social organizations at any level are constituted by the individuals who participate in them and whose behavior and ideas are influenced by them; sub-systems and organizations through which the actions of the organization are implemented; and the material traces through which the policies, memories, and acts of decision are imposed on the environment: buildings, archives, roads, etc. All features of the organization are embodied in the actors and institutional arrangements that carry the organization at a given time. At each point we are invited to ask the question: what are the social mechanisms through which this institution or organization exerts influence on other organizations and on agents' behavior?

Methodological localism has numerous intellectual advantages. It avoids the reification of the social that is characteristic of holism and structuralism, it abjures social "action at a distance," and it establishes the intellectual basis for understanding the non-availability of strong laws of nature among social phenomena. It is possible to offer numerous examples of social research underway today that illustrate the perspective of methodological localism; in fact, almost all rigorous social theorizing and research can be accommodated to the assumptions of methodological localism.[9]

4.2   The Centrality of Causal Mechanisms in Social Explanations

A second major component of this philosophy of social science has to do with the most fundamental assumptions we make about the nature of explanation. I maintain that social explanation requires discovery of the underlying causal mechanisms that give rise to outcomes of interest.[10]

---

[8] Hacking (1999) offers a critique of misuses of the concept of social construction. This use is not vulnerable to his criticisms, however.

[9] Examples of theories and analyses contained in current comparative and historical social science research may be found in McDonald (1996) and Mahoney and Rueschemeyer (2003). Many of these examples illustrate the fecundity of the approach to social analysis that emphasizes the "socially constructed individual within a concrete set of social relations" as the molecule of social action. See also Sewell (2005) for a treatment of historical change that is compatible with this approach.

[10] Little (1991, 1998). Important recent exponents of the centrality of causal mechanisms in social explanation include Hedström and Swedberg (1998); McAdam, Tarrow, and Tilly (2001) and George and Bennett (2005). George and Bennett offer careful analytical treatment of how the causal mechanisms approach can be developed into specific research strategies in the social and historical sciences. Volume 34, Nos 2 and 3 of *Philosophy of the Social Sciences* (2004) contains a handful of articles devoted to the logic of social causal mechanisms, focused on the writings of Mario Bunge. Steel (2004) provides a philosophically rigorous assessment of several features of the mechanisms approach and directs particular attention to the claim that discovery of

On this perspective, explanation of a phenomenon or regularity involves identifying the causal processes and causal relations that underlie this phenomenon or regularity, and causal mechanisms are heterogeneous.[11] Social mechanisms are concrete social processes in which a set of social conditions, constraints, or circumstances combine to bring about a given outcome.[12] On this approach, social explanation does not take the form of inductive discovery of laws; rather, the generalizations that may be discovered in the course of social science research are subordinate to the more fundamental search for causal mechanisms and pathways in individual outcomes and sets of outcomes.[13] This approach casts doubt on the search for generalizable theories across numerous societies. It looks instead for specific causal influence and variation. The approach emphasizes variety, contingency, and the availability of alternative pathways leading to an outcome, rather than expecting to find a small number of common patterns of development or change.[14] The contingency

mechanisms is necessary for successful causal explanation. One specific interpretation of the idea of a social mechanism is formulated by Tyler Cowen: "I interpret social mechanisms ... as rational-choice accounts of how a specified combination of preferences and constraints can give rise to more complex social outcomes" (Cowen 1998: 125). The account offered here is not limited to rational choice mechanisms, however.

[11] Other philosophers of science have argued for a similar position in the past decade. See particularly Salmon (1984) and Cartwright (1983, 1989).

[12] The recent literature on causal mechanisms provides a number of related definitions: "We define causal mechanisms as ultimately unobservable physical, social, or psychological processes through which agents with causal capacities operate, but only in specific contexts or conditions, to transfer energy, information, or matter to other entities" (George and Bennett 2005). "Mechanisms are a delimited class of events that alter relations among specified sets of elements in identical or closely similar ways over a variety of situations" (McAdam *et al.* 2001: 24). "Mechanisms ... are analytical constructs that provide hypothetical links between observable events" (Hedström and Swedberg 1998: 13).

[13] Authors who have urged the centrality of causal mechanisms or powers for social explanation include Harré and Secord (1972), Cartwright (1983), Salmon (1984), Cartwright (1989), Dessler (1991), Varela and Harré (1996), and Sørensen (2001). In their important volume devoted to this topic, Hedström and Swedberg write, "The main message of this book is that the advancement of social theory calls for an analytical approach that systematically seeks to explicate the social mechanisms that generate and explain observed associations between events" (Hedström and Swedberg 1998: 1). Jack Goldstone expresses this method in his analysis of the explanation of revolutions: "In making my argument ... that population growth in agrarian bureaucracies led to revolution, I did not proceed by showing that in a large sample of such states there was a statistically significant relationship between population growth and revolution ... Rather, I sought to trace out and document the links in the causal chain connecting population growth to revolutionary conflict" (Goldstone 2003: 48).

[14] An important expression of this approach to social and historical explanation is offered by Charles Tilly: "Analysts of large-scale political processes frequently invoke invariant models that feature self-contained and self-motivating social units. Few actual political processes conform to such models. Revolutions provide an important example of such reasoning and of its pitfalls. Better models rest on plausible ontologies, specify

of particular pathways derives from several factors, including the local circumstances of individual agency and the across-case variation in the specifics of institutional arrangements – giving rise to significant variation in higher-level processes and outcomes.[15]

This approach places central focus on the idea of a causal mechanism. To identify a causal relation between two kinds of events or conditions, we need to identify the typical causal mechanism through which the first kind brings about the second kind. What, though, is the nature of the social linkages that constitute causal mechanisms among social phenomena? I argued above for a microfoundational approach to social causation: the causal properties of social entities derive from the structured circumstances of agency of the individuals who make up social entities – institutions, organizations, states, economies, and the like. So this approach advances a general ontological stance and research strategy: the causal mechanisms that create causal relations among social phenomena are compounded from the structured circumstances of choice and behavior of socially constructed and socially situated agents. And hypotheses about social mechanisms will generally find their grip on the social world through identifying features of agency, features of institutional setting, or both.

The general answers I propose as a theory of social causation flow from the assumption that the "molecule" of social causation is the deliberative agency of the socially situated agent – what is referred to here as "methodological localism." This perspective affirms that there is such a thing as social causation. Social structures and institutions have causal properties and effects that play an important role within historical change (the social causation thesis). They exercise their causal powers through their influence on individual actions, beliefs, values, and choices (the microfoundations thesis). Structures are themselves constituted by individuals, so social causation and agency represent an ongoing iterative process (the agency–structure thesis). There are no causal powers at work within the domain of the social that do not proceed through structured individual agency. And hypotheses concerning social and historical causation can be rigorously formulated, criticized,

fields of variation for the phenomena in question, reconstruct causal sequences, and concentrate explanation on links within those sequences" (Tilly 1995: 1594).

[15] McAdam *et al.* describe their approach to the study of social contention in these terms: "We employ mechanisms and processes as our workhorses of explanation, episodes as our workhorses of description. We therefore make a bet on how the social world works: that big structures and sequences never repeat themselves, but result from differing combinations and sequences of mechanisms with very general scope. Even within a single episode, we will find multiform, changing, and self-constructing actors, identities, forms of action and interaction" (McAdam *et al.* 2001: 30).

and defended using a variety of tools: case-study methodology, comparative study, statistical study, and application of social theory.[16]

Is there such a thing as "macro–macro" causation – that is, causal relations between higher-level social structures? Yes – but only as mediated through "micro-foundations" at the level of structured human agency. State institutions affect economic variables such as "levels of investment," "levels of unemployment," or "infant mortality rates." But large institutions only wield causal powers by changing the opportunities, incentives, powers, and constraints that confront agents. So the hard question is not: "Do institutions and structures exercise autonomous and supra-individual causal primacy?," since we know that they do not. Instead, the question is, "To what extent and through what sorts of mechanisms do structures and institutions exert causal influence on individuals and other structures?"

On this approach, the causal capacities of social entities are to be explained in terms of the structuring of preferences, world views, information, incentives, and opportunities for agents. The causal powers or capacities of a social entity inhere in its power to affect individuals' behavior through incentives, preference-formation, belief-acquisition, or powers and opportunities. The micro-mechanism that conveys cause to effect is supplied by an account of the actions of agents with specific goals, beliefs, and powers, within a specified set of institutions, rules, and normative constraints.

4.3   Generalizations and Predictions

A final large issue for the philosophy of social science is the question of the availability of general laws that might be said to govern social processes and outcomes. Is the social world law-governed? Are there social laws, analogous to laws of nature, that give rise to social phenomena? And is it a central task of the social sciences to discover law-like regularities among social phenomena? Some social scientists believe that the scientific credentials of their disciplines rise or fall on the strength of the law-like generalizations and regularities that they are able to identify.[17] The task of social science research is to discover the laws that govern social processes. I disagree with this view (Little 1993). The general view I defend is that there are regularities to be found within the social

---

[16] See Little (1998) for further exposition of these ideas.
[17] Social scientists who have taken this view include Adelman and Morris (1967), Huntington (1968), Zuckerman (1991), and King *et al.* (1994). Philosophers supporting the idea that law-like generalizations must undergird social explanations include Mill (1879), Hempel (1942), Thomas (1979), Kincaid (1990), and McIntyre (1996).

sciences, at a variety of levels of social description; that these regularities derive from features of individual agency in the context of specific social arrangements; and that the discovery of such regularities is one important goal of social science research. But I also maintain that these regularities are substantially weaker than those that obtain among natural phenomena. Moreover, I maintain that these regularities have a much more limited role within good social explanations than they are often thought to do. The upshot of these arguments is that we should have little confidence in the projectability of social regularities as a basis for prediction, and must therefore pay more attention to the specifics of the social and individual-level mechanisms that produce the regularities as well as the exceptions.

Much of the impulse toward emphasizing the explanatory importance of regularities in the social sciences derives from a misleading analogy with the natural sciences. The successes of the natural sciences have given natural scientists confidence that natural systems operate in accordance with a strict set of laws, that these laws may be given precise mathematical formulation, that they derive from the underlying real properties of constituent physical entities, and, finally, that these facts entail that the future behavior of physical systems is in principle (though perhaps not in practice) predictable. And this conception of the nature of physical systems in turn gave rise to a paradigm of scientific explanation: to explain a phenomenon is to derive the explanandum from a set of general laws and a description of the initial conditions of the system. Prediction and explanation go hand in hand, and both depend on the availability of empirically supportable general laws.[18] However, I do not believe that this is a good way of understanding social phenomena.

We need always to keep in mind that social phenomena are the consequence of intentional human actions (sometimes in vast numbers). Some social regularities are the intended consequence of individual action (e.g., the revenue-maximizing property of typical states follows from the interest that government officials have in maximizing the power and income of the state). Others are the unintended consequence of purposive individual action; e.g., the falling rate of profit is the unintended consequence of profit-maximizing strategies by large numbers of individual capitalists (according to Marx). But in either case the regularity is strictly derivative from the constitution and powers of the individuals whose actions give rise to the social phenomenon in question. And individuals are not homogeneous or predictable.

---

[18] Note, however, the powerful arguments against this general view put forward by Nancy Cartwright in *How the Laws of Physics Lie* (1983).

It might be thought that the emphasis on causal mechanisms in the previous section implies that there ought to be discernable regularities among the phenomena to which these mechanisms give rise.[19] After all, if we have discovered a causal mechanism working in one set of circumstances, then it should have the "same" effect when working in a relevantly similar set of circumstances. Causal mechanisms are robust across settings, so they should be expected to support generalizations across cases under the rare circumstance that they are operating singly or dominantly; therefore there ought be a regularity across those relevantly similar circumstances. This is so in the abstract; but circumstances in the social world are rarely sufficiently similar to give rise to such macro-level regularities. The problem is that causal mechanisms combine with each other, and interact with contingent initial conditions, in ways that make their effects difficult to observe in macro–macro comparisons. Causal mechanisms are multiple and conjunctural; it is very unusual to find multiple settings embodying just one causal mechanism. Macro-states initiate and embody multiple dynamic causal mechanisms. So there are few macro–macro regularities. Making the situation even worse, many macro-concepts cover a causally heterogeneous set of social phenomena. "States" and "riots" enclose multiple types of social action. So, once again, we should be cautious about the possibility of discovering macro–macro generalizations; it seems likely that the best we can hope for at the macro–macro level are weak and exception-laden generalizations.

Consider an example. Are there laws of the modern state as an institution? Social scientists have discerned a variety of regularities concerning the state: states maximize revenues (Levi 1988); state crises cause revolutions (Skocpol 1979); states create entrenched bureaucracies; and the like. These statements represent generalizations across a number of cases, and they are intended to have counterfactual import. But it is plain that these generalizations are subordinate to our hypotheses about the underlying institutional and individual-level circumstances that give rise to the identified regularities of state behavior. States conform to regularities because they are the product of a number of agents whose purposes, powers, and opportunities are similar in many different social contexts; this leads to a regularity of state behavior. But there is no intrinsic "inner nature" to all states that would constitute the ontological basis for strong regularities governing state behavior.

---

[19] Comments at the Herdecke conference by Jack Knight, David-Hillel Ruben, and James Alt underlined this observation.

An important consequence of this analysis is that the predictive capacity of the social sciences is very limited. It is certainly possible to make predictions based on microfoundational reasoning, discovery of causal mechanisms, and the like, and it is reasonable to do so for a variety of purposes. However, we should not have a great deal of confidence in the resulting predictions. Causal analysis is conditioned by *ceteris paribus* clauses, incomplete causal fields, and other problems – with the result that predicted outcomes of a given analysis may well fail to obtain because the conditions are violated. And crude inductive generalizations (e.g., "recessions during election years are usually followed by change of party in office") have limited applicability to particular cases because of the complexity and conditionality of the causal circumstances that underlie them. So I conclude that the search for law-like generalizations that permit confident predictions is *not* one of the most central tasks for the social sciences.

So I argue for a position that is neither positivist nor anti-positivist; instead, call it *post-positivist*. Against the main line of positivist philosophy of science, I dispute the idea that law-like generalizations are fundamental to successful scientific explanation. But against current anti-positivist criticisms among some social scientists, I argue for causal realism in social explanation: causal explanation is at the core of much social research, and causal hypotheses depend on appropriate stand ards of empirical confirmation for their acceptability. Finally, successful causal analysis permits us to arrive at statements of social regularities – this time, however, based on an understanding of the underlying processes that give rise to them. And this understanding permits us to assess the limits and conditions of the regularities we affirm, and the likelihood of failure of these regularities in various social circumstances.

This leads to a fairly clear conclusion. Social regularities emerge rather than govern. The governing regularities are regularities of individual agency: the principles of rational choice theory or the findings of social psychology. Social regularities are strictly consequent, not governing. They obtain because of the lower-level regularities; they have no independent force (unlike a common interpretation of the force of the laws of gravitation). Descriptive regularities among social phenomena can be discovered. But these are distinctly phenomenal laws rather than governing regularities; they have little explanatory import; and they are not particularly reliable as a basis for prediction. Better on both counts is a theory of the underlying causal mechanisms that produce them. These theories in turn need to be supported empirically. We can be most confident in statements of law like regularities when we have an account of the mechanisms that underlie them.

5   Conclusions

I believe that the social sciences need to be framed out of consideration of a better understanding of the nature of the social – a better *social ontology*. The social world is not a system of law-governed processes; it is instead a mix of different sorts of institutions, forms of human behavior, natural and environmental constraints, and contingent events. The entities that make up the social world at a given time and place have no particular ontological stability; they do not fall into "natural kinds"; and there is no reason to expect deep similarity across a number of ostensibly similar institutions – states, for example, or labor unions. So the rule for the social world is heterogeneity, contingency, and plasticity. And the metaphysics associated with our thinking about the natural world – laws of nature; common, unchanging structures; and predictable processes of change – do not provide appropriate metaphors for our understandings and expectations of the social world; nor do they suggest the right kinds of social science theories and constructs.

Instead of naturalism, I advocate an approach to social science theorizing that could be described as *post-positivist*. It recognizes that there is a degree of pattern in social life – but emphasizes that these patterns fall far short of the regularities associated with laws of nature. It emphasizes contingency of social processes and outcomes. It insists upon the importance and legitimacy of the eclectic use of social theory: the processes are heterogeneous, so it is appropriate to appeal to different types of social theories as we explain social processes. It emphasizes the importance of path-dependence in social outcomes. It suggests that the most valid scientific statements in the social sciences have to do with the discovery of concrete social-causal mechanisms, through which some types of social outcomes come about. And finally, it highlights what I call "methodological localism": the insight that the foundation of social action and outcome is the local, socially located and socially constructed individual person. The individual is *socially constructed*, in that her modes of behavior, thought, and reasoning are created through a specific set of prior social interactions. And her actions are *socially situated*, in the sense that they are responsive to the institutional setting in which she chooses to act. Purposive individuals, embodied with powers and constraints, pursue their goals in specific institutional settings; and regularities of social outcome often result.

My most central conclusion, however, is ontological. We ought not think of the social world as a *system* of phenomena in which we can expect to find a strong underlying order. Instead, social phenomena are highly diverse, subject to many different and cross-cutting forms of causation. So the result is that the very strongest regularities that will be ever be discerned will remain the exception-laden phenomenal regularities described here and the highly qualified predictions of regularities that derive from structure–agent analyses. There is no more fundamental description of the social world in which strong governing regularities drive events and processes.

It is highly relevant that the examples discussed above of innovative work on the most foundational questions in the social sciences today are very consistent with the broad thrust of this philosophy of social science. The emphasis on discovering concrete social causal mechanisms; the insistence on the contingency and variability of social outcomes; the emphasis on the path-dependent character of many social outcomes; the recognition of the centrality of socially situated actors as the substance of social relationships; the recognition of the pliability and plasticity of important social institutions and structures – all of these provide intellectual support for the main thrusts of the philosophy of social science offered here.

These points are highly relevant to the theme of the future of the social sciences. If social scientists are captivated by the scientific prestige of positivism and quantitative social science, they will be led to social science research that looks quite different from what would result from a view that emphasizes contingency and causal mechanisms. And if there are strong, engaging examples of other ways of conducting social research that can come into broad exposure in social science – then there is a greater likelihood for the emergence of a genuinely innovative and imaginative approach to the problem of social knowledge.

REFERENCES

Abbott, A. D. 1988. *The System of Professions: An Essay on the Division of Expert Labor*. Chicago, IL: University of Chicago Press.
    1999. *Department & Discipline: Chicago Sociology at One Hundred*. Chicago, IL: University of Chicago Press.
    2001. *Chaos of Disciplines*. Chicago, IL: University of Chicago Press.
Adams, J., E. S. Clemens, and A. S. Orloff (eds.) 2005. *Remaking Modernity: Politics, History, and Sociology (Politics, History, and Culture)*. Durham, NC: Duke University Press.
Adelman, I. and C. T. Morris 1967. *Society, Politics, & Economic Development: A Quantitative Approach*. Baltimore: Johns Hopkins Press.
Brinton, M. C. and V. Nee (eds.) 1998. *The New Institutionalism in Sociology*. New York, NY: Russell Sage Foundation.
Cartwright, N. 1983. *How the Laws of Physics Lie*. Oxford: Oxford University Press.
    1989. *Nature's Capacities and their Measurement*. Oxford: Oxford University Press.

Cowen, T. 1998. "Do Economists Use Social Mechanisms to Explain?" In P. Hedström and R. Swedberg (eds.) *Social Mechanisms: An Analytical Approach to Social Theory.*

Darnton, R. 1984. *The Great Cat Massacre and Other Episodes in French Cultural History.* New York, NY: Basic Books.

Dessler, D. 1991. "Beyond correlations: Toward a causal theory of war," *International Studies Quarterly* 35: 337–355.

Dobbin, F. 1994. *Forging Industrial Policy: The United States, Britain, and France in the Railway Age.* Cambridge, New York, NY: Cambridge University Press.

Ensminger, J. 1992. *Making a Market: The Institutional Transformation of an African Society, The Political Economy of Institutions and Decisions.* Cambridge, New York, NY: Cambridge University Press.

George, A. L. and A. Bennett 2005. *Case Studies and Theory Development in the Social Sciences, BCSIA Studies in International Security.* Cambridge, MA: MIT Press.

Goertz, G. and H. Starr (eds.) 2003. *Necessary Conditions: Theory, Methodology, and Applications.* Boulder, CO: Rowman & Littlefield.

Goldstone, J. A. 1991. *Revolution and Rebellion in the Early Modern World.* Berkeley, CA: University of California Press.

    2003. "Comparative Historical Analysis and Knowledge Accumulation in the Study of Revolutions." In J. Mahoney and D. Rueschemeyer (eds.) *Comparative Historical Analysis in the Social Sciences.*

Gouldner, A. W. 1970. *The Coming Crisis of Western Sociology.* New York, NY: Basic Books.

Gulbenkian Commission on the Restructuring of the Social Sciences 1996. *Open the Social Sciences:Report of the Gulbenkian Commission on the Restructuring of the Social Sciences.* Stanford, CA: Stanford University Press.

Hacking, I. 1999. *The Social Construction of What?* Cambridge, MA: Harvard University Press.

Harré, R. and P. F. Secord 1972. *The Explanation of Social Behaviour.* Oxford: Blackwell.

Hedström, P. and R. Swedberg (eds.) 1998. *Social Mechanisms: An Analytical Approach to Social Theory, Studies in Rationality and Social Change.* Cambridge, New York, NY: Cambridge University Press.

Hempel, C. 1942. "The Function of General Laws in History." In C. Hempel (ed.) *Aspects of Scientific Explanation.*

Huntington, S. P. 1968. *Political Order in Changing Societies.* New Haven, CT: Yale University Press.

Kincaid, H. 1990. "Defending laws in the social sciences," *Philosophy of the Social Sciences* 20(1): 56–83.

King, G., R. O. Keohane, and S. Verba 1994. *Designing Social Inquiry: Scientific Inference in Qualitative Research.* Princeton, NJ: Princeton University Press.

Klitgaard, R. E. 1988. *Controlling Corruption.* Berkeley, CA: University of California Press.

Knight, J. 1992. *Institutions and Social Conflict (Political Economy of Institutions and Decisions).* Cambridge, New York, NY: Cambridge University Press.

Knight, J. and J. Ensminger 1998. "Conflict over Changing Social Norms: Bargaining, Ideology, and Enforcement." In M. C. Brinton and V. Nee (eds.) *The New Institutionalism in Sociology.*

Levi, M. 1988. *Of Rule and Revenue.* Berkeley, CA: University of California Press.

Li, H. 2005. *Village Governance in North China, 1875–1936.* Stanford, CA: Stanford University Press.

Lieberson, S. 1963. *Ethnic Patterns in American Cities.* New York, NY: Free Press of Glencoe.

1980. *A Piece of the Pie: Blacks and White Immigrants since 1880.* Berkeley, CA: University of California Press.

1987. *Making It Count: The Improvement of Social Research and Theory.* Berkeley, CA: University of California Press.

1997. "Modeling social processes: Some lessons from sports," *Sociological Forum* **12**(1): 11–35.

2000. *Matter of Taste: How Names, Fashions, and Culture Change.* New Haven, CT: Yale University Press.

Lieberson, S. and F. B Lynn. 2002. "Barking up the wrong branch: Scientific alternatives to the current model of sociological science," *Annual Review of Sociology* **28**(1): 1–19.

Lieberson, S. and M. C. Waters 1988. *From Many Strands: Ethnic and Racial Groups in Contemporary America (The Population of the United States in the 1980s).* New York, NY: Russell Sage Foundation.

Little, D. 1991. *Varieties of Social Explanation: An Introduction to the Philosophy of Social Science.* Boulder, CO: Westview Press.

1993. "On the scope and limits of generalizations in the social sciences," *Synthese* **97**: 183–207.

1998. *Microfoundations, Method and Causation: On the Philosophy of the Social Sciences.* New Brunswick, NJ: Transaction Publishers.

2006. "Levels of the Social." In S. Turner and M. Risjord (eds.) *Handbook for Philosophy of Anthropology and Sociology.* Amsterdam, New York, NY. Elsevier.

Mahoney, J. and D. Rueschemeyer 2003. *Comparative Historical Analysis in the Social Sciences (Cambridge Studies in Comparative Politics).* Cambridge, New York, NY: Cambridge University Press

McAdam, D., J. D. McCarthy, and M. N. Zald 1996. *Comparative Perspectives on Social Movements: Political Opportunities, Mobilizing Structures, and Cultural Framings (Cambridge Studies in Comparative Politics).* Cambridge, New York, NY: Cambridge University Press.

McAdam, D., S. G. Tarrow, and C. Tilly 2001. *Dynamics of Contention (Cambridge Studies in Contentious Politics).* New York, NY: Cambridge University Press.

McDonald, T. J. (ed.) 1996. *The Historic Turn in the Human Sciences.* Ann Arbor, MI: University of Michigan Press.

McIntyre, L. C. 1996. *Laws and Explanation in the Social Sciences: Defending a Science of Human Behavior.* Boulder, CO: Westview Press.

Mill, J. S. 1879. *A System of Logic, Ratiocinative and Inductive, Being a Connected View of the Principles of Evidence and the Methods of Scientific Investigation* (10th edn). London: Longmans Green and Co.

North, D. C. 1990. *Institutions, Institutional Change and Economic Performance.* Cambridge: Cambridge University Press.

1998. "Economic Performance through Time." In M. C. Brinton and V. Nee (eds.) *The New Institutionalism in Sociology.*

Ortner, S. B. (ed.) 1999. *The Fate of "Culture": Geertz and Beyond.* Berkeley, CA: University of California Press.

Salmon, W. C. 1984. *Scientific Explanation and the Causal Structure of the World.* Princeton, NJ: Princeton University Press.

Schelling, T. C. 1978. *Micromotives and Macrobehavior.* New York, NY: Norton.

Sewell, W. H. 2005. *Logics of History: Social Theory and Social Transformation (Chicago Studies in Practices of Meaning).* Chicago, IL: University of Chicago Press.

Skocpol, T. 1979. *States and Social Revolutions: A Comparative Analysis of France, Russia, and China.* Cambridge, New York, NY: Cambridge University Press.

Sørensen, A. B. 2001. "Careers, Wealth and Employment Relations." In A. L. Kalleberg and I. Berg (eds.) *Source Book on Labor Markets: Evolving Structures and Processes.* New York, NY: Plenum Press.

Steel, D. 2004. "Social mechanisms and causal inference," *Philosophy of the Social Sciences* **34**(1): 55–78.

Steinmetz, G. (ed.) 2005. *The Politics of Method in the Human Sciences: Positivism and its Epistemological Others, Politics, History, and Culture.* Durham, NC: Duke University Press.

Thelen, K. A. 2004. *How Institutions Evolve: The Political Economy of Skills in Germany, Britain, the United States, and Japan (Cambridge Studies in Comparative Politics).* Cambridge, New York, NY: Cambridge University Press.

Thomas, D. 1979. *Naturalism and Social Science: A Post-empiricist Philosophy of Social Science (Themes in the Social sciences).* Cambridge, New York, NY: Cambridge University Press.

Tilly, C. 1995. "To explain political processes," *American Journal of Sociology* **100**: 1594–1610.

2003. *The Politics of Collective Violence (Cambridge Studies in Contentious Politics).* Cambridge, New York, NY: Cambridge University Press.

Turner, S. and M. Risjord 2006. *Handbook for Philosophy of Anthropology and Sociology.* In D. Gabbay, P. Thagard, and J. Woods (eds.) Vol. 15, *Handbook of the Philosophy of Science.* Amsterdam; New York, NY: Elsevier Science.

Varela, C. R. and R. Harré 1996. "Conflicting varieties of realism: Causal powers and the problem of social structure," *Journal for the Theory of Social Behaviour* **26**(3): 313–325.

Wallerstein, I. M. 1999. *The End of the World As We Know It: Social Science for the Twenty-first Century.* Minneapolis, MN: University of Minnesota Press.

Zuckerman, A. S. 1991. *Doing Political Science: An Introduction to Political Analysis.* Boulder, CO: Westview Press.

6 – Comment
Causal Mechanisms and Generalizations

Jack Knight

In his provocative chapter Daniel Little sets out to address an important set of questions about the ability of the social sciences to explain contemporary social change. In doing so he raises some important challenges to the dominant approaches in the social sciences today. Much of what he proposes, both as critique and as recommended reform, is thoughtful and persuasive. However, in this brief response I want to raise an important question about his new "post-positivist" alternative.

Little characterizes social change in the following ways: "momentous and globally unprecedented," "complex and large," with causal connections that "are obscure," and effects that "are large and unpredictable." He argues that it is thus a mistake to think that it is possible to create comprehensive social theories to explain such change. In support of this he emphasizes that "social development is a contingent, multi-threaded social fabric" that involves "unpredictable twists and turns" and "myriad ... contingent and variegated forms of social processes."

This characterization of social change seems about right. As I see it, Little has accurately highlighted the complex nature of the trajectory of modern social development and change. And so, like many other contemporary scholars, Little cautions us against embarking on a fool's venture down the path of comprehensive theory-building. This debate about the relative merits of comprehensive vs partial theories of social history is an ongoing one. Although I think that there is considerable merit in the claims of those like Little who doubt the explanatory value of grand macro theory, this is not the aspect of Little's argument that I want to address here. For there is a more challenging methodological critique of contemporary social science research that Little makes the primary focus of this chapter. I should note that it is not clear to me whether or not Little thinks that this methodological critique is conceptually related to the rejection of comprehensive theory. However, since I do not see that the one necessarily follows from the other (i.e., one could favor or oppose grand theory without taking a position on the methodological aspects of the critique), I will treat them separately for purposes of this response.

As the main focus of his chapter Little pursues a critique of what he calls "a tempting response to this sociological challenge" of explaining complex social change. The methodological approach that he wants to reject is "positivist quantitative social science," large n-studies of social phenomena organized either as intra-society or inter-society analyses. Little is clearly opposed to such an approach. He tries to temper his opposition by acknowledging that there is a need for "powerful quantitative tools," but he is not able to find a significant place for them once he begins to offer his impressive survey of good work in the social science research on social change. It seems to me that one can agree, as I do, with Little's basic claim that "we should not imagine that these tools are appropriate to every research question in the social sciences" while at the same time acknowledging that there is an important role for this methodology, even for the task of explaining long-term social change.

From my perspective the most challenging critique that Little offers to contemporary social scientists goes to the question of generalization. Little draws a provocative lesson from his analysis of the complexity of social change: "the 'systematizing' or universalizing impulse that has guided the development of so many of the sciences is less valid in the social sciences." This is an important theme throughout this chapter. He says, in a discussion of the fundamental importance of causal mechanisms, that on the approach that he advocates (more on this below) "social explanation does not take the form of inductive discovery of laws ... This approach casts doubt on the search for generalizable theories across numerous societies." In a subsequent discussion of the specific question of the implications of his approach for our ability to generalize across cases, Little argues that whatever regularities we are able to identify in the social sciences "are substantially weaker than those that obtain among natural phenomena." He concludes that "these regularities have a much more limited role within good social explanations than they are often thought to do."

This claim about regularities and their lack of significant generalizability strikes me as the single most encompassing challenge that Little makes for contemporary social scientists. The vast majority of social scientists who understand their research as grounded in the standard tenets of the positivist tradition will quite obviously reject this challenge to the importance of generalization. However, many scholars would agree with Little's claim that universal laws should not be the basis of social scientific explanations. Similarly, many would agree that prediction is overrated as the criterion for successful research in the social sciences. Nonetheless, many of these same scholars would likely question efforts to undermine the pursuit of generalizations in their research, including many of the fine scholars whom Little rightly recognizes in his survey of good work in the social sciences. So, it might be worthwhile to consider whether the conceptual and methodological approach that Little offers as an alternative, what he calls a post-positivist approach, is really inconsistent with a significant role for generalizations.

Fundamental to Little's proposed research approach is the pursuit of causal mechanisms: "social explanation requires discovery of the underlying causal mechanisms that give rise to outcomes of interest." He endorses a particular conceptual and methodological approach to the discovery of these mechanisms, methodological localism (ML). There is much to recommend about this approach. I will not re-describe all of the details here, but I will emphasize certain features of ML that I find especially appealing as a conceptual framework for social scientific explanation. First and foremost, Little places the importance of micro-foundations at the centerpiece of the approach. He emphasizes that "individuals are the bearers of social structure and causes." Second, he insists on the "social-ness" of individual actors. There are two relevant senses of this social-ness in Little's account. In one sense "social-ness" entails that individuals are socially constituted actors "formed by locally embodied social facts." In another sense "social-ness" requires us to study the actions of these individual actors in the context of the culture and social institutions in which we find them. But, having emphasized that individuals and their actions must be understood in terms of their "social-ness," Little appropriately insists on the "important point that all social facts, social structures, and social causal properties depend ultimately on facts about individuals within socially defined circumstances."

As characterized, ML has the following implication for how we should identify the relevant and appropriate causal mechanisms employed in the explanations of social phenomena: "the causal mechanisms that create causal relations among social phenomena are compounded from the structured circumstances of choice and behavior of socially constructed and socially situated agents. And hypotheses about social mechanisms will generally find their grip on the social world through identifying features of agency, features of institutional setting, or both." I find this approach compelling in its emphasis on both agency and institutional context. It is the basic conceptual approach that I would recommend for the social sciences. And it characterizes, I think, much of the work that is presently at the forefront of the types of sociological research that Little extols as well as much of the best work, properly understood, in the field of the new institutional social sciences as well as related areas like comparative political economy and political development.

However, from this point of substantial agreement I do not see the obvious connection to the de-emphasis of generalization as part of the basic task of the social scientist. As Little readily acknowledges in some of the citations in his footnotes, there has been an ongoing debate in the social science literature over the relative merits of mechanism-based, as opposed to law-based, explanations. Much of the early emphasis on mechanisms built on some early work by Robert Merton. Merton (1968) advocated an approach to social-scientific research that adopted a middle ground between grand theorizing and description of specific events. The task was to identify "social processes having designated consequences for designated parts of the social structure." In this early formulation Merton allowed for the possibility that mechanism-based research was generalizable across contexts.

More recent advocates of this middle-range approach have in various guises shared an emphasis reflected in the alternative, but related, definitions offered by Thomas Schelling (1998): "a social mechanism is a plausible hypothesis, or a set of hypotheses, that could be the explanation of some phenomenon, the explanation being in terms of interactions between individuals and some other individuals, or between individuals and some social aggregate. ... Alternatively, a social mechanism is an *interpretation*, in terms of individual behavior, of a model that abstractly reproduces the phenomenon that needs explaining."

A helpful typology of social mechanisms has been proposed by Hedström and Swedberg (1998). Building on a scheme developed by Coleman (1990), Hedström and Swedberg identify three types of social mechanisms: (1) situational mechanisms which focus on the effects of social context on individual actors, (2) action-formation mechanisms which focus on "how a specific combination of individual desires, beliefs, and action opportunities generate a specific action" and (3) transformational mechanisms which focus on how the interactions of various individuals produce collective outcomes. The common feature of each type of mechanism is that an explanation in terms of social mechanisms requires an account of how the social phenomena to be explained are related to the actions of individuals. As I understand Little, his proposed approach is quite consistent with the typology offered by Hedström and Swedberg.

And Hedström and Swedberg share the distrust of explanations purportedly grounded on general laws:

The covering law model provides justification for the use of "black-box" explanations in the social sciences because it does not stipulate that the mechanism linking *explanans* and *explanandum* must be specified in order for an acceptable explanation to be at hand. This omission has given leeway for sloppy scholarship, and a major advantage of the mechanism-based approach is that it provides (or encourages) deeper, more direct, and more fine-grained explanations. The search for generative mechanisms consequently helps us distinguish between genuine causality and coincidental association, and it increases the understanding of why we observe what we observe (1998: 8–9).

But here it is important to note that the more modest aspirations of the mechanism-based approach as advocated by Hedström and Swedberg do not rule out generalization across cases. On their account, the mechanism-based approach is beneficial precisely because it focuses our attention on basic mechanisms that are prevalent in a wide variety of social situations. In fact some advocates of mechanism-based explanations go even farther. See, for example, Petersen (1999), who incorporates the generality requirement into his definition of a mechanism: "specific causal patterns which explain individual actions over a wide range of settings."

Now the Petersen extension clearly seems to go farther than Little would be willing to accommodate. But the Hedström and Swedberg approach seems to have considerable affinity with what Little wants to recommend for the social sciences. And cast in this way it would appear to be acceptable to an ever-growing number of scholars in the social sciences today. And therefore I come back to the question, do we really need to de-emphasize generalization in order to reap the considerable benefits his proposed focus on causal mechanisms and methodological localism would produce? And similarly, would we be unwise to do so?

In the end I am not sure how Dan Little would answer these questions. In large part this comes down to a question of whether his opposition to generalization is theoretical and conceptual or practical and empirical. If it is the former, then the answers would be "yes we need to do so" and "we don't really have a choice." But it is not clear from this chapter that this is his position. And it is also not clear that this is a position that can be persuasively defended at the conceptual level. There is evidence in the chapter that his opposition is more practical. In one important passage Little argues that his "approach emphasizes variety, contingency, and the availability of alternative pathways leading to an outcome, rather than expecting to find a small number of common patterns of development or change. *The contingency of particular pathways derives from several factors, including the local circumstances of individual agency and the across-case variation in the specifics of institutional arrangements – giving rise to significant variation in higher-level processes and outcomes*" (emphasis added).

As I read this passage, Little's opposition is empirically based, that it is a fact that the complexity of social circumstances is such that we will not be able to find sufficient similarity to justify a claim to generality at the mechanism-level. And, if this is empirically the case, then it might support a recommendation that social scientists should de-emphasize the search for generalization across cases because it will be time poorly spent, time better spent in the pursuit of better explanations about particular cases.

But here we should not lose sight of the fact that this kind of claim about variety and complexity is just another empirical claim open to question. It may or may not be the case in any particular situation and thus it should be a hypothesis open to investigation. On this account, generalization itself would, on an approach grounded in methodological localism and committed to the discovery of causal mechanisms, be an appropriate subject of social scientific analysis.

REFERENCES

Coleman, J. 1990. *Foundations of Social Theory*. Cambridge, MA: Harvard University Press.
Hedström, P. and R. Swedberg 1998. "Social Mechanisms: An Introductory Essay." In P. Hedström and R. Swedberg (eds.) *Social Mechanisms*. Cambridge: Cambridge University Press, pp. 1–31.
Merton, R. 1968. *Social Theory and Social Structure*. Enlarged edition. New York, NY: The Free Press.
Petersen, R. 1999. "Mechanisms and Structures in Comparisons." In J. Bowen and R. Petersen (eds.) *Critical Comparisons in Politics and Culture*. Cambridge: Cambridge University Press, pp. 61–77.
Schelling, T. 1998. "Social Mechanisms and Social Dynamics." In P. Hedström and R. Swedberg (eds.) *Social Mechanisms*. Cambridge: Cambridge University Press, pp. 32–44.

7   What Is This Thing Called "Efficacy"?[*]

Nancy Cartwright

1   The Topic

This chapter is about efficacy, effectiveness, the need for theory to join the two, and the tragedies of exporting the Cochrane medical-inspired ideology to social policy. Loosely, *efficacy* is what is established about causes in RCTs – randomized controlled trials. *Effectiveness* is what a cause does "in the field." The theory, like that describing forces in mechanics, underwrites the assumption that the cause contributes the same effect in the field as in the experiment. The tragedies are multiple and snowball on from one another. On conventional Cochrane Collaboration doctrine, following the model of testing pharmaceuticals, the RCT is the gold standard for evidence of effectiveness in evidence-based policy. The first tragedy is that on dominant characterizations of "efficacy," including, especially, many that try hard to be scientific, it does not make sense to suppose that efficacies make any difference outside experiments. The second tragedy is that once "efficacy" is characterized so that it does make sense, the RCT can hardly be a gold standard since it goes no way toward establishing the theory, or more loosely the story or account, that it takes to get out of the experiment and into the field. The third tragedy is that much of the teaching about evidence-based policy pays little attention to the need for such theories or accounts. Indeed there is often the suggestion that RCTs should replace such accounts since the accounts are almost always controversial. The worst tragedy is that we offer advice that lets policy down, wasting the powerful knowledge that could be provided by RCTs. We pay heavily to measure efficacies in RCTs, but efficacy is not evidence at all for effectiveness without the right kind of account or theory to make it so. Yet we provide hardly any guidance about how to manage when accounts are dicey.

[*] I would like to thank Chris Thompson for his help with the research and production of this chapter and the Spencer Foundation for support of the research.

2   Why It Matters

Evidence-based or evidence-supported policy is all the rage now, man-dated throughout the US and the UK and increasingly in continental Europe, at the international, national, and local levels. But mandates need policing and policing here requires a theory of evidence. To know whether a genuine attempt has been made to support decisions about pol-icy by evidence that the proposed policy will be effective, some notions are required of what effectiveness is and what counts as evidence for it.

Much serious work has been done to supply these, a great deal of which has been inspired by the Cochrane Collaboration, a "volunteer" organ-ization that has been very instrumental in the evidence-based medicine movement. The relatively new cousin of the Cochrane Collaboration, the Campbell Collaboration, takes it as one of its chief tasks to provide an account of evidence for evidence-based social science and policy. Much of the current debate in the Campbell Collaboration is about the extent to which the Cochrane methodology can be taken over and espe-cially about the prevalent view that RCTs should be the gold standard, providing the very best evidence in favour of policy and, sometimes it is argued, the only reliable evidence. Consider for example the advice of the US Department of Education: Good evidence for a new educational policy is two successful RCTs for that policy in "typical" schools that are "like yours."

Part of these efforts are driven by the laudable desire to produce advice that on the one hand is relatively sound and on the other can realistically be expected to be put to use by policy makers who may wish to do the best they can but are untrained in both natural and social science and in the handling and evaluation of evidence and also are pressed for time and resources. It is understandable in these circumstances if the advice does not meet either goal to a very high standard. A lot of compromise should be expected and satisficing rather than optimizing should be the standard for success.

This chapter looks at one standard way of thinking about the use of RCTs as evidence for policy: RCTs establish the efficacy of a cause and that provides one central piece of evidence about how effective the cause will be in live settings outside the experiment. The most stand-ard assumption in practice seems to be that a cause that is efficacious in the experimental setting will be efficacious outside unless its effects are swamped by opposing causes. Consider the widely discussed California class-size reduction program of 1996 which is generally taken not to have produced the good results hoped for. The general ver-dict is that well-conducted trials in Tennessee provided good evidence for the efficacy of class-size reduction in improving academic achievement (Nye *et al.* 2000). What went wrong in California was not that this evidence is faulty. Rather, the program was rapidly rolled out in California, with little time for schools to prepare. As a result there was a big increase in demand for teachers and for classrooms, a demand that well exceeded supply. Implementation lagged in schools serving minority and low income students, in part because they lacked adequate classroom space. As a consequence most of the unqualified teachers ended up in the schools with the most disadvantaged students (Bohrnstedt and Stecher 2002).

This argument seems to suppose that class-size reduction is an efficacious cause whose tendency to produce good effects was overwhelmed in California by the actions of causes that operate in the opposite direction. There are other ways of thinking about efficacy, or indeed of thinking about what exactly it is that is established in RCTs and what use can be made of experimental results, some of which I outline in section five. But this is one very standard and widespread way of treating RCTs and their lessons and it is the one I shall concentrate on in this chapter. My conclusions are unfortunate. If it is to be used in the way just described, efficacy takes far more evidence than RCTs to establish, and evidence that is very different in kind. What the RCT can do is to measure *how much* efficacy a cause has, *given* that there is efficacy there to be measured.

Besides explicitly addressing issues of efficacy, evidence, RCTs, and policy, this chapter has a subsidiary purpose. Scientists, including social scientists, are often dismissive of philosophy. Philosophy, it is said, is too abstract, too fussy, and too taken up with its own problems to matter to real practice. With the issues discussed here I think that the opposite is the case. Bad practice, I maintain, is being recommended without intention and without sufficient notice in part because prissy issues that philosophy fusses about are being ignored, issues like what counts as a proper definition and whether an argument has been laid out with all the necessary premises.

3   Efficacy and Effectiveness

In 1978 the US Office of Technology Assessment (OTA) made a distinction between efficacy and effectiveness that is widely cited. The definition and the summary statement given in the OTA report are themselves a little at odds with each other, but here is the kind of lesson that practitioners take away from it: *Efficacy* is "the ability of a treatment to produce benefit if applied ideally" and *effectiveness* is "the benefit that actually occurs when a treatment is used in practice" (Andrews 1999: 317). How shall we understand "ideally" here? This author also maintains that "a good treatment" should "be shown to be clinically and statistically better than that due to an ineffective or placebo treatment in randomised controlled trials" (1999: 316). The OTA itself explained in its 1978 report, "For efficacy the conditions of use are considered to be ideal, or, as a substitute, experimental research findings" (p. 16). It is laudable to try to be more precise and to make distinctions that help prevent mistaken inferences, such as the easy slide from experimental results to predictions that just the same results will obtain in the field. Nevertheless it is this characterization of efficacy in terms of experimental results that makes the first set of troubles I want to highlight. The more recent report of 1994, "Identifying Health Techniques that Work," concurs with the identification of efficacy with experimental results: "Traditionally, RCTs have been the tool associated with narrowly defined efficacy studies" (p. 17).

The same kind of characterization of efficacy in terms of experimental conditions is common among careful statisticians, and for good reason. Efficacies are average effects and averages are relative to distributions. But distributions and averages make no sense without reference to the population of units over which the distribution is defined and the chance set-up that is supposed to generate it; for an average it must be clear whether it is conditional or unconditional and what it is conditional on.

The canonical source here is probably D. R. Cox's 1958 *The Planning of Experiments*, which defines the notion of "true treatment effects": These effects are defined as effects in an experiment. One of the most standard ways of defining efficacy nowadays is that championed by Paul Holland and Donald Rubin. Although there have been a variety of sophistications and improvements, the fundamental point I want to make still holds. Since my point is most easily illustrated with the original characterization, I will stick with that.

In an early careful account Rubin defines the formal cousin of efficacy: "the average causal effect in P of one treatment relative to another" (where P is a "population of N experimental units" (Rubin 1978: 36).) This is the difference in *conditional* means for the outcome it would experience given exposure to the first treatment and the outcome a unit would experience given exposure to the second (Rubin 1978: 43).

A.P. Dawid objects to the fact that the means are taken for subjunctive conditional variables: The values of these two variables cannot possibly both be observed but at most only one or the other (Dawid 2000, 2007). I am sympathetic to his worries but they are not mine here. Rather I want to point out what these means are conditioned on: observed values of treatment outcomes, of covariates, a missing-indicator variable, and, what matters for my discussion here, a "treatment assignment variable" W. The treatment assignment variable picks up two kinds of information. One is "the treatment assignment mechanism which determines the sample experimental units to receive each treatment" (Rubin 1978: 37) – for instance the random assignment mechanism of the RCT. The other is "the sampling mechanism which determines the experimental units to be studied" (i.e., to be exposed to a treatment). Together these fix whether or not the means concern units in an experimental setting and exactly how the assignment to treatment occurs for units in the experiment. Dawid does a similar thing in his own decision-theory account, which takes means only over actual, not counterfactual, outcomes, for Dawid conditions his means on a treatment assignment variable, F, that can take three values: receives treatment 1 by assignment; receives treatment 2 by assignment; and is not assigned but takes values for either treatment naturally (i.e., Dawid conditions on a variable that determines whether one is in an experiment or not).

In one sense this is exactly as it should be. We should not be defining means without specifying the intended population, chance set-up, and conditioning factors. But in another, it makes for a very odd sense of "efficacy." For the causal effect is now always *defined only relative to a particular experimental design*. That means that if we take this as our formal definition of the efficacy of a cause, it *does not make sense* to talk about the cause being efficacious in any other setting. The concept only applies to those situations in which the experimental design obtains, and in no others.[1]

It is important to note that this is not the familiar problem of external validity. In the problem of external validity we consider a quantity that *could* hold across a variety of situations; we observe the value of that quantity in one situation, like an RCT; and we ask if the very same value of that quantity holds everywhere or in some other specified situation. But with the definitions discussed here, the quantity in question could not possibly take a value anywhere except in the experiment with respect to which it has been defined. By way of analogy, the International Adult Literacy Survey (IALS) is used to measure and compare the functional

---

[1] Alternative to choosing a specific population and assignment mechanism for defining the efficacy of a cause C is to define a concept with an open variable "for any population P and assignment mechanism W, the P.W efficacy of C is ... ". One could then suppose that finding the RCT- efficacy of C in any population P would give one information about the P.W efficacy of C for any target P and W. This alternative raises exactly the same demands as those adumbrated in section 8, for the capacity alternative I focus on.

literacy of adults across the Organization for Economic Cooperation and Development (OECD). It is reasonable to ask whether the test of literacy used by the IALS is applicable to populations outside the wealthy countries that make up the OECD. But it is inappropriate to ask whether the IALS can tell us anything about the numeracy rates in different countries.

We can take this problem to be the result of over operationalizing, and that is the approach I shall discuss here. That is, we mistakenly define a quantity that can hold across a variety of different situations in terms of one way of measuring the quantity in one particular setting. The problems that arise from operationalizing are exacerbated by the drive to provide definitions in terms of probabilities so that the techniques of statistics can be brought to bear. The 1978 OTA report for instance asserts (without defence or comment), "Efficacy is best expressed in probabilities" (p. 4). But probabilities are always defined relative to populations, chance set-ups, and conditioning variables. So definitions expressed in terms of probabilities will always characterize concepts that *ipso facto* have a very limited range of applicability.

4   Capacities

So then suppose that we take the approach that RCTs provide evidence for the effectiveness of causes in real-life settings because the RCT provides information about some fact about what a cause can do that can obtain outside the RCT, in the real-life setting. I have always called these facts about what causes do across a broad range of settings, facts about the *capacity* associated with the cause, where capacity is modeled on J. S. Mill's notion of a *tendency*. Indeed, in the book where I discussed capacities at length (*Nature's Capacities and their Measurement*) one of the prime motivations for positing capacities was to make sense of the highly lauded scientific technique of using the effects that causes produce in very special circumstances – RCTs, ideal Galilean experiments or when all other causes are held fixed – to teach something about what will happen when those causes are present elsewhere. The idea is that:

1. Causes have, or have associated with them, relatively enduring capacities.
2. We can learn about the natural effects of these capacities in various nice situations (like Galilean experiments, RCTs, or when all other causes are held fixed at one set of values).
3. What we learn in these very special situations has some systematic relation to the effect that is produced when the cause obtains elsewhere.

This seems to mesh neatly with the talk about efficacy that we frequently see in practice. Recall the California class-size reduction program. The Tennessee trials provide evidence about how efficacious small class sizes can be on academic achievement; the effect that actually occurred in California was taken to be the resultant of that effect damped down and modified in an intelligible way by other causes, themselves conceived of as being efficacious in opposite, or at least different, directions.

This leads to the following scenario. The efficacy of a cause T for an outcome O (or the efficacy of T for O relative to alternative Ts) is a relatively enduring feature associated with T. The magnitude of T's efficacy for O can be measured more or less accurately in a variety of ways and that magnitude contributes in some systematic or intelligible way to what happens whenever T is present. The ideal RCT is then taken to be a good way to measure aspects of the efficacy of T.[2]

This is a good scenario but philosophers of science and metaphysicians generally do not like it, and these are people who spend their lives thinking about what kinds of categories it makes sense to use to describe reality. I can hardly refer to them as an objection to the hypothesis that efficacy is a capacity as I have spent considerable effort arguing that their point of view is mistaken. It is only fair to note for practicing social scientists, however, that my claim that the notion of capacity makes sense and that it plays an important role throughout the natural, social, and biomedical sciences, and especially in their link with technology and policy, is by far a minority view in philosophy. Here I want to raise my own problems with deploying it for evidence-based policy. To do so it will help to rehearse some conventional lessons from philosophy of science, especially in the venerable discussion about the realism and acceptability of scientific concepts, which differs from discussions of construct validity that social scientists may be more familiar with. Before doing so, however, I shall give a very brief review of some other methods on offer for avoiding the problems raised by the kinds of definitions of efficacy described in section 3. This should help us to understand better what thinking in terms of capacities involves and what its advantages are.

---

[2] I say "aspects of" because one would seldom expect the actual mean differences to be the very effect the cause *contributes* in this capacity sense. At the very least, we know that this difference is an average over the contribution it makes in various subpopulations represented in the population enrolled in the experiment. It is a complicated – and context-dependent – question of exactly what information an RCT yields about the contribution of a cause on the capacity interpretation. I ignore the issue here in order to avoid excess complication.

5   **Alternatives to Capacities**

Let me recap the problem. We want to know what facts about a cause can be established in an RCT that could allow positive results in an RCT to be evidence – albeit possibly only partial and defeasible evidence – for the effectiveness of the cause elsewhere. One answer, the one I look at in detail here, is that the RCT can provide information about the size and direction in which a (relatively enduring) capacity operates. Other strategies propose the following:

5.1   *Causal Claims and Inductions from Them*

Given some natural assumptions about causality, an increase in probability of an effect, E, in the "treatment" arm of an ideal RCT over that in the "placebo" arm shows that for at least some individuals in the population enrolled in the experiment, the treatment, T, *caused* E. The defence of this as a valid conclusion to draw from a successful RCT depends on how one characterizes an ideal RCT and on what assumptions one makes about singular causal claims. This is a long story that I shall not review. What matters for the issues here is how establishing this kind of causal conclusion about T for the population enrolled in the experiment can provide evidence for the effectiveness of T outside the experiment. How does establishing that T causes E in some individuals in the experiment give reason to think that T causes E outside?

The answer is that it does not, without a great deal more ado. Here we do encounter the problem of external validity, and writ large. In this case it makes sense to ascribe the same feature inside and outside the experimental set-up, but there is no reason yet to do so. We can attempt an induction: Some singular causal events that happened once somewhere will happen in the target population. I do not mind inductions, but wild-eyed ones, on a wing and a prayer, are to be avoided in evidence-based policy. We are only entitled to an induction on some given feature when there is reason to think that the new case is like the original with respect to that feature. We properly induce the color of the camellia buds that have not yet opened on the plant at the bottom of my garden from the colour of those that have, but we do not induce the color of the flowers my geranium will produce from the flowers on my camellia. That is why the US Department of Education remarks that evidence for efficacy of a program in your school should be positive results in RCTs in two schools *like* yours. But it is notorious that the kinds of similarities that make for a good basis for induction are seldom transparent.

So, what features could a cause have that make induction reasonable? Once one begins to lay out the kinds of assumptions presupposed in making inductions from experimental outcomes to real-life situations, the enterprise begins to look a lot like the postulation of a capacity. Why should we think that the cause T will do for some targeted individuals what it did for some individuals in the experiment? If we have reason to ascribe an enduring capacity to T, we have a reason: T caused E in the observed individuals because it had the capacity (power, tendency) to do so, and having that capacity, it will do so for other individuals unless too strongly interfered with. So it may not be that this alternative differs significantly from the capacity idea after all. At the very least the postulation of capacities and the use of induction from singular causal conclusions established in RCTs have in common the concern I am worried about in this chapter: The RCT itself goes no way to underwriting the assumption crucial for taking it as evidence for effectiveness outside the setting, whether this be seen as evidence that the cause T is associated with a relatively enduring capacity or some kind of evidence that there is a proper base for an induction. In particular, analogues for the three requirements I outline in section 8 must be satisfied just as much for inductions as for the use of the logic of capacities.

5.2   A Casual Law and its Analogues

Ideal RCTs can also establish what I call *causal laws*. Again, the proof depends on what one counts as an ideal RCT, what one takes a causal law to be and what features one takes systems of causal laws to have. Causal laws as I characterize them – and this is the characterization that is employed in the proof that ideal RCTs can establish causal laws – are always relative to a "test population." These are populations that are homogeneous with respect to "all other" causes for the same effect. So, what the positive RCT can establish is that there is some test population, δ, which is a subset of the population enrolled in the experiment, for which "T causes E in δ" is true.

Laying aside problems with policies that shift the underlying causal structure, this means that the causal law that T causes E holds for any population that satisfies δ. Two immediate problems face employing this nice result as evidence for effectiveness. First, we usually do not know what δ is. The ideal RCT tells us that there is such a subpopulation but not how to pick it out. Second, we could seldom expect that our target population is a δ population, or has δ as a subpopulation within it. So the information would be useless – unless we suppose that the causal law that holds in one population can be relied on to hold in others, which we often do assume. We do not expect exactly the same effect to appear in other kinds of populations since what actually happens in a given population depends on the combined efforts of all the causes at work in it. But sometimes we expect that the contribution guaranteed by one causal law will be the same for different kinds of populations, and often with very good reason. But again this returns the focus to capacities and their justification: The logic that supports this assumption is the logic of capacities, or something very akin to it.

5.3   Probabilities and Inductions from Them

A third alternative is to abjure causal and efficacy talk altogether and just stick with the probabilities. Take the conclusion simply to be the difference in means between the two experimental groups. Even better, take each mean separately since, without the causal connotations, the difference clearly gives less information than the two means separately, each of which can be used independently in decision-theoretic analyses, as Dawid suggests. Again the question arises, how do the means in the experiment provide evidence about anything outside? Again the answer is that without a lot more ado, they cannot. The problem in this case is usually worse than with external validity of causal conclusions. In general the mean in two populations is the same only if the probability distribution over confounding factors is the same in the two. There may be some cases where the population enrolled in the experiment can be seen as a really good sample from the target population so that the distributions should be the same. But we know this is impractically rare. Indeed this is why J. S. Mill thought that political economy could never be an inductive science. The probabilities with which a given kind of event occurs depend on the complex mix of causal factors that bring it about and there are generally a very large number of these, each itself brought about by yet an earlier large mix of factors, and so on. So the probabilities for an outcome should be expected to be in constant flux; they are the very thing we would want not to do inductions on. The contribution of a cause (in the technical "capacity" sense of "contribution") at least is the kind of thing that we have seen to be relatively stable for a variety of different causes across a variety of different fields – like the gravitational action of one mass on another. Mill for one took the fundamental social causes to be among these. If he is right that social causes are often associated with relatively stable capacities, then there is some hope that RCTs can be of use for social policy. For pure probabilities, by contrast, we almost always have good reason to suppose change rather than stability.

5.4   Econometrics

This alternative does not rely on RCTs as evidence at all. There are a number of different formal treatments of efficacy on offer in econometrics. These treatments define efficacy in a population-relative way, not however relative to an experimental population but rather to the target population. These then provide schema for modelling the causal structure of the population and investigate econometric techniques for estimating efficacy from data. As with the treatments in section 3, it is still the case that by definition a cause can only be efficacious in the population relative to which it is defined, but now it is defined where the policy-maker needs it. These techniques can provide powerful evidence directly relevant for policy. But they have a big drawback, which is just the flipside of this virtue. They deliver results about the target population directly, but then they require both data and careful modeling on that population, and this can frequently be too expensive and too time consuming to help in policy considerations. Better to have results concerning capacities, "off-the-shelf" results that do not have to be established anew for each population – if only we can get them.

Let us turn then to considerations about what it takes to establish capacity claims and what role RCTs can play in doing so.

6   What Makes a Concept Legitimate?

Contemporary science studies features that can be ascribed to real systems in the world: the charge of an electron, the structure of a DNA string, the fitness of a population, or – in the case of interest here – the efficacy of a treatment for a given effect, like the efficacy of class-size reduction to improve academic achievement or of phonic awareness to reading development. Very often, as with charge and efficacy, the features are *quantities*   they have magnitudes that measure the amount of the quantity and allow for comparisons of amounts of the quantity that obtain in different systems. With respect to quantities three different notions need to be distinguished.

1. The quantity itself (e.g., *negative charge*).
2. The scientific representation of it (e.g., negative charge is represented as a discrete-valued function that maps physical systems into $-n$ x *1.6022* x *10^{-19} coulombs* where n is any integer including 0, and -1.6022 x 10^{-19} coulombs is the charge of the electron, the smallest unit of negative charge).
3. Reliable operations for measuring the quantity (e.g., measuring the charge of a particle by observing the deflection in its trajectory produced by a known electromagnetic field).

I rehearse these three because they are often conflated, sometimes harmlessly, sometimes not. For instance, "Fuel poverty is defined as circumstances where a household has to spend more than 10 per cent of its income on fuel to maintain an adequate standard of warmth" (DWP 2006: 70) conflates 1. and 3. Or consider debates about measures of economic freedom (e.g., how adequate is a sheer cardinality measure, which measures the economic freedom of individuals by the size of their set of options?). These are debates about the best scientific representations of economic freedom, but they can often look like debates about 1. – What is economic freedom? – or 3. – What operations will reliably measure the amount of economic freedom? Physics too is rife with conflations. "The quantum state is evolved deterministically by the quantum Hamiltonian" is typical. The quantum state is supposed to be a real feature of systems in the world. The quantum Hamiltonian is definitely a piece of mathematics, which cannot evolve anything in the real world. Either some real feature of systems in the world must be pointed to that is represented by the Hamiltonian or the claim must be read as a purely mathematical claim about the relation of the two representations, with "deterministically" read as some appropriate mathematical characteristic. From the point of view of capacities, the efficacy definitions from section 3 conflate 1. and 3. (Some of the econometric approaches discussed in section 5 make efficacy model-relative, which conflates 1. and 2.)

Which of these three matters for establishing that a scientific concept stands for something real, or at least for licensing the concept as legitimate? All three have their advocates. But I would wish to argue that in the ideal for a good scientific concept we need all three and the three need to mesh tightly. The account of what the quantity is must dovetail with how it is represented, where the canonical way to show they dovetail is via representation theorems, like those involving probabilities, preferences, and utilities (where expressed preferences are considered a function of probabilities and preferences) or those for "measures" of economic freedom. These theorems show that the features picked out to characterize the quantity are appropriately captured by the mathematics used to represent it. (For instance, lengths do not get represented by negative numbers and temperatures do not get represented in a way that insists that twice the numerical value means twice as hot.) Similarly, both the characterizations of the quantity and the representation of it must fit with how we actually measure it in the world. The measurement procedures must be justified as good ways of finding out about the very thing we have characterized and represented. The justification may well be elaborate and indirect, relying on a large number of background assumptions, as in the justification for using the deflection of the trajectory of a particle in an electromagnetic field to measure its charge. But the justification needs to be there if we are to place trust in our measurement procedures.

7   Return to Efficacies as Capacities

From this ideal point of view what is wrong with taking efficacies as enduring capacities measured by RCTs? Generally far too much is missing. We could conceive of the real-life experiment as a way of measuring, but measuring what, represented how? We could additionally think of definitions in terms of mean effects in ideal RCTs relative to well specified populations, like the ones described in section 3, as laying out the requisite representations for some quantity, with real experiments as the real-life measurement procedures. Then we would at least have a transparent connection between the representation and the measurement procedures. But we would still lack an account of what the quantity is that is being represented and measured. The efficacy itself, conceived as a quantity that can be measured in one setting and relied on for prediction in another, is a mere shadow. We have a procedure for measuring, measuring something, but we do not have an account of what it is. We can of course provide such an account by reading off a definition of the quantity from the representation or the measurement procedure. But then we are back where we started in section 3. We would have a good, scientifically well-formulated definition of efficacy. But causes could never be efficacious outside the experimental settings which enter into our definition.

The problem I note here is in no way peculiar to "efficacy" but is widespread throughout the social sciences. To legitimate a quantity concept we need to characterize the quantity itself, its representation and its measurement procedures. Then we need to tie these three together – in a way that can be justified. And we need to be explicit, precise, and rigorous. We naturally face problems here, but which problems depends on which end we start from in tying the bundle together. What generally happens is that the account of representation slides close to one side or the other, making the distance on the far side difficult if not impossible to traverse. We can build a representation very close to our measurement procedures, in which case the bond between those two is transparent and easy to justify. But then, how to reach over to an account of the quantity itself becomes a problem. As we saw with the representations for "efficacy" described in section 3, we sometimes tie representation so closely to measurement that the gulf to the other side where the quantity dwells is unbridgeable.

On the other hand the representation can flow naturally from the account of what the quantity is. Consider the representation of charge referred to above, as a discrete-valued quantity. This is readily justified by the principles of electromagnetic theory that help explain what negative charge is, especially the principles that explain that electrons carry the smallest unit of negative charge, that this charge is indivisible and that the charge of an electron is -1.6022×10^{-19} coulombs. In this case we have the converse problem to that with efficacy: It is very difficult to justify our measurement procedures – Why look at the deflection of a particle's trajectory in an electromagnetic field in order to measure its charge? – and doing so will demand a large number of auxiliary empirical assumptions. It is no surprise then that those who distrust almost all claims to knowledge in the human sciences err in the other direction and tie representation too closely to measurement.

If the situation is so bad, you may wonder why I have defended efficacies as capacities for so long. The reason depends on noting the difference between the abstract description of a category and a concrete instance of something that fits the category. I have defended capacities as an appropriate abstract category of concepts for use in science and I have defended specific efficacies as concrete instances of that category and I have done the first largely on account of the second. That's because I see specific efficacy concepts at work in science, concepts that are well defined, appropriately represented and properly measurable.

The paradigm is Mill's own – forces in physics. First of all, "force" is defined implicitly via the laws in which it participates. Importantly for counting "force" as a capacity term, among these laws is the law of composition of forces, which fixes what it means to say that a given cause, like gravity or charge, "contributes" its canonical effect even when other forces are at work as well. This then gives sense to the idea, central to the concept of efficacy as capacity, that the cause should contribute in some "systematic or intelligible way to what happens whenever it is present." Second, force has a well-understood mathematical representation, roughly as a vector in 3-space whose components are non-negative real numbers. Third, it is measured in a vast variety of ways that can be defended as appropriate given the force laws in conjunction with acceptable auxiliary assumptions.

To make vivid the gaps that beset the concept of efficacy, consider the force of gravity. The effect of one mass on another is not defined as the force or acceleration that is measured in a very carefully controlled experiment, like those Galileo aimed for. It is defined instead by the role laid out for it in the laws of motion, laws that maintain that the mass contributes the same effect universally, not just in Galileo's experiment, laws whose implications in this regard have been widely confirmed. These well-confirmed laws also show why Galileo's experiments are good for measuring the effect of one mass on another – these are experiments in which the pull due to the mass operates almost on its own, with no other forces to interfere. They tell us in addition exactly what effect to take away from Galileo's experiment and, as I stressed above, they tell us just what it means to say that this effect is contributed in other situations outside the ideal measurement setting.

8   Speaking Realistically

The take-home lesson from the considerations in section 7 is that, for making sure our concepts are good ones, theory matters, and so do auxiliary empirical assumptions, good well-confirmed theory and sound, reliable empirical assumptions. But it is well known that policy cannot wait for the advance of theory. So we had best consider what is minimally required if we are to take a well-conducted RCT as evidence for effectiveness, even if only very partial evidence. Perhaps there are aspects of the ideal we can get along without. For instance, we may not need a thick theory to tell us just what the capacity involved is and under just what laws it operates. But the very logic involved in using the idea of enduring capacities as a rationale for taking RCTs as evidence bearing on the effectiveness of the cause outside the experiment makes three clear demands. Meeting these three demands provides the three ingredients necessary to turn the efficacy measured in an RCT into evidence for effectiveness in a new setting. It's like making pancakes. RCTs are the baking powder. But the baking powder is useless without the flour, milk, and eggs.

1. We need good reason to think that the effect produced in the experiment is an enduring one, reason to think that when we see differential effects with the cause present versus absent in the RCT setting, we are seeing the effects of a genuine capacity, one that can reliably be expected to operate in various new settings and in new populations.
2. We should have good reason to think that the proper effect has been identified, that the effect we focus on in the experimental setting is not piggybacking in a misleading way on the true, generalizable effect. "For example, let us say that an experiment [RCT] is conducted to increase security and reduce theft in two schools through the introduction of closed circuit television [CCTV]. The effect is a reduction in theft in the experimental school. Exactly what is the cause here? It may be that potential offenders are deterred from theft,

or it might be that offenders are caught more frequently, or it might be that the presence of the CCTV renders teachers and students more vigilant ... " (Morrison 2001: 72)

3. We need some sense of what it means for the observed effect to be *contributed* in new cases. In formal theories this is supplied by the *rules for composition*. But there are different methods in different theories. The vector addition of forces is different from simple addition. Sometimes we introduce new, intermediate words to describe how the cause contributes even though we don't know how to predict the ultimate outcome. Consider: The magnet always *pulls on* the pin even when we cannot predict whether the pin will be set free. Or the commitment about contribution can be couched in a cautious, modalized form: Eating Wheaties *can* (or "may") improve your heart health. And of course threshold effects are a notorious problem.

To illustrate the importance of these three basic ingredients for using efficacies to help predict effectiveness let us look in more detail at the third and consider – in caricature – the difference between two well-known examples, simultaneous equation models in economics and forces in physics. These different notions of how causes contribute give radically different predictions about what happens when the cause is present in new circumstances.

Economics first, where supply and demand jointly determine quantity exchanged, via two equations:

Supply: $q_s = \alpha p + \mu$
Demand: $q_d = -\beta p + v$

The theory supposes that in equilibrium $q_s = q_d$ and that *both equations are satisfied at once*. This last is a rule of composition. The supply equation describes the *contribution* to the quantity exchanged from the combined supply-side causes (here price, p, and other unnamed causes, μ) but it does not select what the actual value will be because price and quantity are taken to be fixed simultaneously by both equations of the model. Similarly, the demand equation describes the contribution to the quantity exchanged from the combined demand-side causes (price, p, and other unnamed causes, v) and again it does not select any actual value. The rule of composition describes how these two separately-contributed effects combine to fix the value that the quantity exchanged actually takes: Both equations must be satisfied at once.

Now forces. A particle of negative charge q_1 hovers in mid-air, pulled down by the mass M of the earth and upwards by the pull of a positive charge q_2. The equations that govern its acceleration are from gravitational and electromagnetic theory respectively:

$$\text{acc}_g = gM/R^2$$
$$\text{acc}_{em} = \varepsilon q_1 q_2 / R'^2$$

Here R is the vector representing the particle's distance from the earth; R', its vector distance from the second charge; g, the gravitational constant and ε, the electromagnetic constant. As is well known, the acceleration the particle actually experiences – which in this example is zero – is given by vector addition.

Compare. There is a sense in which the separate capacity claims in the economics model are more informative than those in the physics case. That's because the effect described is bound to happen no matter what particular form the contribution from the other side takes since both equations must be satisfied. Even though the effect is not fixed precisely, there is information about the quantity exchanged that is bound to hold no matter – the information that it lies on the line described. This information may turn out to be of great use or it may be of little use in evaluating the efficacy of a policy that proposes to tinker with demand or supply-side causes. That depends on the policy setting. But (as the theory has it) knowing the demand equation puts the quantity on the demand line and that can be relied on even if we know nothing about the supply equation, and vice versa.

The economics rule for calculating the net effect of different contributions is in sharp contrast with the physics case. With vector addition, knowledge of either contribution separately does not provide any information about the actual value of the resulting acceleration. The acceleration is not constrained in any way by the presence of gravitational or of electromagnetic causes alone whereas the quantity exchanged is constrained to lie along given lines by both the supply-side and the demand-side causes separately. For physics the best we get are counterfactual constraints: If the other causes at work (whatever they are) were to stay fixed, adding the pull of gravity would change the acceleration by the vector addition of $\text{acc}_g = gM/R^2$; similarly, adding the pull of another charge would change the acceleration by the vector addition of $\text{acc}_{em} = \varepsilon q_1 q_2 / R'^2$. These strongly *ceteris paribus* counterfactuals are of limited use in situations where other changes may get introduced by the policy as well.

For most cases of social policy the causes to be worried about are not described by nice theories like these with nicely articulated rules of combination, nor do they have nicely articulated accounts supporting the claim that the causes are associated with relatively enduring capacities that guarantee effects that contribute in some systematic way across new situations. What do we do then?

In the case of combination, we often assume simple linear addition: If the cause produced an improvement of size x in the RCT, it will produce that size improvement elsewhere. But of course we know better and normal intelligence dictates caution. Sometimes the effects taper off at the margins, sometimes there are clearly drawn thresholds, sometimes, without the appropriate helping factors, the effect cannot be produced at all. Even if, unlike what happened in California, other factors are held fixed, smaller classes may make very little difference where reading scores are already high and they will make none at all in classes of children who do not have the capacity to read.

What legitimates the assumption of one kind or another of additivity and where do cautions like these about the assumption come from? They come from some story, account, or theory of what the capacities of small class size are and how they work. And this is where they must come from. We need – we always need – an account that supplies the three basic ingredients if we are to turn efficacies into evidence for effectiveness. We may well not have very good accounts to supply these basic needs but that does not mean we can pretend we do not need them.

If we are not to be led astray, sometimes far astray, by introducing RCT results into policy considerations, it is not enough to be told that the RCT was well conducted. These three requirements must be met in some reasonable way as well. You can't make pancakes without flour, milk, and eggs no matter how much baking powder you pour into the bowl. By itself the RCT is not evidence for efficacy in a new setting. It is evidence only *conditional* on these other three ingredients. If we have little idea about how to supply these basic ingredients, we need to figure out how to cope with that problem, and ignoring it is not coping. We can make bets in situations of ignorance but for policy we should make intelligent bets. The probability that the RCT provides evidence and exactly what it provides evidence for depends on what the other three ingredients might be like that will turn it into evidence. These issues need to be thought about, not swept under the rug; and in the end, as with most cases where bets are dicey, perhaps we need to hedge them heavily.

These remarks are not meant to be of serious practical help. They are rather a warning and a call for refocusing attention in evidence-based policy. Clearly the possibility of establishing efficacies in well-conducted RCTs is a great boon, no matter which definition of efficacy is settled on. But they are only a help when the other ingredients are there to make them so. So, far more effort needs to be focused on how to secure the

other ingredients and how to cope when our knowledge of them is insecure, as it so often is.

9   A Brief but Important Warning

I say that we need a theory – a story, or an account – strong enough to
supply the three basic ingredients of section 8 if efficacy is to contribute
evidence for effectiveness. It is a major problem for policy that often the
theory is missing. What is important to keep in mind is that in many
cases our problem may not be due to the fact that we have not yet found
the appropriate theory but rather that there is no theory to be found.
Much of science works by postulating capacities, by the use of the analytic method – we study the components separately then make predictions by "adding" their effects in the appropriate manner. This works
for forces and for biological mechanisms (such as chemical transmission
between neurons) (Machamer *et al.* 2000) and Mill was convinced it
would work for political economy. But we should be wary. There is no
necessity that what a cause produces in one setting will have any systematic relation with what it does in another. Much of social phenomena may be too holistic to yield to the analytic method. Indeed both
Anna Alexandrova and Julian Reiss argue that a lot of the recent work in
experimental economics suggests that the analytic method is not working so well in political economy as Mill – and a great many more recent
economists – had hoped. In many cases evidence is mounting up against
Mill's hope that the important economic causes are associated with stable tendencies (or, in the vocabulary used here, "capacities"). And if this
is true for economic causes, it seems all the more likely to be true for the
social causes outside the domain of economics.

What then of RCTs? When the analytic method fails they may be
suggestive of what to look for in new cases, but they can hardly count
as evidence at all. Inferring from efficacy to effectiveness is induction
on a wing and a prayer. I doubt, however, that serious social scientists
conduct many RCTs without good reason to think the causes tested are
associated with relatively stable capacities. The reason though is generally bound up with theory, which is despised by many of those who
advocate taking over Cochrane doctrines for policing social evidence. It
is despised because the relevant theories are controversial or ill-formed
or poorly supported or all three. This returns us to the finishing point of
section 8: we need theoretical assumptions, and assumptions of just the
right kind, or else all our very careful efforts at experimentation go to
waste. So we had best not despise dicey theories but learn how to manage their uncertainties.

10   In Sum

The purpose of this chapter is not to attack RCTs or any other way to inform ourselves about efficacy, nor to promote theory above or to the exclusion of everything else in deciding policy. Its aim is merely to argue that in our attempts to get clear and rigorous about what we mean by evidence, we have concentrated too much on getting the methodology of RCTs straight and too little on getting clearer what theories, or accounts, or stories we need to make an RCT evidence at all. For without some theory about the capacities of the causes we are studying, the evidentiary value of an RCT is not just weakened, but, I argue, made empty, zero.

The contemporary theory and practice of the RCT is a brilliant intellectual achievement. But a good RCT – and there are many of them – is only one component of a set of ingredients necessary to create a piece of evidence that policy will be effective. I say "a piece of evidence" deliberately. Even where we have a very good, very full theory, one measurement by one team in one environment is not very strong evidence. (It certainly never suffices in physics, where the theory is often exceedingly strong – though a result from an experimental team that is known to be exceedingly good can carry considerable weight.) This may in the end be all we have – and lucky to have it at that. But that does not make it strong evidence. Again it is important to recognize this and to factor in the large uncertainty that is left, rather than avert our gaze and pretend it does not stalk our policy. The efficacy that we claim to have established by an RCT to test a particular policy is an empty notion with zero evidentiary power if we have no theory to supply the missing ingredients that turn the efficacy into evidence for effectiveness. This is a more fundamental concern about RCTs than the worry that in very many cases there may be difficulties in eliminating bias, controlling for confounding factors, getting the right sample, and so on. It says that even if all these technical problems are absent, there is no read across from efficacy to effectiveness without some theory of some sort, and efficacy as a notion has no power beyond that defined for it by the statistical procedures themselves. Its connection with effectiveness is no more than sharing its first three letters.

I use words like "theory" or "account", or "story" and "some sort of" advisedly. We all agree that in the social sciences, as compared with in the natural sciences, our theories are typically quite poor. We wish it otherwise, and work to make it so. But we are where we are. And the consequence of recognizing this poverty is not that we should abandon theory until it performs better, because without some sort of a theory, some non-statistical idea of what may be going on, we have no evidence from RCTs at all. We just have words and numbers written on a page. The theory is not some fancy add-on which can be bypassed by going straight to the evidence. It is what gives the RCT life.

It is easy to see how RCTs have earned so much attention. Consider the position of policymakers – not methodologists – wanting conscientiously and intelligently to decide what to do. Their raw material will include some theory, about what, say, makes juveniles offend; some factual evidence (e.g., if they are lucky an RCT); and a good deal of anecdote and folk wisdom. (Forget the politics.) When they talk to serious academics and similar, they learn to neglect the anecdotes, as not scientific enough to be evidence. They also learn that there are a lot of theories about juvenile offending, from the sociological to the medical to the psychological, but none of them looks conclusive, and they conflict.

So they say: I will just look at the evidence, at what works. And advocates of RCTs say – and this chapter agrees with them – that RCTs represent an extraordinarily powerful engine for testing efficacy, and hence – and this chapter does not agree – for identifying effectiveness. At the very worst, the policy makers may then think: We are at least looking at that part of the raw material that is telling us something precise and relatively certain and uncontroversial. It would be nice to have a good theory of juvenile criminal behavior, as the physicists have of gravity. But we don't. Theory plus evidence may add up to the gold standard for policy determination. But we have to use what we actually have – the evidence of what works. And maybe that is all we need. Why do we need theory if we have incontrovertible evidence of efficacy?

My argument is that if you don't have both you don't just have half or whatever of what you should have, but that you have nothing. We all recognize that theory without evidence to support it leads to no conclusions. The reverse is true as well.

REFERENCES

Andrews, G. (1999). "Efficacy, effectiveness, and efficiency in mental health service delivery," *Australian and New Zealand Journal of Psychiatry*, **33**: 316–322.
Bohrnstedt, G.W. and B.M. Stecher (eds.) 2002. "What We Have Learned About Class Size Reduction in California," *California Department of Education*.
Dawid, A.P. 2000. "Inference without counterfactuals," *Journal of the American Statistical Association* **95**(450): 407–424.
   2007. "Counterfactuals, Hypotheticals and Potential Responses: A Philosophical Examination of Statistical Causality." In F. Russo and

J. Williamson (eds.) *Causality and Probability in the Sciences.* London: College Publications, Texts in Philosophy Series 5: 503–532.

DWP (Department of Work and Pensions) 2006. *Opportunity for All.* Eighth Annual Report 2006, Strategy Document.

Machamer, P., L. Darden, and C. F. Craver 2000. "Thinking about mechanisms," *Philosophy of Science* **67**(1): 1–25.

Morrison, K. 2001. "Randomised controlled trials for evidence-based education: Some problems in judging 'what works'," *Evaluation and Research in Education* **15**(2): 69–83.

Nye, B., L. V. Hedges, and S. Konstantopolous 2000. "The effects of small classes on academic achievement: The results of the Tennessee class size experiment," *American Educational Research Journal* **37**(1): 123–151.

Rubin, D. 1978. "Bayesian inference for causal effects: The role of randomization," *The Annals of Statistics* **6**(1):34–58.

7 – Comment
Randomized Controlled Trials
and Public Policy

Gerd Gigerenzer

In an inspiring chapter, Nancy Cartwright argues that there is a gap between *efficacy* and *effectiveness*. That is, if a randomized controlled trial (RCT) demonstrates that a cause has an effect (efficacy), this does not imply that the same cause–effect relation holds in the field (effectiveness). The term "field" refers to settings outside those of the original RCT. Bridging this gap, she argues, requires theory, or at least good reasons to think that the proper effect has been identified and is an enduring one, among others. Sound policymaking, she argues, depends on both RCTs and theory, and, without at least some theory, the evidential value of an RCT is made empty, that is, zero.

In this comment, I will apply her philosophical analysis to the "field" of health care. Screening programs for breast cancer have been established in many countries with the aim of saving lives. RCTs represent the scientific basis for these programs. How is the efficacy–effectiveness gap being dealt with in practice? I will argue that, in the world of cancer screening, the logical problem of induction as well as the lack of theory is pressing, and that the most recent Cochrane review ignores both, consistent with Cartwright's analysis. Moreover, I argue that health and government organizations do not seem to be interested to bridge the gap in the first place, but largely ignore the evidence provided by RCTs because of conflicts of interest, among others. In screening practice, experimental evidence does not count for much, and many health organizations rely on framing and omission to make sure that the public is likely to misunderstand what the evidence actually is. Many doctors do not know the results of the RCTs, and few citizens ask for evidence, but rather want to trust the technology, their doctor, and the effectiveness of early detection. Given this collective lack of interest in the results of RCTs, I end up with a slightly more positive view of the agenda of the Cochrane Society, including their positivistic approach toward reporting data without addressing the problem of effectiveness, than suggested by Cartwright's thoughtful analysis.

1   Randomized Controlled Trials on the Efficacy of Early Detection

A recent Cochrane Review of screening for breast cancer with mammography lists seven RCTs with a total of about 500,000 women (Gøtzsche and Nielsen 2006). About half of these women underwent regular mammography (the screening group), the others did not (control group). Screening checks women who have no symptoms of breast cancer, and its purpose is to save lives through early detection and treatment. These RCTs address the question of how large the effect of early detection is, that is, the difference in the means of the two groups – the number of women who die after a specified time. Mortality reduction is the possible effect, and early detection plus subsequent early treatment represents possible causes. The efficacy of mammography screening is measured by benefits – the breast-cancer-specific mortality reduction and the total mortality reduction – and harms, such as unnecessary mastectomies or lumpectomies (overtreatment) for the women in the screening group.

For women aged fifty and older, and an average thirteen years follow up, the results across the seven trials were:

1. The breast-cancer-specific mortality reduction was one in every 2,000 women. That is, out of every 2,000 women in the control group, about ten died of breast cancer after an average of thirteen years, while this number was about nine in the screening group. Thus, there was evidence that screening reduced deaths from breast cancer, albeit the effect was very small.
2. No total reduction in mortality (all causes) was found, and no reduction in death from cancer (all cancers). That is, there was no evidence that screening saved lives or reduced total cancer deaths.
3. Out of every 2,000 women in the screening group, ten healthy women, who would not have been diagnosed if there had not been screening, were diagnosed as breast cancer patients and treated unnecessarily, undergoing mastectomies and lumpectomies. Thus, overtreatment lowered the quality of life for women in the screening groups.

This short summary does not do justice to the subtleties of the Cochrane Review, which is an immensely careful report. The attention of the Cochrane analysis is directed at the internal validity of the individual trials, such as, 21% of the women with breast cancer who died in the Malmö trial also had two or three types of different cancer. The Review documents carefully that belief in the effectiveness of screening may influence the decision to which kind of cancer the patient's death is attributed, and that there seems to be a bias to attribute death to breast cancer more often when women were not screened. Since cancer mortality is less likely to be subject to misclassification, they suggest that "death by any cancer" may be the more reliable outcome measure than "death by breast cancer." These issues concern the reliability of efficacy measures, not the question of whether the results of the RCTs can be expected to hold outside the experimental setting.

2   Do These Results Generalize to the Field?

Cartwright's question is: Does efficacy imply effectiveness? She suggests that evidence-based medicine, as represented by the Cochrane Collaboration, pays little attention to this question. Is this the case in the present Cochrane Review? In fact, the question of whether the results of the RCTs apply to women not taking part in the controlled studies is neither posed nor addressed in Gøtzsche and Nielsen's Cochrane Review. But omission is not sufficient evidence for overlooking the logical gap; the authors may simply not deal with this question. Yet, the Review provides indirect linguistic cues. If one assumes that the results of RCTs generalize to the field, then one might easily slip from past tense to future tense, that is, from what the RCT actually showed (1 in 2,000 women *did not* die of breast cancer) to what will happen in the future (1 in 2,000 women *will not* die of breast cancer). In fact, this jump from past to future tense occurs throughout the report. For instance, under the heading "implications for practice," the authors write: "the chance, that a woman *will* benefit from attending screening is very small, and considerably smaller than she *will* experience harm. It is thus not clear whether screening does more good than harm" (p. 13 [italics added]).

The authors of the Review might conjecture that their (implicit) conclusions about the future effectiveness are justified, since there were seven trials, not just one RCT. These were conducted in Canada, New York, Malmö, Göteborg, Edinburgh, Stockholm, and Kopparberg and Östergötland. Furthermore, the results in these seven locations were similar, except that the suboptimally randomized trials reported slightly higher breast-cancer-specific mortality reductions than the two adequately randomized trials (in Canada and Malmö). Yet, this empirical defense will not convince those who insist on the logical gap and the necessity of theory to bridge it. The logical argument is that, in order to argue from an RCT situation to a similar effectiveness in a field situation, both the RCT and the field situation must be random samples from the same population. The term population refers here not only to women but also to the entire set of relevant causes, such as the potential effect of participating in an RCT. The authors of the Cochrane Review might respond that such a situation virtually never exists in the world of health care, only in the imagination of statistical theory, and that, therefore, a generalization from seven RCTs that show similar results is the best evidence for effectiveness one can ask for. Replication compensates lack of random samples. Yet, the philosopher will not be satisfied by this response.

3   Is There a Theory to Bridge the Gap?

Causal theories are strikingly absent in the Cochrane Review. It is about effect sizes, biases, blindness, randomization, post-randomization, meta-analyses, and the reliability of outcome measures. But how would early detection cause a reduction of mortality, or what prevents early detection from doing so? The classical theory of breast cancer, proposed by surgeon William Halsted, assumes that cancer emerges at some point in the body and spreads from there to other areas of the body, leading to death. Halsted accepted the German pathologist Rudolf Virchow's doctrine that all cellular changes, including neoplasms, were caused locally. From the 1890s until about 1975, in the footsteps of this account, the standard treatment for breast cancer was mastectomy, which involves complete removal of the breast, surrounding tissues, and lymph nodes. On this standard account, cancer is a localized phenomenon and early detection followed by removal of the cancer cures the patient. This standard story, however, can hardly build a bridge over the efficacy–effectiveness gap, since it is not supported by the results of the RCTs. Early detection led to a minimal reduction of breast-cancer-specific deaths, and to no reduction in deaths from all cancers and overall mortality. In contrast, the recent concept of nonprogressive cancer assumes a continuum between cancer and no cancer and biological mechanisms that halt the progression of cancer (e.g., Mooi and Peeper 2006). In this view, the term cancer encompasses a broad spectrum of conditions, making the cancer/no cancer distinction suspicious. This concept can explain that screening detects cancers that would have never become clinically relevant during the lifetime of patients (overdiagnosis), and that the patients with nonprogressive cancer will, by definition, not die from the disease. This can explain the unnecessary mastectomies and lumpectomies, but it cannot provide the theoretical bridge between RCTs and the field. In the case of screening for breast cancer, there is presently no causal theory that can fill the logical gap.

4   Do Policymakers Pay Attention to RCTs?

RCTs sometimes guide policy, as the Californian class-size reduction program illustrates. Cartwright's argument is that policymakers may be too quick in assuming that efficacy implies effectiveness in a different setting. I want to draw attention to the opposite problem, which I believe is the rule rather than the exception in cancer screening. Rather than ignoring the gap, policymakers ignore the RCTs.

I begin with the observation that the above three results of the seven RCTs, discussed in medical journals for more than a decade, are unknown to most women in Western democracies, including those who participate in screening, and are also unfamiliar to many physicians including gynecologists (Gigerenzer *et al.* 2007). Amazingly, few people know the results of these RCTs. Health policy is guided by other goals than information, such as increasing the participation rates in screening programs. Transparent information about the results of the RCTs would conflict with this interest and likely decrease participation rates. When screening advocates and public policy organizations disseminate the results of the RCTs to the trusting public, then the benefits are typically exaggerated and the harms are downplayed. I illustrate this resistance to evidence with the reaction of German and US health organizations and governmental agencies.

5   **Evidence Not Wanted: Germany**

In 2001, the German government proposed mammography screening for all women between the ages of fifty and sixty-nine: "Mammography screening could reduce mortality from breast cancer by 30%, that means, every year about 3,500 deaths could be prevented, ca. 10/day" (cited in Gigerenzer *et al.* 2007). Note first that the public is not informed about the benefit in a transparent way, such as an absolute reduction from about ten to nine breast cancer deaths in every 2,000 women. Rather, the result is communicated in the form of a relative risk reduction, which is a large number and looks impressive. According to the Cochrane Review, the relative risk reduction is in the order of 10% (from about ten to nine deaths), with their precise estimate being 15%. Since the results of the seven trials vary somewhat, the German government seems to have picked the most optimistic one. My point is that most citizens do not understand the difference between an absolute and a relative risk reduction, and believe that 30% means that many lives are saved. This was the way in which the public learned about the first result concerning cancer-specific mortality reduction.

The figure of 3,500 deaths that could be prevented is an estimate for deaths from breast cancer, but it is presented so that it appears to be the total number of people saved. Nowhere does the government report mention that there is no reduction in the total number of deaths, or deaths from any cancer. This was the government's way of presenting the second result concerning total mortality. The harms of overtreatment were not mentioned at all.

The Berlin Chamber of Physicians protested in a 2002 press release against a general screening program on the grounds that there is no scientific evidence that the potential benefits of screening are greater than its harms, whereas the parliament's health committee overstated benefits and downplayed harms. Two days later, the German Minister of Health, Ulla Schmidt, responded in a press release that RCTs provide sufficient evidence in favor of screening because "there is an up to 35% reduction in breast cancer mortality" (Gigerenzer *et al.* 2007). Note once again the use of relative risk reduction. When I clarified what this number means in an interview in the German weekly *Die Zeit*, the advisor of the Secretary of Health, Professor Karl Lauterbach, defended the use of relative risk reduction by responding that "In justifying the programs, the Secretary of Health does not inform individual women, but the public. If an individual doctor advises patients, he should, like Mr. Gigerenzer, state the absolute risk and its reduction" (Lauterbach 2002: 16). According to this logic, transparency is for individual women, not for the public.

In Germany, as in most other countries, the organizations that are in charge of the screening programs are the same as those who are in charge of the dissemination of the information about cancer screening. In their first role, their goal is to increase participation rates, which conflicts with the goal of informing patients about benefits and harms in a transparent way.

6   Evidence Not Wanted: United States

In 1997, the National Institutes of Health Consensus Development Conference on Breast Cancer Screening for Women Aged 40 to 49 was convened at the request of the Director of the National Cancer Institute (NCI). The expert panel reviewed the medical studies and concluded with a ten to two vote that there is insufficient evidence to recommend screening for this age group, and that "a woman should have access to the best possible relevant information regarding both benefits and risks, presented in an understandable and usable form" (National Institutes of Health Consensus Development Panel 1997: 1015). At the news conference, Richard Klausner, Director of the NCI, said he was "shocked" by this evidence, and that night, a national television program began its news coverage with an apology to American women for the Panel's report. Eventually, the Senate voted ninety-eight to zero for a nonbinding solution in favor of mammography for women in their forties. The Director of the NCI asked the Advisory Board to review the Panel's report, a request that they first declined, but, in March 1997, the Board voted seventeen to one that the NCI should recommend mammography screening every one or two years for women in this age group – against the conclusion of its own expert panel (Fletcher 1997). The voting members of the NCI Advisory Board are appointed by the US President, not by the medical experts in the field, and are under great pressure to recommend cancer screening. In 2002, new studies became available that again indicated that the benefits of mammograms may not outweigh the harms, and Donald Berry, Chairman of the Department of Biostatistics at the MD Anderson Cancer Center explained this result to the Senate, but to no avail. The Bush Administration restated the recommendation, and Andrew von Eschenbach, the Director of the NCI at that time, announced that women in their forties should have mammograms (Stolberg 2002).

7   Evidence-based Medicine as Science Fiction

In this comment, I took off where Nancy Cartwright left us: The problem of how to make an argument from efficacy – what a well-conducted RCT shows – to effectiveness in the field. I respect her argument that there is a logical gap when we generalize from one population to another, and from the controlled conditions in an RCT to the unknown features of the wild. I looked at screening for breast cancer with mammography and a recent Cochrane Review as a concrete illustration of her philosophical issue. The Review pays no attention to the efficacy–effectiveness gap, and, moreover, the neglect of the problem is implied by the use of future terms instead of past terms when discussing the results. There is no theory that could fill the logical gap, but the Cochrane reviewers might, instead, point to the convergent empirical evidence in the seven trials. Cartwright observed that the results of RCTs are often quickly assumed to hold true in the field, and are relied upon uncritically for policymaking. In this comment, I drew attention to the opposite observation in cancer screening: RCTs are mostly ignored by policymakers in Germany and the United States, results are distorted rather than followed, and policymaking is shaped by conflicting interests rather than evidence. Despite the neglect of the logical gap between efficacy and effectiveness, there is good cause to be made for the positivistic enterprise of the Cochrane Collaboration. In cancer screening policy experimental evidence plays little role. Few patients ask for evidence, few doctors understand it, and paternalism reigns.

REFERENCES

Fletcher, S. W. 1997. "Whither scientific deliberation in health policy recommendations? Alice in the Wonderland of breast cancer screening," *New England Journal of Medicine* **336**: 1180–1183.

Gigerenzer, G., W. Gaissmaier, E. Kurz-Milcke, L. M. Schwartz, and S. W. Woloshin 2007. "Helping doctors and patients make sense of health statistics," *Psychological Science in the Public Interest*, **8**: 53–96.

Gøtzsche, P. C. and M. Nielsen 2006. "Screening for breast cancer with mammography," *Cochrane Database of Systematic Reviews* (4), Article CD001877. DOI: 001810.001002/14651858.CD14001877.pub14651852.

Lauterbach, K. W. 2002, August 28. "100 000 überflüssige Operationen" [Letter to the editor]. *Die Zeit* p. 16.

Mooi, W. J. and D. S. Peeper 2006. "Oncogene-induced cell senescence – Halting on the road to cancer," *New England Journal of Medicine* **355**: 1037–1046.

National Institutes of Health Consensus Development Panel 1997. "National Institutes of Health Consensus Development Conference statement: Breast cancer screening for women ages 40–49," *Journal of the National Cancer Institute* **89**: 1015–1020.

Stolberg, S. G. 2002, February 6. "Study says clinical guides often hide ties of doctors," *New York Times*.

How Philosophy and the Social Sciences Can Enrich Each Other: Three Examples

Part III of the book deals with three quite diverse problems: the problem of cooperation in society; the problem of virtuous behavior; and the problem of the hermeneutic circle. The problems are carefully selected in order to address issues of fundamental importance for the philosophy of the social sciences, which are, however, very diverse in scope. The aim is to show that these problems can be most fruitfully addressed, and at least partly solved, if both philosophical and scientific resources are used. This volume takes an anti-foundationalist view: that is, it rejects the notion that philosophy can or should provide any kind of foundations for the social sciences; and it takes the stance that philosophy and the social sciences are equally important partners in the enterprise of solving problems. In this process the resources of both philosophy and the social sciences are mutually enriched and their boundaries are continually shifting. The three examples provided in this part thus show *How Philosophy and the Social Sciences Can Enrich Each Other*.

The so-called "Hobbesian problem of social order" is at the heart of political philosophy and the social sciences. How is it possible, given the self-interested nature of human beings, that a stable social order can emerge and prevail, and cooperation can take place? In centuries of discussion of the proposed solutions to this fundamental problem, different arguments have been advanced and different methods have been applied. *James Woodward* addresses the question, "Why do people cooperate as much as they do?" and uses the resources of game theory and experimental economics in trying to explain the degree of cooperative behavior shown by human agents. The upshot of Woodward's discussion is that actors are best understood as conditional cooperators or, as they are sometimes called, "strong reciprocators." The preferences of conditional cooperators (and their behavior) are very complex and context sensitive, and Woodward suggests that one important role of social norms is to deal with the indeterminacy and unpredictability that would otherwise exist in the interactions of conditional cooperators with one another and with purely selfish actors. *Werner Güth and Hartmut Kliemt,* in "Putting the Problem of Social Order into Perspective," remind us first of Hume's dictum that "[t]hough men be much governed by interest, yet even interest itself, and all human affairs, are entirely governed by opinion." In modern parlance, Hume and the other British Moralists thought that, without referring to the role of accepted rules and standards in guiding human behavior, the emergence of social order and inter-individual cooperation cannot be explained. Though Güth and Kliemt do have disagreements concerning the way Woodward mixes rational choice, behavioral and evolutionary arguments, they positively point out that the classical controversy is center stage in his discussion: Can all forms of cooperation be explained as instrumental to the pursuit of some non-cooperative interest or must there be intrinsically motivated cooperative inclinations including rule-following?

In recent years philosophers have mounted a sustained attack on the use of human virtues and character in ethics (and more recently in epistemology). The attack is based on various results in social psychology. It is argued that virtues and character are explanatorily inert, and indeed theoretical illusions altogether. In "Situations Against Virtues. The Situationist Attack on Virtue Theory" *Ernest Sosa* discusses this attack on the theoretical integrity of virtues and character and on their use in normative disciplines and in the explanation of human behavior. He defends virtue ethics by focusing on competence, more specifically on an analogy between moral competence and driving competence. In his comment, *Steven Lukes* shows that the dispute between situationism and virtue theory has the appearance of a duck–rabbit problem – the problem being that you cannot see both the duck and the rabbit at one and the same time. Lukes shows that Sosa is right and that this appearance is an illusion. The dilemma that either there are character traits or there are no character traits, but only situations are relevant in determining behavior, is detected as illusionary and due to the way that situationists pose the issue.

The hermeneutic circle serves as a standard argument for many of those who raise a claim to the autonomy of the human sciences or propagate an alternative methodology for the human sciences. In the last chapter of the book I check the soundness of this argument. Trying to answer the question "What Kind of Problem is the Hermeneutic Circle?" it is argued that the hermeneutic circle is neither a genuine ontological problem nor a logical nor methodological problem, and consequently that it cannot lend legitimacy to the claim of the autonomy of the human sciences. Rather, it is shown to be an empirical problem, which has long been studied using the tools of the empirical sciences. *David-Hillel Ruben*, who as the sole exception in this book, provides a comment from

the philosophical point of view rather than from the viewpoint of a practicing scientist, contends that there is no plausible distinction between
understanding and explanation in advance of a thesis about the irreducible differences between knowledge in, or the methodology of, the natural and the social sciences. Without assuming that understanding and
explanation are different ideas, Ruben situates the hermeneutic circle in
a typology for circles that he develops, taking into consideration that a
circle can be a circle of particulars, concepts, or facts.

8 Why Do People Cooperate as Much as They Do?*

James Woodward

1 Introduction

Why do humans cooperate as much as they do? Stated at this level of generality, it seems obvious that this question has no single answer. The same person may cooperate for very different reasons in different situations, depending on the circumstances in which the cooperation occurs, the behavior, and motivation of others with whom they cooperate and so on. Moreover, both casual observation and (as we shall see) experimental evidence suggest that people differ in type, with self-interest playing a larger or more exclusive role in explaining cooperative (and non-cooperative) behavior among some people, and non-self-interested motives (which may themselves take a variety of different forms but often involve dispositions toward conditional rather than unconditional cooperation) playing a larger role among other subjects.

Nonetheless, one finds in philosophy, social science, and evolutionary biology a number of different stylized accounts of why cooperation occurs, often presented as mutually exclusive alternatives. In what follows I explore some of these, with an eye to assessing their explanatory adequacy. I begin with the familiar idea that cooperative behavior can be understood simply as an equilibrium in a repeated game played by self-interested players. I suggest there are substantial empirical and theoretical problems with this idea and that the empirical evidence predominantly supports the view that cooperation often involves interactions among players of different types, some of whom are self-interested in the sense that they care only about their own monetary payoffs and others of whom are best understood as conditional cooperators of one kind or another, or, as they are sometimes called, *strong reciprocators*; e.g., Fehr and Fischbacher (2003), Henrich *et al.* (2004), and the essays in Gintis, Bowles, Boyd, and Fehr (2005). Mantzavinos (2001: 101ff.)

* Many thanks to Ken Binmore, Francesco Guala, Dan Hausman, Philip Hoffman, Chrys Mantzavinos, Toby Page, and Jiji Zhang for helpful discussions of the issues addressed in this chapter.

also emphasizes the importance of conditional cooperation. Conditional cooperators (CCs) differ among themselves but for our purposes we may think of them as having at least one of the following two features: 1. In some circumstances CCs cooperate with others who have cooperated or who they believe will cooperate with them, even when the choice to cooperate is not payoff maximizing for the CCs; and 2. In some circumstances CCs do not cooperate with (and may punish or sanction) others who behave in a non-cooperative way (or who they expect to behave non-cooperatively) even when non-cooperation is not payoff maximizing for the CCs. Typically, CCs respond not just to the payoffs they receive as a result of others' behavior, but to the intentions and beliefs which they believe lie behind others' behavior (presumably in part because these have implications for the possibility of future cooperation).

This has the consequence that the preferences of conditional cooperators (and their behavior) is very complex, and context sensitive – we cannot, for example, adequately predict their behavior across different contexts simply by supposing that they have, in addition to self-interested preferences, stable "other regarding" preferences regarding payoffs to others that are independent of their beliefs about the behavior and intentions of those others. I suggest that one important role of *social norms* (and more generally social and moral rules, and institutions) that one often finds governing real life human cooperative interactions is to deal with the indeterminacy and unpredictability that would otherwise be present (i.e., in the absence of norms) in the interactions of conditional cooperators with one another and with more purely selfish types. Thus rather than thinking of appeals to preferences for reciprocation and conditional cooperation on the one hand, and appeals to social norms on the other, as competing explanations of cooperative behavior, I instead see these as complementary – at least in many cases, conditional cooperators need norms if they are to cooperate effectively and norms require conditional cooperators with a willingness to sanction for their successful enforcement. Put slightly differently, the high levels of cooperation sometimes achieved by humans depends not just on our possession of social preferences of an appropriate sort but on our capacity for what Alan Gibbard (1990) calls "normative governance" – our ability to be guided by or to conform our behavior to norms. Moreover, different people are able to achieve common conformity to the same norms despite being (in other respects) motivationally fairly heterogeneous. This capacity is crucial to the ability to achieve cooperative solutions to the social dilemmas human beings face – social preferences (whether for conditional or unconditional cooperation) by themselves are not enough.

1.1   Self-interest in Repeated Games

One familiar account of human cooperation – call it the "self-interest in repeated games" (SIRG) account – goes like this. Assume first that we are dealing with subjects who are "rational" (their preferences satisfy the usual axioms of expected utility theory) and are entirely "self-interested." ("Self-interest" is a far from transparent notion, but for present purposes assume this implies that people care only about their own material payoffs – their monetary rewards in the experiments discussed below.) Assume also that interactions are two-person or bilateral and that in particular two players find themselves in a "social dilemma": there are gains for both players (in comparison to the base-line outcome in which each chooses a non-cooperative strategy) which can be obtained from mutual choice of cooperative strategies, but a non-cooperative choice is individually rational for each if they interact only once. Assume, however, that their interaction is repeated – in particular, that it has the structure of a repeated two-person game of indefinite length, and that certain other conditions are met – the players care about their future sufficiently (do not discount too heavily) and they have accurate information about the choices of the other player. For many games meeting these conditions, the mutual choice of a cooperative strategy is a Nash equilibrium of the repeated game.[1] To take what is perhaps the best known example, mutual choice of tit for tat is a Nash equilibrium of a repeated prisoner's dilemma of indefinite length, assuming that the players have sufficiently high discount rates and accurate information about one another's play. This (it is claimed) "explains" why cooperation occurs (when and to the extent it does) among self-interested players.

An important part of the appeal of this approach, in the eyes of its proponents, is that it is "parsimonious" (Binmore 2007) in the sense of not requiring the postulation of "other regarding" preferences to explain cooperation. Instead, cooperative behavior is explained just in terms of

---

[1] That is, it is a Nash equilibrium, assuming that the players care only about the monetary payoffs to themselves. In what follows, when I speak of various games as having Nash equilibria, I will mean (unless explicitly indicated otherwise) equilibria that are Nash on the assumption that the players care only about their own monetary payoffs. I will also largely ignore various refinements of the Nash equilibrium concept (subgame perfect equilibria etc.) in what follows because I think that in most games of any complexity there is little empirical evidence that subjects actually choose in accordance with such equilibrium notions.

That a mutual choice of cooperative strategy is a Nash equilibrium of the repeated game is an implication of the so-called "fork theorem" which says, roughly, that under the conditions envisioned above any possible outcome (including cooperative outcomes) in which each player's minimax condition is satisfied (each player minimizes his maximum loss) is a Nash equilibrium of the repeated game.

self-interested preferences which we already know are at work in many other contexts.

I do not wish to deny that this SIRG account isolates one important set of considerations which help to explain cooperation. (See below for more on this.) Nonetheless, when taken as a full or exclusive explanation of cooperative behavior the approach is subject to a number of criticisms, both theoretical and empirical. In what follows, I explore and assess some of these.

1.2   Theory

At the theoretical level, while it is true that cooperative strategies are among the Nash equilibria of games like the repeated prisoner's dilemma, it is also true that a very large number of non-cooperative strategies (like mutual defection on every move) are among the Nash equilibria of the repeated game. Moreover, philosophical folklore to the contrary, it is simply not true that among these multiple Nash equilibria, those involving highly cooperative strategies are somehow singled out or favored on some other set of grounds. For example, it is false that tit for tat is an evolutionarily stable strategy (ESS) in a repeated PD, still less that it is the only such ESS. As Binmore (1994: 198ff.) points out, in Axelrod's well known computer simulation, it is not true that tit for tat "won" in the sense of being much more highly represented than less nice competitors. In fact, the actual outcome in Axelrod's tournament approximates a mixed-strategy Nash equilibrium in which tit for tat is the most common strategy, but is only used a relatively small portion (15%) of the time, with other less cooperative strategies also being frequently used. When paired against other mixes of strategies besides those used by Axelrod, tit for tat is sometimes even less successful.[2] Thus what SIRG accounts (at least of the sort we have been considering) seem to show at best is that it is *possible* for cooperative play to be sustained in a repeated two-person social dilemma (or alternatively, they explain why, *if* the players are already at a cooperative equilibrium, they will continue to cooperate rather than deviating from this equilibrium). However, given the large number of non-cooperative equilibria in many repeated games involving social dilemmas, SIRG analyses do not by themselves tell us much about why or when cooperative (rather than non-cooperative) outcomes occur in the first place.

---

[2] For example, as described by Binmore (1994) in a simulation by Linster, the Grim strategy of defecting forever in response to a single defection takes over more than half of the population.

In addition, many social dilemmas are not bilateral but rather involve a much larger number of players (e.g., they may take the form of n person public goods games for large n). Theoretical analysis (e.g., Boyd and Richerson 1988) suggests that it is much more difficult for cooperative behavior to emerge among purely self-interested players in such multi-person games (in comparison with two-person games) when such games are repeated. This is so not only because information about the play of all the players is less likely to be public, but also because the strategy that is used to enforce cooperation in the two-person game (playing non-cooperatively when the other player has played non-cooperatively on a previous move) is a much more blunt or less targeted instrument in games involving larger numbers of players: in such games choosing to play non-cooperatively will impose costs on all players, rather than just those who have been uncooperative. For example, non-contribution to a public good in order to sanction the non-contribution of other players in a previous round of play will harm the other contributors as well as non-contributors and will presumably encourage contributors to be less cooperative in subsequent rounds. Indeed, in agreement with this analysis, under the conditions that we are presently envisioning (players cannot sanction non-cooperators, or avoid interacting with them) it is found as an empirical matter (both in experiments and in observation of behavior in the field) that cooperation in repeated public goods games declines to relatively low levels under repeated play. As we shall see, defenders of SIRG accounts often contend that this decline supports their claim that people are self-interested – their idea being that the decline occurs because self-interested subjects do not initially realize that non-cooperation is pay-off maximizing, but then gradually learn this under repeated play of the game. On the other hand, it is also clear from both experiment and field observation that substantial levels of cooperation can be sustained under repetition in public goods games when the right sort of additional structure is present (e.g., arrangements that allow the more cooperative to identify and interact preferentially with one another and to avoid free riders, or arrangements which allow for the sanctioning of non-cooperators (see below)). Thus, although this moral is not usually drawn by defenders of the SIRG analysis, what the empirical evidence seems to suggest is that whether or not people are self-interested, cooperative outcomes in social dilemmas are typically *not* automatically sustained just on the basis of repeated interactions in groups of substantial size. This again suggests that when cooperation occurs in such circumstances (as it does with some frequency) it is not well explained just by pointing to results about the existence of cooperative equilibria in repeated games, involving selfish subjects. More is involved.

Yet another theoretical issue for the SIRG approach concerns the structure of the game being played. There is a strong tendency in much of the literature to focus on what Philip Kitcher calls "compulsory" repeated games (Kitcher 1993) – that is, games in which there is no move available which allows the subject to opt out of the repeated interaction. The repeated prisoner's dilemma has this character, as does the repeated *n* person public goods game envisioned above – subjects *must* interact with one another, with the only question being whether they will interact cooperatively or instead defect. Yet in many real life situations subjects have a choice about whether to enter into an interaction which presents the possibility of cooperation/defection or instead to go solo and avoid the interaction.[3] For example, they may either hunt cooperatively in a group or hunt alone. Relatedly, subjects also have a choice about who they select as a potential partner for cooperation and who they avoid. This difference between mandatory and optional games matters for several reasons. First, insofar as the situation allows for the possibility of opting out of any form of repeated interaction, we need to explain why (rather than just assuming that) repeated interaction occurs (to the extent that it does). If we simply assume a game in which there is repeated interaction, we may be assuming away a large part of the explanatory problem we face, which is why a repeated game with a certain interaction structure occurs in the first place.[4] Second, and relatedly, both theoretical analysis and experimental evidence (some of which is described below) suggest that when players are allowed to decide whether to interact and with whom to interact, this can considerably boost the incidence of cooperation.

1.3   Empirical Evidence

I turn now to a discussion of some empirical evidence bearing on the adequacy of the SIRG account – mainly but not exclusively experimental evidence.

---

[3] Similar points are also made in Tullock (1985) and Vanberg and Congleton (1992). Mantzavinos (2001) also emphasizes the importance of considering games with an exit option.

[4] Of course there are cases in which the representation of an interaction as a mandatory repeated game is completely appropriate. If you and I are marooned on a very small desert island, with only one limited source of water, we face a forced interaction over the distribution of the water and one which is repeated as long as both of us remain alive. My point is simply that not all cases in which cooperation occurs have this structure. Sometimes whether there is repeated interaction at all should be regarded as endogenously determined.

In some two-person repeated games such as repeated prisoner's dilemma and trust games, subjects often achieve relatively high levels of cooperation, although cooperation declines fairly steeply when subjects know that they are in the final rounds. For example, Andreoni and Miller (1993) report results from a repeated PD of ten rounds in which roughly 60% of subjects begin by cooperating, with this proportion dropping gradually in earlier rounds (50% cooperation in the fifth round) and then steeply to roughly 10% by the tenth round. In repeated public goods games with a substantial number of players, the generic result again is that many subjects begin with a relatively high level of contribution (40–60% of subjects in Western countries contribute something) but then contributions decline substantially as the game is repeated, eventually reaching a level at which a core of only about 10% of subjects continue to contribute.

These results do not of course tell us *why* subjects behave as they do in repeated games. However, it is commonly argued (e.g., Binmore 2007 and Samuelson 2005) that this pattern is broadly consistent with the SIRG account – in games like a repeated prisoner's dilemma, the high levels of cooperation observed in the early and middle periods of the game are just what one would expect from self-interested players and the sharp declines in cooperation at the end of the game are just the sort of "end game" effects one would expect among self-interested players who recognize that their choices will no longer influence whether others play cooperatively. The initially high levels of contribution in public goods games are attributed to "confusion" (failure to understand the game that they are playing), and the decline in cooperative behavior as the game is repeated is attributed to the players gradually learning the structure of the game and coming to realize that defection is the self-interested strategy.

1.4   Social Preferences?

Others (Camerer and Fehr 2004 and Gintis 2006) have challenged this interpretation. One popular line of argument goes like this: if players are purely self-interested (and if they are also fully rational and understand the strategic structure of the situation they face) then while (as we have seen) they may cooperate in repeated games, they should not do so in one-shot games in which the Nash equilibria all involve non-cooperative choices, at least when care is taken to ensure that the players believe that the possibility of future interactions and reputational effects are absent. However, this is contrary to the rather high levels of cooperation that are observed experimentally in many one-shot games involving social dilemmas. For example, in one-shot PDs 40–60% of players in developed countries cooperate and in one-shot public goods games in developed countries, subjects contribute on average about half of their endowment, although again there is a great deal of variation, with a number of subjects contributing nothing. In one-shot trust games in developed countries, on average subjects transfer around 0.4–0.6 of their stake and trustors return approximately the amount transferred (when the self-interested choice is to return nothing), although there is once more considerable individual variation. In one-shot ultimatum games in developed countries, mean offers of proposers are about 0.4 of the stake and offers below 0.2 are rejected half the time. Offers of 0.5 are also common. Of course, it also appears to be true that people behave cooperatively in one-shot interactions in real life in which non-cooperation is the self-interested strategy: they leave tips in restaurants to which they think they will never return, they give accurate directions to strangers when this requires some effort and so on.

Also relevant in this connection are the results of a fascinating series of cross-cultural experiments (described in Henrich *et al.* 2004) in which various experimental games were played in a number of different small-scale societies in Asia, Africa, and South America. In these games, there was considerably more variation in results than in developed countries like the contemporary US. For example, among the Machiguenga in Peru, one-shot ultimatum game offers had a mean of 0.26 and a mode of 0.15 (far lower than mean and modal offers in developed societies) and almost no offers were rejected. In contrast, among the Lamerela in Indonesia offers in excess of 0.50 (hyper-fair offers) were common and there was frequent rejection, even of hyper-fair offers. A similar variation was found in public goods games and trust games.

Moreover, in at least some cases these results appear to correlate with features of social life in the societies in question. For example, the Machiguenga are described as "socially disconnected" by anthropologists, with economic life centering on the individual family and little opportunity for anonymous transactions. By contrast, the Lamerala are a whaling society in which there is a high degree of cooperative hunting and food sharing and in which it is common for people to reject gifts out of concern that this will place them under a strong obligation to recip-rocate. The Orma in Kenya make relatively high contributions in public goods games (both in comparison to many other small-scale societies and the contemporary US: 0.58 of their endowments). Participants in this game associated it with a harambee, "a Swahili word for the institu-tion of village level contributions for public goods projects such as build-ing a school" (Ensminger 2004: 376) and in fact a harambee collection was going on at the time the public goods game was played. It seems plausible that the relatively high level of contribution was at least in part a reflection of the willingness of the participants to think of it as an instance of this practice.

A number of writers take such observations to show that in addition to whatever self-interested preferences people may possess, many people also have non-self-interested (or "social") preferences. (These may include preferences to sanction or reciprocate negatively as well as preferences that benefit others.) For example, many of the contributors to the Henrich *et al.* volume (e.g., Camerer and Fehr 2004) take themselves to be using one-shot experimental games to "measure" social preferences that exist in real-life social interactions even though these take the form of repeated games in the societies under study. They believe that the use of one-shot games allows them to disentangle such social preferences from the influence that purely self-interested preferences exert on cooperative behavior in repeated games. However, this interpretation of the experimental evidence concerning one-shot games has itself been challenged by several defenders of SIRG. Since this challenge has important implications for how we should best explain cooperation it will be worth considering in some detail.

1.5   Challenges to the Existence of Social Preferences

The challenge begins with the claim that most real-life social interactions are best modeled as moves in a repeated game in which accurate information about previous play is available to most of the players. In other words, it is claimed that genuinely one-shot games are rare in real life. This is particularly likely to be true (it is claimed) in the small-scale societies or groups in which human beings lived through most of prehistory since these are groups with relatively stable membership in which the same individuals interact repeatedly, face similar dilemmas again and again, and generally have accurate information about one another's behavior. (It is sometimes also suggested that as a consequence of this, our cognition and motivation have been shaped by natural selection as adaptations for repeated interactions and that we are not well adapted to one-shot interactions or that we inevitably construe them in terms appropriate to repeated interactions.[5])

---

[5] The extent to which it is true empirically that virtually all interactions in small scale or "primitive" societies are non-anonymous and repeated is contested (see e.g., Gintis 2006). There is also disagreement about how common genuinely one-shot interactions are in contemporary large-scale societies.

It is further suggested that the upshot of all this is that when subjects play a one-shot game in the laboratory, it is likely that their choices either (a) will be confused (because of their unfamiliarity with the game they are playing) and/or (b) that their choices will reflect the importation of patterns of behavior or heuristics, or norms that derive from their experience in playing some repeated game that looks similar to the laboratory game. (For later reference, I will call claim (b) the importation thesis.) In either case, a subject's behavior cannot be taken as evidence for their possession of non-self-interested preferences. Under (a) their choices will not reflect stable, coherent preferences of any kind, self-interested or non-self-interested. Under (b), cooperative behavior in the one-shot game also cannot be taken as evidence for their possession of non-self-interested preferences, since it will merely reflect the importation of strategies that are appropriate for self-interested players in the repeated game. As an illustration, suppose subjects are presented with a one-shot game in which monetary payoffs to each player have the structure of a PD and that the subjects behave cooperatively. If subjects are purely self-interested, but model this interaction as a repeated game and import cooperative strategies like tit for tat that are equilibria of this repeated game (for self-interested players), then we have an explanation of why they cooperate that does not require appeal to non-self-interested preferences. It is true, of course, that experimenters go to considerable lengths to make sure that subjects understand (i.e., at a conscious level) that they are playing a one-shot game and to try to control for concerns about reputational effects that might lead subjects to treat the game as repeated. Thus for the importation thesis (b) to be credible, it should be interpreted as claiming that despite their intellectual recognition that the game is one-shot, subjects nonetheless import patterns of play from similar looking repeated games (as a result of something like unconscious processing, implicit learning, or perhaps the triggering of innate dispositions).

An additional prediction that is at least suggested (if not strictly entailed) by this portion of the SIRG analysis is this: Suppose we have a one-shot game in which subjects cooperate only because they import cooperative strategies supported by self-interest from a similar looking repeated game. Then, if the one-shot version of this game were played over and over again, but with different partners, one would likely see a decline (and probably a very steep decline) in cooperation. This would be because (on the SIRG account) one would expect that subjects would gradually learn that (i.e., come to model the interaction as though) they were playing a one-shot game and the behavior which is appropriate to the repeated game would gradually be replaced with behavior that is appropriate to the one-shot game (e. g., defection in a one-shot PD, assuming the players are motivated only by self interest).[6]

Binmore (2007) in fact claims that one does see just this pattern in many experimental games and he takes this to support the SIRG account. He points out that subjects who repeatedly play one-shot PDs against different partner become more likely to defect as the game goes on: the defection rate goes from an initial rate of 40–60% to nearly 90% defection. A similar pattern is observed with subjects who play a rotating one-shot public goods game (a so-called stranger treatment, with the group composition changed every time period). Here also contributions decline over time, although, interestingly, it is controversial and the evidence equivocal whether the decline is more or less steep than the decline in contributions in repeated public goods games with the same subject (so-called partner treatment). Binmore takes this to be evidence that subjects gradually learn the structure of the one-shot game with repetition and converge on the Nash equilibrium (mutual non-cooperation), which is just the behavior that the SIRG account predicts.

2   Conditional Cooperation

2.1   Conditional Cooperation as an Alternative to Self-interest

The extent of the support that these observations provide for the SIRG account depends of course on the extent to which there are alternative, non-SIRG accounts of subjects' motivation that predict the same behavior – that is, the extent to which it is only the SIRG account that predicts this behavior. In fact, there is an alternative account (or really, a family of accounts) that, as we shall see, has independent evidential support. This is that a substantial portion of the subjects are conditional cooperators of one sort or another. For the purposes of this chapter, conditional cooperators are people with the following features: (1) they will cooperate even when this is not the payoff maximizing choice if (but only if) enough others with whom they are interacting also cooperate or are expected (by the conditional cooperators) to cooperate. For example, they will cooperate in a one-shot PD or on what is known to be the last move of a repeated PD. (2) They will impose costs on non-cooperators (either through failure to cooperate or through sanctions of some other sort) when this is not payoff maximizing. For example, they will reject

---

[6] Or at least one would expect this pattern unless one were prepared to claim that subjects are so fixed and inflexible in their behavior that they cannot unlearn strategies that have once been learned or that it takes an enormously long time for this to happen – assumptions that game theorists tend to reject in other contexts.

low offers in one-shot ultimatum games and will punish free riders at a cost to themselves in the last round of a public goods game. The general category of conditional cooperator covers a variety of more specific possibilities: some subjects may begin by cooperating and defect only against those who have previously defected, as in the standard tit-for-tat strategy. Other subjects may begin by cooperating, but then defect more indiscriminately (even against those with whom they are interacting for the first time) whenever they have experienced enough defections by others. (There is experimental evidence that a significant number of subjects behave in this way.) Still other subjects may begin by defecting (or by not interacting at all, if that is an option) and only begin cooperating if they observe a sufficient amount of cooperative play by other subjects. Similarly different subjects may have different thresholds for sanctioning. Note that conditional cooperators differ not only from subjects whose preferences are entirely self-interested, but also from subjects whose preferences are unconditionally altruistic (i.e., subjects who have preferences to benefit others regardless of the behavior of those others). For example, unlike CCs, unconditional altruists will never reject positive offers in a one-shot UG in which reputational concerns about possible future interactions etc. are not an issue since this makes both players worse off than acceptance. Similarly, unlike CCs, unconditional altruists who maximize some function of both players' payoffs may cooperate in a PD in which the second party defects, provided that the gain to the second party is large enough.

An alternative explanation (besides the SIRG account) for the decline in cooperation when repeated one-shot games are played is that this is due to the interaction between two different types of subjects, both present in significant numbers: subjects who, because they have purely self-interested preferences (or for some other reason), are prepared to defect or free-ride when others cooperate and subjects who will make non-payoff maximizing cooperative choices in some circumstances, but who will decline to cooperate (even with new partners) after they have experienced a sufficient number of interactions in which their partners have behaved non-cooperatively (perhaps especially when they themselves have behaved cooperatively). With such a combination of subjects, the rate of cooperation among conditional cooperators will decline over a series of one-shot interactions as they encounter more and more defectors.

2.2   *Evidence for Conditional Cooperation*

There are a number of converging pieces of evidence that suggest that there is at least some truth in this alternative account. Ahn, Ostrom, and

Walker (2003) asked subjects to state their preferences regarding outcomes in one-shot double-blind prisoner's dilemmas, both involving simultaneous and sequential moves. In the simultaneous move version of the game, 10% of subjects ranked the mutually cooperative outcome (C, C) higher than the outcome in which they defect and their opponent cooperates (D, C). Another 19% said they were indifferent between these two outcomes, despite the fact that the latter (D, C) outcome is payoff maximizing for them. In the sequential move game, 40% of subjects ranked (C, C) higher than (D, C) and another 27% said they were indifferent between these two outcomes. Lest these be dismissed as mere survey results, Ahn *et al.* also report results from behavioral experiments that are broadly consistent: 36% of subjects cooperated in the simultaneous move game, and in the sequential game, 56% of first movers cooperated and 61% of second movers cooperated when first movers cooperated, despite the fact that the latter is not the payoff maximizing choice. Ostrom comments elsewhere that these results "confirm that not all subjects enter a collective action situation as pure forward looking rational egoists" and that "some bring with them a set of norms and values that can support cooperation" (2005: 129). However, she also adds that

> Preferences based in these [cooperative] norms can be altered by bad experiences. One set of 72 subjects played 12 rounds of a finitely repeated Prisoner's Dilemma game where we randomly matched partners before each round. Rates of cooperation were very low. Many players experienced multiple instances where partners declined to cooperate (Ahn, Ostrom, and Walker 2003). In light of these unfortunate experiences, only 19 percent of the respondents now ranked (C, C) above (D, C) while 17 percent were indifferent (ibid.). In this uncooperative setting, the norms supporting cooperation and reciprocity were diminished by experience, but not eliminated. (2005: 129)

This comports with my suggestion above: a number of subjects are conditional cooperators, but of the sort that will defect even against a new partner if they have sufficient experience of defection themselves.

A number of other experimental studies suggest a similar picture of subjects' preferences.

2.3   Heterogeneity of Types and Endogenous Grouping

Further support for this general picture is provided by a series of experiments involving grouping effects in repeated public goods games. A representative example[7] is an experiment conducted by Page *et al.* (2005)

---

[7] There are a number of other experiments reporting results that are consonant with those reported by Page *et al.* For example, Gunnthorsdottir, McCabe, and Houser (2000) sorted players into cooperators or free riders based on their play in an initial

in which subjects were given information every three rounds about the contributions of other subjects in previous rounds and also an opportunity to express preferences about future partners. Subjects whose mutual rankings were the lowest (with a low ranking meaning that a partner is most preferred for cooperation) were then organized into a first group; those whose rankings were intermediate were organized into a second group and so on, for a total of four groups. In other words, the result was that those who had contributed most in previous rounds played the next three rounds among themselves, those who had contributed at an intermediate level were organized into a second group and so on, down to a fourth group that consisted very largely of free riders. The result of this "endogenous grouping treatment" was that the average level of contribution across all groups (70%) was much higher than the average level of contribution in a baseline treatment in which there is no regrouping (38%). Moreover, average levels of contribution varied greatly across the groups, with this following the order of group formation – that is, contributions were highest and continued to be so in the first group formed of high cooperators and were lowest in the fourth group: in the first group, 50% contributed their entire endowment in the final period (with the corresponding numbers from the second, third, and fourth groups being 43, 18, and 0).

public goods game. Players were then either (1) divided into groups depending on their contributions (with the highest contributors grouped together, the next highest contributors grouped together etc.) and then played a repeated public goods game or (2) alternatively, played repeated public goods games under a random grouping condition in which there is no sorting. The public goods games are played under a variety of MPCR (marginal per-capital return) conditions – that is, different rates at which money is distributed to group members as a function of individual contributions. The authors find that, in the random grouping condition and for all MCPRs in every round, cooperators (as typed by the initial sorting) contributed more on average than free riders, suggesting that the initial sorting captured some stability of type in subjects' behavior. Under the random grouping treatment, when cooperators and free riders interacted, Gunnthorsdottir *et al.* found the characteristic decay in aggregate contributions to the public good over time that is reported in other experiments. They also found that "Almost all of this decay can be attributed to decay in cooperators' contributions." By contrast, there were much slower rates of decay under the treatments in which there was sorting into groups and cooperators met free riders less frequently, although in these cases, decay rates are (unsurprisingly) lower under high MPCRs. Indeed, Gunnthorsdottir *et al.* found that "by sufficiently reducing cooperators' interactions with free-riders ... at least when the MPCR is not too low, cooperators' public contributions were sustained." The natural interpretation of these results is again that cooperators respond to interactions with free riders by reducing contributions but that such reductions do not occur if interactions with other cooperators are sufficiently frequent. Other experiments supporting a similar picture in which a substantial subset of players condition their contributions on their beliefs about the contributions of other players are reported in Burlando and Guala (2005) and Croson (2007).

Like the experiments reported by Ahn *et al.*, this experiment also suggests that subjects are heterogeneous and differ in type, with some concerned only to maximize their personal payoffs and others appearing to be some form of conditional cooperator.[8] Evidence for this comes from the observation that across all four groups, 59% contributed at least something in the last period, even though the dominant strategy for payoff maximizing is to contribute nothing in the last period.[9] Evidence for subject heterogeneity and the presence of a significant number of conditional cooperators is also provided by the great differences in the levels of cooperation across the four groups – if there are no differences of some kind, it is hard to see why the grouping treatment should be so effective in fostering different levels of cooperation.

A second general conclusion that is suggested by this experiment is that understanding the interaction between different types appears to be important to understanding when and why cooperation occurs. If conditional cooperators are forced to play with people who are prepared to free-ride (and there is no punishment option or other treatment which will encourage more cooperative behavior – see below), then cooperation will decay.[10] If instead, conditional cooperators are able to interact only with others who have been cooperative in the past, much higher levels of cooperation can be sustained. Presumably this happens for several reasons. To begin with, this grouping procedure allows conditional cooperators to interact preferentially with each other. Second, it provides an incentive for self-interested players to mimic conditional cooperators (at least until the final round of play) or to signal that they will play as if they are cooperators and in this way obtain the benefits of interacting with them. This confirms the point, noted earlier, that it matters a great deal, in understanding cooperation, whether repeated interaction is forced or whether people have a choice over who their potential partners will

---

[8] A major tradition in social psychology emphasizes the extent to which behavior is influenced by situational or environmental variables rather than fixed personality traits. The observation that there is also evidence for some consistency of type at the level of individual behavior when the same or very similar games are played is not in conflict with these claims, but does constitute an interesting addendum to them.

[9] Of course it might be argued (in line with the SIRG account) that contributions in the last period are instead due self-interested players who are "confused" or unthinkingly import previous patterns of play into the last period. Nonetheless, it is interesting that these same players are at least *stably* confused etc. since they behave in a similarly cooperative way the first round.

[10] Page *et al.* draw attention to an experiment study of grouping a public goods game by Ehrhart and Keser (1999) in which, unlike the Page *et al.* experiment, subjects were allowed to move unilaterally from one group to another without the agreement of the group being joined. Unsurprisingly, free riders attempt to join groups containing high contributors, with the result that contributions decline over time.

be. More generally, the structure of the interactions between different types of players and hence the institutions that mediate this have a major influence on the extent to which cooperation occurs.

2.4   *Repeated Public Goods Games with Punishment*

When a costly punishment option is introduced which allows subjects in a repeated public goods game to specifically punish individual noncontributors but at a cost to themselves, a number of subjects will punish, even in the final round, in which punishment cannot influence future behavior. Moreover, introduction of this option prevents the decline in contributions with repeated play that is seen in an ordinary repeated public goods game. Indeed, humans are willing to engage in so-called third-party punishment, in which an outsider or third party punishes (at a cost to himself) uncooperative or unfair behavior involving other parties, even though this behavior does not disadvantage the punisher. For example, third-party observers will punish unfair or uncooperative behavior in a DG or PD involving two other players (Fehr and Fischbacher 2004).

This suggests several points. First, note that willingness to punish, particularly in the final round, does not have an obvious interpretation in terms of purely self-interested play – instead it suggests a disposition to engage in negative reciprocity in response to non-cooperative play by others. Second, the results from games with punishment again seem broadly consonant with the alternative explanation considered above (in terms of heterogeneous types, one of which is conditional cooperators) of the decline in contributions in ordinary public goods games without punishment. This is that the decline occurs because the subject population contains both conditional cooperators and free riders. When the former behave cooperatively and the latter free-ride, the only way the conditional cooperators can retaliate is by withholding contributions themselves, but when they do so, this of course indiscriminately punishes all other cooperators as well, leading to an unraveling of cooperation. What the introduction of the punishment option does is to make it possible for retaliation to be directed specifically at non-contributors and this both encourages potential non-contributors to cooperate and makes non-contribution less attractive as a retaliatory measure.

2.5   *Restart Effects*

Suppose that a repeated public goods game is "restarted" – that is, play is stopped (e.g., after ten rounds) and subjects are then told that a new ten-round repeated game will begin (with the same partners)

(Andreoni 1988). During the first ten rounds, contributions decline just as they do in an ordinary repeated public goods game. However, the effect of the restart is to temporarily boost contributions – that is, the average level of contribution is much higher in the eleventh round (which begins the "new" game after the restart) than it is in the tenth round which ends the old game. Note that this is again *prima facie* inconsistent with an interpretation according to which subjects are gradually learning through repeated play that zero contribution is their best strategy. If they have learned this fairly well by the end of the tenth round, why do they suddenly forget it and begin contributing at a higher level when the game is restarted?

An alternative interpretation of the restart effect is that there is some substantial population of people who view the restart as an opportunity for the group to make a fresh start and break out of a bad pattern of declining contributions. In other words, they contribute when the restart begins as part of an attempt to signal anew their willingness to contribute if others are willing to do so as well; their increased level of contribution is an attempt to get others to change their behavior and no longer free-ride. Of course this explanation assumes a particular kind of conditional cooperator: one who is willing to begin cooperating anew with the restart.[11]

2.6   Field Studies

The picture suggested by these experimental results is also supported by field observation. In her classic study, Ostrom (1990) suggests a number of features shared by successful cooperative arrangements involving Common Pool Resources (CPRs). Two of these features are particularly relevant to our discussion. First, it is crucial that the CPR be governed by rules that clearly define who has access rights and which allow for the exclusion of opportunistic outsiders. Second, there must be a system which allows for the monitoring and sanction of those who breach the rules governing use of the CPR. These features correspond to the experimental observations that cooperation will decay under repetition if there is no way for cooperators to exclude or avoid interaction with free riders and that cooperation is much more likely to be sustained if non-cooperative behavior is detected and sanctioned. The general picture that is suggested by Ostrom's study is again one in which there is heterogeneity of types and in which sustaining cooperation requires rules and institutions that protect the more cooperative types from free

[11] Thanks to Francesco Guala for helpful correspondence regarding this effect.

riders. The explanation of sustained cooperation thus requires reference to these rules and institutions and not just to the presence of repeated interaction among self-interested types.

2.7   The Role of Intentions

The experiments considered so far suggest that some significant number of subjects are conditional cooperators in the sense that they respond to cooperative behavior with cooperative behavior and uncooperative behavior with non-cooperative behavior even when it is not payoff-maximizing to do so. Other experiments suggest that the subjects do not just reciprocate other's behavior but respond to (what they believe to be) the intentions with which others act.

Consider the following two versions of a trust game, which are taken from McCabe, Rigdon, and Smith (2003).

In Figure 8.1, player 1 has an outside option – she can move across, ending the game and ensuring a payoff of (20, 20) for herself and her partner. Alternatively, she can move down, in which case her partner has a choice of moving across with payoffs (25, 25), or moving down with payoffs (15, 30). Player 1's choice to move down is thus a "trusting" move in the sense that she forgoes a certain payoff in order to give her partner an opportunity to choose an alternative that is better for both, but also trusting that her partner will not choose down, producing an outcome that is better for player 2 but worse for player 1. Thus by choosing down instead of across, player 1 communicates to player 2 her intention to trust player 2 to choose across. A natural hypothesis about interactions of this type, due to Philip Pettit (1995), is to suppose that under the right conditions player 1's trust can help to induce player 2 to

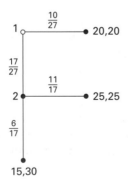

Figure 8.1

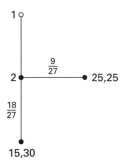

Figure 8.2

behave in a trustworthy manner and that when player 1 moves down, she acts with this intention. That is, player 1 hopes or expects that her intention to trust player 2 (expressed by moving down) will be recognized by player 2 and that as a result of recognizing this intention, player 2 will be more likely to move across than player 2 would be in the absence of the communication of this intention.

Now compare this with the game in Figure 8.2, which is just like the game in Figure 8.1, except that the outside option has been removed – player 1's only choice is to move down. Because of this, player 1's move down cannot communicate to player 2 an intention to trust player 2. Supposing that something like Pettit's analysis is correct, this removes *one* of player 2's motives to move across: the motive of reciprocating the trust shown by player 1. As a result, one would expect that player 2 would be more likely to choose down in the version of the game without an outside option than in the version of the game with an outside option. As the numbers attached to the game trees (which give the proportions of players choosing each alternative at each node) show, this is indeed what is found – the difference in choices of down vs across by player 2 across the two versions of the game is significant at the 0.01 level.

Notice that this difference in behavior across the two games is not predicted by models of subjects' preferences that attribute their behavior to unconditional dispositions to cooperate or benefit others or by models that take subjects to care only about the outcomes they receive. Such models predict that there should be no difference in the proportion of player 2s choosing across versus down across the two games since the outcomes for both players under these two choices are the same across both games.

Additional evidence for the role of perceived intention in influencing behavior in contexts involving the possibility of cooperation comes from experiments conducted by Falk *et al.* (2003), although here it is negative reciprocity which is illustrated. These authors had subjects simultaneously enter choices for each of four mini-ultimatum games (MUGs)– these are ultimatum games in which the proposer faces a binary choice between two alternatives which the responder may either accept or reject, with both players getting nothing in the latter case. In each game, one of the choices for the proposer is (8, 2) but the other option varies among (10, 0), (8, 2), (5, 5), or (2, 8). Unsurprisingly, Falk *et al.* observe that responders' behavior is strongly influenced by the alternatives available to the proposers, just as second player behavior in the trust game described above is influenced by perception of the alternatives available to the first player – presumably this is because these alternatives influence the intentions or motives with which the proposer is perceived to act. For example, rejections are much more common when the proposer chooses (8, 2) when the alternative is (5, 5) than when the alternatives are (10, 0) or (2, 8). This is exactly what one would expect if second movers detect and respond to first movers' intentions, assuming that intentions and motives depend on alternatives that are not chosen as well as those that are chosen: most responders will regard a choice of 8, 2 over 10, 0 as revealing a kinder or less hostile intention on the part of the proposer than a choice of 8, 2 over 5, 5. Again, by contrast, if responders care only about outcomes, they should reject the choice of an 8, 2 split at a constant rate across these games.[12]

---

[12] Why should conditional cooperators (or, for that matter, self-interested types) care about the intentions of those with whom they interact, rather than just the outcomes they receive? A natural hypothesis is that at least part of the explanation is that intentions provide information about the likelihood of future cooperative interactions: someone who not only benefits me but intends that I receive this benefit or chooses the benefit in part because it will benefit me (rather than providing it inadvertently or as a mere byproduct of some purely self-interested choice) reveals a kind or favorable disposition toward me that may manifest itself in future favorable treatment. Someone who makes a choice that reveals a hostile intention toward me is typically more likely to make choices that harm me in the future than someone whose behavior has a negative impact on me, but who exhibits no hostile intention. My conjecture is that this information about future cooperativeness figures in our psychic economy in *both* a calculative, instrumental way and in a non-instrumental "reactive" way. In the former role, we react positively to kind intentions because we consciously recognize that they signal an increased probability of future cooperative benefits. However, independently of this, it is also likely that we have been wired up by some combination of natural selection and past learning to react in a non-instrumental way to the intentions others display toward us: this selection/learning has occurred because intentions convey information about future behavior, but we don't have to go through a process of conscious calculation when we react in this way. Instead, we have fast, automatic "reactive attitudes" (in the sense of Strawson) toward others' revealed intentions that influence our utility and our responses to others, independently of our evaluation of outcomes.
    It is worth noting that it's a feature of deontological moral theories that they assign an independent significance to the intentions with which actors act, in addition to the

3   Models of Cooperative Behavior

3.1   Introduction

At this point I want to step back a bit and ask the following question: If
people exhibit the kinds of behavior described above (if at least some are
conditional cooperators) how should we model their behavior? Since the
structure of this portion of my argument will be somewhat complex, I
begin with an overview of what I attempt to show. One obvious strategy,
explored by a number of economists, for modeling cooperative behavior is to assume that people have "social preferences" (i.e., preferences
that are not purely self-interested in the sense that payoffs to others figure in people's utility functions, but that are like self-interested preferences in the sense that they are well behaved and stable at the individual
level across different situations and contexts). I focus in 3.2, for illustra-
tive purposes, on what is perhaps the best known proposal along these
lines   the inequity aversion model of Fehr and Schmidt (1999). I will
argue that this whole approach suffers from certain fundamental limitations: as an empirical matter, people's utility functions, at least when
parameterized in the way that Fehr and Schmidt suggest, are not stable
across different games. Moreover, one important reason for this is that
the purely outcome-oriented or consequentialist character of the utility
function they propose is inadequate to capture the reciprocal behavior
of conditional cooperators.

Sections 3.2–3.4 then explore the question of what would be required
to adequately capture the preference structure of conditional cooperators, assuming that a purely outcome-oriented model is inadequate. My
intention here is not to propose an original positive theory, but rather
to show how complex any successful model must be and how much
information it must incorporate. In particular, I will contend, following
Rabin (1993) and others, that empirically adequate models of reciprocal
behavior must incorporate information about the beliefs and intentions of other reciprocators into the subjects' utility functions; that is, to
adequately explain their behavior, we must assume that the utility that
they get from various interactions depends on, among other things, the
intentions and beliefs with which other players act. This in turn has the

outcomes they produce. Many consequentialists claim that there is something unreasonable about this. The experimental evidence described above seems to suggest
that deontological theories are better descriptions of what people actually care about
than purely consequentialist theories. Moreover, the evidence described above gives
consequentialists a reason to care about the intentions with which people (including
themselves) act. People's perceptions of others' intentions influence how they react to
others' choices and hence the overall consequences that are produced.

further consequence that very specific background information (e.g., about baselines with respect to which kind or hostile intentions can be identified) must be shared among the players – in the absence of such shared information, conditional cooperators will find it very difficult to predict one another's behavior and hence to successfully cooperate. As I then argue in Section 4, one role of shared norms and institutions is to supply this background information. This raises an explanatory quandary which is explored in Section 5: On the one hand, norms and institutions are themselves the product of human cooperation; one would like an account framed in terms of the beliefs and preferences of individual agents which explains how these emerge and are sustained. On the other hand, if those agents are conditional cooperators, it appears that to adequately model their behavior we require reference to preferences and beliefs that already embody information about norms and institutions.

3.2   Social Preferences: Inequity Aversion as an Illustration

The overall strategy of Fehr and Schmidt (1999) is to assume that people have stable utility functions in which payoffs to others as well as to self figure as arguments and then use this assumption to explain other regarding behavior. Suppose that $X = x_1 . x_n$ represents the monetary allocation among each of n players. Fehr and Schmidt propose that the utility of the ith player for this allocation is

$$U_i(X) = x_i - a_i/n - 1 \sum \max[x_j - x_i, 0] - b_i/n - 1 \sum \max[x_i - x_j, 0]$$

where the summations are for $j \neq i$ and $b_i \leq a_i$ and $0 \leq b_i \leq 1$. The utility of the ith player is thus a function of the monetary payoff he receives and two other terms, the first reflecting how much i dislikes disadvantageous inequality and the second reflecting how much i dislikes advantageous inequality, the assumption being that i dislikes the former more than the latter. The parameters a_i and b_i measure how sensitive i is to each form of inequality and are assumed to be relatively stable characteristics of individuals.

Adopting the assumption that subjects have stable utility functions of this form, Fehr and Schmidt try to "measure" or estimate subjects' a_is and b_is from their behavior in one set of games (ordinary ultimatum games) and then use this information to predict behavior in other games, such as an ultimatum game with responder competition, and public goods games with and without punishment. Such prediction should be possible if subjects have stable social preferences. Although they claim some success, their project suffers from two serious limitations, both of which are instructive for our discussion.

First, contrary to what Fehr and Schmidt claim, there is little evidence for coefficient stability even across the limited class of games they consider. The data from UGs on which they rely does not allow for reliable point estimation of the values of the coefficients a_i and b_i and it is only by making additional and very *ad hoc* assumptions about the values and distribution of these parameters that Fehr and Schmidt are able to produce results that are consistent with (but not really predictive of) play across different games. More natural assumptions suggest that the coefficients change in value across games.[13]

Second, as Fehr and Schmidt explicitly recognize, their utility function assumes that subjects care only about outcomes (for themselves and others), even though it is clear from the experimental evidence described above that subjects also care about the intentions and previous play of the other subjects with whom they are interacting. For example, according to the Fehr/Schmidt model, trustees should behave in exactly the same way in both versions of the trust game devised by McCabe, Rigdon, and Smith (see Figures 8.1 and 8.2 above). Indeed, trustees should also choose in the same way in a dictator game in which there is no previous move by the trustor and the trustee must allocate 3x between himself and the trustor in circumstances in which it is known that the trustor already has n-x, since according to the Fehr/Schmidt model, the trustee's allocation should depend only on his degree of inequality aversion, and not (independently of this) on the trustor's previous choice. Both casual observation and experimental results show that many subjects do not treat these situations as equivalent: because they are conditional cooperators, subjects are more generous to others who have allocated money to them via a previous move in a trust game than to subjects who have not made such a move. Presumably, it is at least in part because many subjects are conditional cooperators (and hence care about intentions, previous play etc.) that purely outcome-based utility functions like Fehr/Schmidt's are predictively inadequate.

It is worth observing that similar limitations infect some of the most popular normative theories of other-regarding preferences in the philosophical literature, at least insofar as these are construed as descriptive of people's actual preferences. Like the Fehr/Schmidt model, many of these theories are also purely outcome oriented (and in this respect, consequentialist) – they don't capture the idea that people often care not just about outcomes, but also about the intentions of other players and that, because they are conditional cooperators, they may evaluate the same pattern of payoffs (both for themselves and others) quite differently

---

[13] For details, see Shaked (2007) and for additional discussion, Woodward (2009).

depending on the choices of others that preceded this payoff. This focus on outcomes is characteristic, for example, of versions of utilitarianism according to which utility is measured by monetary payoffs (or by some monotonically increasing function of those payoffs) for those affected. Similarly for a theory that says that people's social preferences should be to maximize the payoff of the worst-off person or group, independently of the intentions they exhibit.[14] When treated as descriptive claims about people's preferences, such theories imply that subjects will (contrary to actual fact) treat McCabe *et al.*'s two versions of the trust game as equivalent to each other and to the trustee dictator game described above. Similarly, such theories recommend acceptance of any positive amount in a one-shot UG (assuming that reputational effects can be neglected), since rejection makes both players worse off. Notice that the problem under discussion is distinct from the problem posed by the fact that such theories also make what are probably unrealistic assumptions about how unselfish people are. Even if we confine ourselves to people's non-self-interested preferences, such theories leave out important components of the structure of such preferences – the components that underlie conditional cooperation. The result is that such theories are descriptively inadequate as accounts of people's social preferences.[15]

[14] This is not intended as a criticism of Rawls' theory since Rawls makes it clear that his difference principle is meant to apply only to the basic structure of society and is not intended as a guide to individual choice. Nonetheless the values and normative commitments that underlie the difference principle seem, *prima facie* at least, to be rather different than those that underlie conditional cooperation. An explicitly conditional version of Rawls' principle of fairness ("To the extent that others contribute to a cooperative scheme and you benefit from this, you are required to contribute as well.") seems to come closer to capturing the commitments of a conditional cooperator.

[15] Of course, this leaves open the possibility that the theories in question may be normatively defensible despite their descriptive inadequacies – that is, that they are adequate theories of how we *ought* to choose or judge even if they are inadequate as theories of how we *do* choose. On the other hand, normative theories are often defended in part on the grounds that they capture or represent what we in fact care about or value (or underlying features of our intuitive judgments). When assessed on these grounds, theories that leave out the kinds of considerations that motivate conditional cooperators seem normatively inadequate. Let me also add that I do not deny that people have preferences over outcomes for others in circumstances in which there is no possibility of conditional cooperation with those others. For example, given some good which is to be divided among other people with whom one is not involved in any sort of reciprocal interaction (e.g., as in a pure case of charitable giving), people will certainly have preferences for how the good is to be divided and it is a further empirical question whether utilitarianism, inequity aversion, maximin and so on provide a descriptively adequate characterization of those preferences. What I deny is that these preferences for unconditional distributions (as we might call them) are good models of the preferences of conditional cooperators or that they explain the facts about conditional cooperation on which I focus in this chapter. People behave very differently (and in accord with different preferences) when they are faced with issues of conditional cooperation than when they are faced with choices among unconditional distributions.

3.3   The Structure of Reciprocity

My suggestion so far (which I take to be illustrated by the failures of the Fehr/Schmidt model) has been that purely outcome-based utility functions are inadequate as models of the preferences of conditional cooperators – instead we require models that incorporate information about subjects' intentions and beliefs. In what follows I want to explore some of the complexities that arise when we try to do this. I emphasize again that my intention is *not* to try to construct an adequate positive model of reciprocal preferences, but instead the much more modest one of arguing that once we attempt to incorporate information about beliefs and intentions into the subject's utility functions, we also find that we apparently must incorporate context-specific information that makes reference to background norms and institutions. The remarks that follow are heavily influenced by Rabin (1993) (which is easily the most sophisticated and psychologically realistic treatment available) but are intended to be generic – they also reflect common assumptions about modeling reciprocity found elsewhere in the literature.

Let us focus for simplicity on a two-person extensive form game and think about it in the following way: player 1 has a choice of different actions or strategies, then player 2 chooses in response to these, with the joint choices a_1, a_2 of 1 and 2 determining the material payoffs Π_1, Π_2 for both players. 1's choice a_1 expresses an intention toward (or sends a signal to) 2, an intention that may be more or less "kind," cooperative or trusting (or alternatively hostile or uncooperative). 2 acquires from this a belief – call it b_2 – about the intention or signal that 1 is attempting to express. This belief in turn influences 2's choice of a_2 in response to a_1 – that is, b_2 is an argument in 2's utility function in addition to 2's material payoff Π_2. Moreover, 1 is aware of this influence of b_2 on a_2 and chooses her own action a_1 with an eye to this influence, forming a belief c_1 about the belief b_2 that will be induced in 2 by her (1's) choice of a_1. Similarly player 2 has beliefs about the beliefs that 1 is attempting to induce by her choice a_1 and so on.

In principle, then, players' assessments of one another's kindness will depend not only on their material payoffs, but also in part on their assessments of one another's intentions and on their expectations about one another, and they choose in accordance with utility functions that represent this information. However, to gain tractability, most analysts begin in one way or another with a purely outcome-based characterization of what it is for one player to behave kindly and then attempt to complicate or extend their analysis to reflect the role of intention and expectations. Proceeding in this way, let us characterize player 1's kindness toward player 2 in choosing a strategy a_1 as follows: first, establish some reference or neutral point which is a possible payoff for 2 – let this be Π_2 (x_0). The kindness of 1 toward 2 will then depend on the difference between the payoff Π_2 (m_{a1}) that is the maximum payoff that 2 can achieve for himself, given that 1 has chosen a_1, and Π_2 (x_0): that is, 1's kindness toward 2 is measured by Π_2 (m_{a1}) $-\Pi_2$ (x_0). In other words, 1's choice of a_1 will be kind or unkind according as to whether it places 2 in a position in which 2's payoff-maximizing response is above or below the reference payoff. [16] (If desired, this expression can be normalized in some way – e.g., by dividing by the difference between 2's maximum and minimum payoffs under a_1.)

This raises the obvious question of how the reference payoff is to be determined. The analysis would be greatly simplified if the reference payoff for player j could be specified as some context-independent function of possible alternative payoffs just to j, where this does not depend on the payoffs and intentions of the other player i, bringing this latter information (which is obviously relevant to how j responds to i's choices) into the analysis in some other way. Rabin (1993) adopts this strategy, suggesting that j's reference payoff might be taken to be the average of the maximum and minimum payoffs for j.[17] However, as many of the above examples suggest, it seems more empirically accurate and psychologically plausible to follow Cox et al. (2007) in allowing the reference point for player i to be influenced by (among other considerations) the alternatives available to j elsewhere in the game. For example, 1's choice of (8, 2) in the 5–5 MUG described above seems much more unkind than 1's choice of (8, 2) in the 10–0 game and this assessment is reflected in the much higher rate of rejection of the 8–2 split in the 5–5 game. Presumably, this is because of the availability of the alternative (5, 5) in the former game as compared with the alternative (10, 0) in the latter game. Similarly, the choice of take in the 2–8 game seems less unkind in than the choice of take in the 5–5 game. What this suggests is that in a realistic, predictively adequate model of reciprocity, the reference payoffs for each player will vary with the details of the structure of the game being played so that choices of the same final allocation

---

[16] Versions of this strategy are followed by Rabin (1993), Cox et al. (2007), and Guala (2006). For example, Cox et al. write: "A natural specification for the reciprocity variable [that is, the variable that measures reciprocity] is $r(x) = m(x) - m_0$, where $m(x)$ is the maximum pay-off that the second mover can guarantee himself given the first mover's choice x and m_0 is m (x_0) where x_0 is neutral in some appropriate sense" (2007: 22–23).

[17] A similar strategy is suggested by Guala (2006) who takes i's kindness toward j to be measured by the difference j receives under i's choice of strategy and some alternative reference payoff that j could have received if i had chosen differently, but does not suggest that this reference payoff for j in turn depends on the alternatives available to i and the payoffs to i associated with these.

across different games will vary in the kindness and unkindness that they express depending on the structure of the rest of the game.

3.4   Property Rights

Another set of considerations that suggest that reference points must be specified in a context-dependent way that depends on the particular game being played has to do with what Cox *et al.* (2007) call status or property rights. For example, when subjects are assigned the role of proposers in ultimatum games (UGs) on the basis of their performance on a pre-game quiz, they make lower offers (in comparison with UGs in which there is no such pre-game treatment) and these offers are accepted more often. Within the reference point framework described above, the effect of the quiz is to shift both players' reference points in such a way that proposers are not seen by responders as behaving unkindly when they keep more for themselves, and as a consequence, responders are less likely to reciprocate negatively by rejecting the offer. Put slightly differently, one function of reference points is to specify what one player may expect of another as a "reasonable sacrifice" that constrains purely self-interested play and this will be determined contextually by prior rights and by the alternatives available to each player – for example, many would suppose that it is "reasonable" to expect the proposer in a UG to sacrifice by forgoing making an offer of (10, 0) when (5, 5) is an alternative but not when (1, 9) is the only alternative.[18]

4   Norms and Institutions

4.1   Implications for Modeling Conditional Cooperators

What does all this imply about attempts to write down utility functions that describe the choices of interacting conditional cooperators? It seems to me that what it suggests is that although rough qualitative prediction may be possible in the absence of detailed information about reference points, and property rights, both theory and experiment imply that precise predictively useful quantitative prediction requires utility functions that incorporate such information. This in turn requires information about background norms and institutions. Of course one might *hope* for a general quantitative theory that allows one to somehow derive all

---

[18] Similar background assumptions about reasonable sacrifice are ubiquitous in the moral domain: it is generally thought reasonable to expect that one will save a drowning child at the cost of soiling one's clothes, but perhaps not if this involves substantial risk to one's own life.

such background information from assumptions that do not presuppose any background information about norms – that is, a theory which explains why (or allows one to predict that) particular reference points etc. emerge and are sustained in different situations. But at present we lack such a theory and there is an obvious puzzle about how we might get it or what form it would take, if my remarks above are correct. The puzzle is that understanding the behavior of conditional cooperators (and hence appealing to that behavior to explain the existence of background norms etc.) already apparently requires at least some of the information about norms and institutions that we were hoping to explain, thus threatening a kind of explanatory circularity. For the moment, however, I want to put aside this issue and simply observe that as far as our present level of understanding goes, it looks as though assumptions about reference points and so on need, as it were, to be put in by hand on a case-by-case basis, reflecting cultural knowledge that seems local or specific to individual games, rather than anything that we can derive from general theory.

This makes it hard for *theorists* to model interactions involving reciprocity or conditional cooperation in a non-*ad hoc* way, but it also has another consequence. This is that, in the absence of the right sort of local information, it may be very difficult for the *players* themselves to predict in detail how other players will behave, even in the heavily idealized case in which each player is some form of conditional cooperator, believes the other players to be conditional cooperators, and so on. Or, to put the point more cautiously, to the extent that the players are able to successfully predict one another's behavior, they are likely to be relying on something more than general beliefs about the extent to which others are motivated by a general taste for reciprocity or conditional cooperation, combined with information about the general type of game they are playing and their own and other players' payoffs. Instead, they must be relying on additional information that is more specific to the individual game they are playing – information about reference points in the above representation. I will suggest below that one function of *social norms* is to supply this game-specific information – that is, it is norms that specify reference points, tell us whether an alternative strategy available to a player involves a reasonable or unreasonable sacrifice and so on.

Another related factor which may make it difficult for players to predict one another's behavior derives from the way in which (as we have seen) conditional cooperators are influenced by their *beliefs* about others' choices, intentions, and beliefs. Suppose that player 1 does not expect player 2 to reciprocate kind behavior and that as a consequence 1 does not behave kindly toward 2. If (a) 2 believes that 1 fails to behave kindly toward him only because 1 believes that 2 will not reciprocate, this may well lead 2 to a different assessment of 1's behavior (and a different response to it) than if (b) 2 believes that 1 believes that 2 would reciprocate if 1 were to behave kindly and also believes that 1 has chosen to behave unkindly despite this belief. In the former case (a) if 2 is a reciprocator he may try to take steps to change 1's beliefs (perhaps by continuing, at least for a while, to behave kindly toward 1); in case (b) 2 may decide to negatively reciprocate 1's unkindness. Similarly, suppose that 1 causes some outcome that benefits 2, but 2 believes that 1 did not choose to produce this outcome because he wished to benefit 2; instead 2 believes that 1 caused the outcome accidentally or that player 1 chose the action while unaware that it would benefit 2 or without caring about this. To the extent that 2 is a reciprocator, he presumably will be less likely to reciprocate 1's behavior with an intentionally kind response in such cases (in comparison with cases in which player 1's kindness is perceived as intentional). Finally, consider cases in which 1 intends to behave kindly toward 2 and chooses an action which he (1) believes will be believed by 2 to be kind or beneficial but that in fact 2 does not regard the action as kind. In this case also, 2's response will depend not just on his beliefs but perhaps also on his beliefs about what 1 believes and so on. For example, in a trust game, the trustor may believe that sending $3 (out of a possible $10) will be viewed as a kind and trusting choice, while the trustee may regard anything less than $5 as indicating distrust. Similarly, a payback (e.g., of slightly more than the amount invested) that is regarded by the trustee as kind, trustworthy, or generous may not be so regarded by the trustor. This in turn will affect the trustor's future behavior toward the trustee.

More generally, to the extent that the players who are conditional cooperators are uncertain about one another's beliefs and expectations, and yet nonetheless understand that these will influence each other's behavior, this introduces an additional level of unpredictability into the whole interaction and makes it even more difficult to fully anticipate exactly how others will behave (at least in the absence of the sort of game specific local knowledge described above). Within Rabin's framework this shows up in the fact that for many games there are a very large number of fairness equilibria and these are often very difficult to compute (indeed the required computations may be intractable).

As a concrete illustration of these points, consider the empirical results from the trust game. Focus first on those trustors who are conditional cooperators and who wish to behave kindly toward the trustee. (These might be defined as, e.g., those who send some positive amount.)

Although this is an issue that needs to be investigated empirically, my conjecture is that at least among contemporary American subjects, there is no specific norm governing the amount of money to be sent by trustors. The absence of such a norm might be revealed by, for example, a lack of consensus or considerable variability in this group in response to questions about what trustors "should" send (or, alternatively, a lack of consensus about the amounts that will lead trustees to regard trustors as cooperative). In other words there will be uncertainty about reference points and what counts as kind behavior. I conjecture also that there is a similar absence of a clear norm about the amounts that should be returned by trustees, which would show up in the answers to parallel questions.[19] In the absence of such norms, we should expect a great deal of variability even among conditional cooperators in the amount sent and a great deal of variability among conditional cooperators in the proportion of this returned. This is consistent with the data (reporting results from the "no history" version of the game) from the original paper in which the trust game was introduced (Berg, Dickhaut, and McCabe 1995) which is reproduced in Figure 8.3.

If, for the sake of argument, we regard any trustor who sends a positive amount as a conditional cooperator, then, in the "no history" form of the game thirty out of thirty two proposers are conditional cooperators, but the amount they send shows a great deal of variation – amounts sent and number of subjects sending that amount are as follows ($10, 5), ($8, 1), ($7, 3), ($6, 5), ($5, 6), ($4, 2), ($3, 4), ($2, 2), ($1, 2). Looking at the response of the trustees, we see that a sizable portion of trustees are free riders or near free riders – of the twenty-eight trustees who received more that $1, 12 returned either $0 or $1. However, in the same group of 28, 11 returned more than they received. Taking the latter to be conditional cooperators of one sort or another,[20] Figure 8.3 shows there

---

[19] Readers might ask themselves how much they think that they "should" send if they were in the position of trustor and how much they should return if they were trustees. In the latter role, is one behaving cooperatively as long as one returns more than one is sent? Or does cooperative behavior instead require a more generous response (e.g., an even split of the amount generated by the investment, so that if the initial endowment is e and the amount invested i, the trustee returns $3i/2$, with the trustor getting $e-i+3i/2=e+i/2$ and the responder $e+3i/2$)? Or should the responder return $2i$, so that each player ends up with an equal amount $(e+i/2)$? Or should the responder return even more, since, after all, the risk associated with the interaction is borne entirely by the trustor? In this connection it is worth observing that in the results reported in the version of the trust game from Berg *et al.* described in Figure 3, out of the eleven trustees who returned more than invested, six returned $2i$ or more, with the remainder returning an amount intermediate between i and $2i$ that bore no obvious relationship to i.

[20] Obviously this is a very permissive stipulation about who counts as a conditional cooperator. My guess is that many trustors will not regard trustees who return only slightly more than they receive as kind or cooperative.

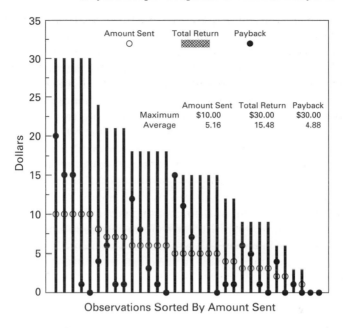

	Amount Sent	Total Return	Payback
Maximum	$10.00	$30.00	$30.00
Average	5.16	15.48	4.88

Observations Sorted By Amount Sent

Figure 8.3

is considerable variation in the amount returned both in absolute terms and as a percentage of the amount invested even among this group. Again I take this to reflect, in part, the absence of any clear norm about what is the right amount to return even among those who wish to behave reciprocally. The presence of a sizable (but not exactly known) proportion of free riders also of course contributes to making it difficult for the players to accurately predict one another's behavior.

A non-experimental illustration of these points is provided by Avner Greif's well-known study of commercial relationships among the Maghribi – a group of Jewish traders in the Mediterranean who flourished in the eleventh century (Greif 2006) These traders employed agents to oversee the sale of their goods in foreign countries, but faced the problem of ensuring that these agents did not act inappropriately with the goods in their care (e.g., by embezzling them or selling them for prices that were too low). Appropriate behavior was enforced by a "grim trigger" strategy: an agreement among the merchants not to ever hire again anyone as an agent who once cheated one of the merchants in the community. However, to be most effective agents needed to have considerable discretion, based on their judgment about local conditions, in selling the goods and this meant that it was by no means obvious in the abstract what sort of actions amounted to cheating or dishonesty. According to Greif, this problem was solved, at least in part, by reference to a body of customary rules – a "social norm" – known as the merchant law, that "defined what actions [on the part of the agents] constituted appropriate conduct" (2006: 59). He also remarks that the effectiveness of the merchants' threat of collective sanctions "depend[ed] critically on a common cognitive system that ascribes meaning to various actions, particularly actions that constitute cheating" (2006: 69). In other words, even if the merchants are conditional cooperators with a general willingness to punish "cheating," in order for this disposition to be effective in enforcing cooperative behavior a common understanding, specified by as set of social norms and shared by both merchants and agents, of what sort of behavior amounts to cheating is required. Similarly, even if agents wish to be cooperative, such norms are required for agents to know what sorts of behavior will be perceived by merchants as cooperative and for merchants to be able to anticipate and assess the behavior of their agents.

4.2   Norms as a Solution to the Problem of Unpredictability

Let us assume that, for the reasons just described, there are circumstances in which (in the absence of norms) it is difficult for groups of conditional cooperators to predict one another's behavior or accurately infer one another's motives and intentions. What are the likely or possible consequences of this unpredictability when conditional cooperators find themselves in such circumstances? It seems obvious that it can easily lead to the decay of cooperation even among those who are well disposed to cooperating. In such circumstances, subjects who intend (by their lights) to behave cooperatively will make choices that strike others as uncooperative (or insufficiently cooperative) prompting non-cooperative responses, failures to coordinate in optimal ways, and so on. This will be particularly likely to the extent that (as we may safely assume is usually the case) even subjects with preferences that incline them toward conditional cooperation will also be influenced by self-interested preferences which will incline them to interpret cooperative requirements in self-serving ways, so that while they recognize, e.g., that if they have received a positive amount in a trust game, they should provide some return, they will find it easy to persuade themselves that it is appropriate to return less rather than more. Similarly, when some subjects behave uncooperatively, conditional cooperators with a disposition to negative reciprocity will impose sanctions of some kind or other, but this doesn't settle the question of what the character or magnitude of those sanctions will be, or exactly when they should be imposed. To the extent that the character and magnitude of sanctions is unpredictable and disagreements arise about their appropriateness, there will be a further danger of decay of cooperation.

An obvious solution to this problem is the establishment of systems of social norms (rules and institutions) to govern cooperative interactions.[21] Such norms will specify the terms of cooperation in various kinds of interactions in a much more detailed and precise way than the very general guidelines about returning cooperation with cooperation and non-cooperation with non-cooperation that characterize conditional cooperators as a qualitative type. For example, such norms will specify how much one should contribute in a particular kind of public goods interaction such as the harambee, and what sanctions should be imposed if someone fails to contribute in accordance with this norm. Similarly, they will specify (within the subject's particular social group) what is a cooperative or kind offer in an ultimatum game and what is not – and hence which offers call for a non-cooperative response. My assumption is that in the sorts of contexts that we have been discussing, the behavior specified by norms will usually involve conditional (rather than unconditional) cooperation and hence will draw on preexisting general preferences for conditional cooperation (among other motivations). However, I also assume, for the reasons given above, that these preferences will be relatively qualitative and indeterminate in the absence of specific norms (i.e., people's preferences for conditional cooperation will not by themselves, and in the absence of any particular norms, dictate anything very specific about how much they should contribute in a particular kind of public goods game, how much they should return in a trust game, and so on). Norms are required to supply this specificity and because of this, rather different systems of norms governing cooperation will arise in different groups, consistently with many of the members of those groups having some qualitative dispositions toward conditional cooperation. Relatedly, even among members of the same group, the norms governing one sort of interaction may be very different from those governing other interactions that may look (to outsiders) rather similar. For example, subjects may conform to food-sharing norms that require a high degree of generosity but exhibit no such generosity in other contexts where the relevant norms do not require sharing (Ensminger 2004: 359).

---

[21] For a detailed study of social norms and their role in fostering cooperation, see Bicchieri (2006). I am much indebted to her discussion.

As noted above, the empirical evidence suggests that, to be effective, norms governing cooperation usually must be accompanied by sanctions for norm violations, which may be either formal or informal. This is closely connected to the fact that norms are regarded by those whose behavior is regulated by them as not mere regularities in behavior, but rather as prescriptions that establish how one ought to behave. The violation of norms thus at least calls for criticism or censure of some kind, in which violators may be directly told they should act differently, may be subject to adverse gossip, damage to their reputations, ridicule, and withdrawal of social contact or support and so on. Of course, it may also lead to other, more formal liabilities and to physical punishment. As a number of writers have observed, the willingness of humans to sanction others even when it is not in their self-interest to do so and even when they are not directly adversely affected by the norm violation appears to play a central role in the maintenance of systems of norms – this is reflected both in the role that punishment plays in maintaining stable levels of cooperation in experimental games and in the fact that most working systems of cooperation observed in the field include sanctions of some kind for free riders. This is not to say, though, that most people conform to norms only because of a desire to avoid sanctions. Instead, many different motivations seem to be at work in norm-compliance – see below. And one important role of sanctions is to reassure conditional cooperators (who are not themselves motivated entirely by fear of sanctions) that their cooperative tendencies will not be taken advantage of by more selfish types. It is also one of the factors that allows for a certain arbitrariness in exactly which norms get established in connection with a given cooperative problem – our general willingness to sanction can be employed to support any one of many different systems of norms.[22]

4.3   Normative Guidance

The picture for which I have been arguing is one according to which the establishment of norms (rules, institutions etc.) and human willingness

---

[22] Indeed some writers (e.g., Boyd and Richerson 1992) claim that punishment can stabilize virtually *any* system of norms. This strikes me as implausible as an empirical claim – stable systems of norms are unlikely to be supported just by a willingness to punish but will also need to fit with and be supported by preexisting human motivations, including self-interested preferences and preferences for conditional cooperation. The latter motivations are not infinitely variable or plastic. There is a reason why socialist norms of economic behavior did not work very well in the former Soviet Union, even when backed by sanctions.

to comply with or be guided by these norms plays an important role in the high levels of cooperation we are able to achieve. General dispositions toward conditional cooperation may be necessary for the achievement of cooperation, but in the absence of guidance by norms, they are far from sufficient. In the absence of norms, we find it difficult to predict one another's behavior or even to gauge accurately whether others are behaving (or are intending to behave) cooperatively or not. We deal with this problem by making our behavior more predictable and interpretable by creating norms, rules, and institutions.[23] Let me conclude by trying to put this idea in more general perspective and also drawing attention to a problem that it creates for the project of trying to explain the emergence and continued existence of norms and institutions in terms of the beliefs and utilities of individual actors and their interactions.

As a point of departure, consider a rough contrast between two types of belief/desire (or expected utility) explanations. The first appeals just to beliefs and desires that can be specified independently of norms.[24] These will include desire for generic goods that are shared by virtually everyone, regardless of cultural background: desire for food (of some kind or another), shelter, avoidance of physical harm, etc. Arguably, many self-interested preferences fall into this category. Norm-independent beliefs will include veridical beliefs about the physical environment that are likely to be acquired by virtually anyone, regardless of cultural background, with normal perceptual and reasoning abilities who is awake and attentive. Also in the category of norm-independent predictions/explanations are those that appeal to desires that are not universally shared and are not supported by a preference to conform to some norm, but are nonetheless fairly stable for a given individual across some range of contexts (e.g., stable individual level personality traits, assuming these exist, such as shyness and novelty-seeking).

---

[23] In other words, rather than thinking of the degree of predictability of human behavior as given or fixed, independently of the presence of rules and institutions, in the way that natural phenomena like planetary motions are, we should think of human predictability as something that we are able to affect by creating appropriate rules and institutions. To put the point more provocatively: rather than thinking of folk (or belief/desire) psychology as a theory that yields reliable predictions of human behavior in general, we should think of it instead as a theory whose predictive success (when present) depends in part on the existence of the right sort of background scaffolding – norms and institutions to which people conform.

[24] The contrast which follows has some affinities with Rawls' contrast between "object dependent" desires and one subgroup of what he calls "principle dependent" desires – the subgroup having to do with "reasonable principles" that "regulate how a plurality of agents ... are to conduct themselves with respect to each other" (Rawls 2000: 45–48).

The belief/desire predictions/explanations just described contrast with those that advert to beliefs and desires that are *norm-dependent* in the sense that either (a) they can't be specified independently of facts about norms, or (b) people are unlikely to possess them in the absence of certain norms, or (c) we lack information that would allow us to reliably ascribe them in the absence of information about norms. For example, if we wish to predict/explain the detailed behavior of a customer and waiter when the customer orders food in a restaurant, this will almost certainly require an appeal to norm-dependent beliefs and desires – shared beliefs and desires about the norms governing such interactions (e.g., specifying how the customer requests certain food by telling the waiter about his order rather than barging into the kitchen, when payment is expected, tipping, and so on). General, norm-independent information (e.g., about the subject's food preferences or the waiter's desire for income) will not allow us to predict or explain the details of their interaction. My suggestion is that to adequately explain the detailed behavior of conditional cooperators, we must similarly appeal to norm-dependent and not just norm-independent beliefs and preferences.

As already remarked, this creates a problem about a kind of explanatory circularity. On the one hand, rules and institutions are the product of people's interactions, both cooperative and competitive. We would like to be able explain why these emerge and are sustained by reference to the activities and choices of individual agents, which are in turn explained by those agents' preferences, beliefs, and other psychological states (e.g., by exhibiting these rules and institutions as equilibria of some repeated game, given suitable psychological assumptions about the agents playing that game). On the other hand, realistic psychological assumptions about individual agents seem to require that many of them be modeled as conditional cooperators. Moreover, as we have seen, to explain or model the cooperative activities of conditional cooperators we seem to be required to ascribe norm-dependent preferences and beliefs to them. But then we are in danger of presupposing, in our characterization of the agents' beliefs and preferences, at least some of the facts about the existence of rules and institutions that we would like to explain. This potential circularity could be avoided if we could explain the behavior of agents in terms of preferences and beliefs that are not themselves principle-dependent, and then show how norms and institutions emerge from the interactions of such agents, but given the arguments made above, it is not obvious that this can be done.

My concern in this section has been simply to argue that conditional cooperators require norms and institutions if they are to cooperate successfully; I have not tried (and am certainly unable) to provide a general explanation of how such rules and institutions emerge and are sustained. So I will not try to resolve the issue described in the previous paragraph, except to remark that insofar as the problem is genuine, it constitutes a challenge only to a very ambitious explanatory enterprise – that of providing a completely general theory of rules and institutions that explains their existence and emergence just in terms of the interactions of agents whose psychology can be characterized in norm-independent terms. More modest explanatory enterprises are left untouched. To take only the most obvious possibility, it may be possible, by assuming the existence of some background norms and institutions, in the form of norm-dependent preferences and beliefs, to use these in conjunction with other assumptions (including assumptions about norm-independent beliefs and preferences) and game theoretic analysis, to explain the emergence or existence of *other* different norms and institutions. This is in effect what Greif does. Cooperative behavior (and solutions to principal agent problems) among the Maghribi merchants is sustained in part by a system of collective sanctions which presumably draw on self-interested motives as well as perhaps a general willingness to punish transgressions, but, as we have seen, these in turn depend upon a background legal system (the merchant law) which characterizes what will count as dishonest behavior. These background norms are in effect presupposed in Greif's explanation of the cooperative institutions established and maintained by the Maghribi. To assume the existence of these background norms is in effect to presuppose that certain patterns of cooperative behavior are already in place, but the resulting explanation is not perniciously circular since the aspects of the cooperative behavior of the Maghribi that Greif seeks to explain are not those that are presupposed in appealing to the existence of the merchant code.[25]

---

[25] Another way of putting the basic point, which I owe to Dan Hausman is this: if one confines oneself to norm-independent beliefs and preferences, then it is not easy to see what explanatory work could possibly be done by appeal to norms. On the one hand, if the norm-independent preferences and beliefs are by themselves sufficient to explain norm conformity (e.g., because this is the equilibrium of some game that is being played) then any explanatory role for norms seems superfluous – norm adherence is an effect that is explained in terms of norm-independent beliefs and preferences, rather than a part of an explanation of behavior. On the other hand, if players' norm-independent preferences and beliefs are not sufficient to explain norm adherence and players do not have any other sorts of beliefs and preferences, it is a mystery why conformity to norms occurs at all. My solution to this dilemma is to suggest that there is another alternative – norm-independent preferences and beliefs are not sufficiently determinate to explain norm conformity by themselves and only become determinate when supplemented by knowledge and acceptance of norms.

4.4   *The Psychology Behind Conformity to Norms*

Implicit in the picture of cooperation that I have been defending is the assumption that people can learn to follow or conform to norms. I will not try to provide a general theory of what this capacity involves, but a few clarifications may be helpful.[26] First, my discussion assumes that people are motivationally quite heterogeneous, both in the sense that different motivations play a role in different people's behavior and that the same individual may be influenced by different motivational factors, both at the same time and in different mixes over time.[27] For example, someone may behave cooperatively in a situation in which others are cooperating as well because she is influenced both by preference for conditional cooperation and by self-interested preferences. The latter may include both purely instrumental considerations having to do with securing future monetary payoffs by signaling a willingness to cooperate now, but it may also be that the subject derives utility from being well-regarded by others or from conforming to what most others are doing. Perhaps also the subject derives a reward from norm compliance per se. It may be unclear both to others and even to the subject herself what the relative role of these various factors is in influencing her behavior.

Second, for the same reasons that predicting people's behavior in cooperative interactions on the basis of their norm-independent preferences and beliefs is extremely difficult, I also assume that it is often difficult to identify interventions or manipulations of people's norm-independent beliefs and preferences that will reliably alter their behavior in cooperative contexts in precise ways.

Third, I assume we are so psychologically constituted that there is a way around this difficulty: we can create norms (either explicitly and

[26] I'm in broad agreement with the general account of guidance by norms provided in Gibbard (1990). This remains one of the best discussions of this notion that I know of.

[27] In some critical comments on an earlier draft, Chrys Mantzavinos asks why I focus so heavily on motivational considerations rather than more purely "cognitive" factors in explaining norm compliance. This is a complex issue to which I cannot really do justice here. I certainly agree that cognitive factors such as shared beliefs or "mental models" are necessary for the effective maintenance of norms – this is apparent in many of the examples discussed above. However, it also seems to me that the strong message of recent work in social cognitive neuroscience is that motivational, desiderative, and affective factors are crucial too and that furthermore, this is a dimension along which people can vary considerably. In particular, facts about human emotional processing and our capacities for emotional regulation seem central to the explanation of norm-governed judgment and behavior. For more on this general theme, see Woodward and Allman (2007).

deliberately, as a result of discussion, negotiation, and agreement or as a result of more informal processes involving the gradual adoption of shared customs and expectations or through some combination of these) and we can get ourselves and others to conform to them. Requiring/expecting that people conform to (or be guided by) a norm thus can function as what John Campbell (forthcoming) has called a *control variable*: it is a particularly effective (and often readily available) point of intervention that allows us to alter our own and others' behavior in precise, targeted ways. Depending on the circumstances, establishing and appealing to norms in order to influence people's behavior can be more effective than trying to influence behavior by appealing to or manipulating norm-independent beliefs and preferences. In invoking norms in this way we in effect say to others (and to ourselves): we don't care about the exact details of the psychological processes by which you accomplish this, but we expect you to get yourself into a psychological state such that you conform to such and such a norm. We may also add that we expect you to influence others to conform to the norm, to adopt a critical attitude to norm violations and so on. Because the intervention asks only that people conform to the norm and exhibit attitudes and behavior that support the norm, it need not require detailed knowledge of preferences and beliefs that exist independently of and prior to the institution of the norm.

At present we have little understanding of what the processes underlying normative compliance and guidance (and the motivational plasticity that apparently accompanies this) involves in mechanistic or neurobiological terms. However, it is a natural conjecture, supported by some evidence (e.g., see Spitzer et al. 2007 and the references therein), that the neural basis for this capacity involves frontal structures like the ventrolateral and dorsolateral cortex (known to be involved in inhibition, impulse control, and emotional regulation) and the ventromedial and orbitofrontal cortex (involved in the processing of complex social emotions) as well as the anterior cingulate (conflict resolution) and the extensive projections forward and back from these frontal structures to form reward processing structures like the dorsal striatum and the caudate nucleus. These projections not only enable reward structures to influence the frontal structures involved in higher level thought, decision making, and conflict resolution, but also allow the latter structures in turn to modulate and alter reward structures, so that we can learn to find even very abstract concerns not based in immediate biological needs (e.g., a trip to Mecca, or conformity to norms mandating non-cheating on taxes) to be rewarding and motivating.

4.5   *Empirical Implications*

Suppose that we accept the idea that people are sometimes guided by norms in cooperative endeavors. What distinctive patterns of behavior should we expect to see when this occurs in laboratory games and in the field? First and most obviously, to the extent that behavior in the laboratory is guided by norms, we should expect to see some correspondence between laboratory behavior and norms that can be recognized as at work in the field, as with the public goods games played by the Orma. When this is the case, it seems appropriate to think of the laboratory game as measuring *in part* the influence of this norm – "in part" because play in the laboratory game will be influenced, at the individual level, by many other factors as well, as evidenced by the fact that in many cases there is considerable individual variability in degree of norm adherence, and norm adherence in general tends to be weaker under conditions like anonymity.

Presumably, when (or to the extent that) a norm is operative, it should be possible to identify it independently of laboratory play; there should not just be a regularity in behavior in the field corresponding to the norm, but subjects should exhibit behavior showing they recognize the norm. (This might show up under inquiries about what people should do in various situations, in the treatment of certain behaviors as deviations from the norm, in criticisms or sanctions of such deviations, and so on.) Claims that norms are at work in guiding behavior will thus be more plausible in some situations than in others. For example, it is a plausible empirical conjecture that among contemporary American populations, a norm is at work in an ordinary UG that proscribes a highly unequal division of the stakes. On the other hand, consider an asymmetric UG in which the stakes are in the form of chips which are worth twice as much to the proposer as to the responder; a fact that is known to the proposer but not to the responder. In such games, on average, proposers offer a roughly equal division of the number of chips and hence a highly unequal monetary division, which responders accept (Bicchieri 2006). Although it is an empirical question whether there is a distinct norm in the contemporary US which permits a very unequal monetary division of the stakes in this asymmetric information version of the game, this seems unlikely. It is more plausible to think instead either that the same norms (relatively equal division) govern play in both an ordinary UG and the asymmetric information version or that no norm at all governs play in the latter game. In either case, the difference between proposer's play in the asymmetric information game and an ordinary UG will be best explained not in terms of conformity to a norm, but rather by some model in which the proposer is influenced by self-interest to take advantage of the informational asymmetry to achieve an unequal monetary division.

Next, to the extent that people's behavior is influenced by norms, it should exhibit some distinctive patterns that will differ from those we would expect if their behavior were primarily influenced by norm-independent social preferences – at least on simple and straightforward versions of the latter approach. Other things being equal, there should be more variability of individual behavior in games in which it is not clear (for the subject pool) what the relevant norm is or in which there are competing norms than in games in which there is a single governing norm. In the latter case, one might expect to see a "spike" in frequency of behavior conforming to the norm, with behaviors that do not conform to the norm generally occurring less frequently, although there may also be another spike at whatever behavior corresponds to self-interested play. When there are a small number of competing norms that govern the interaction, this should show up in a somewhat discontinuous frequency distribution of behaviors that cluster around the competing norms, with lower frequency of behavior in between norms (i.e., bimodal or poly-modal distributions). In general, variability in behavior should decrease as more context is supplied which plausibly can be regarded as cuing a particular norm.

As concrete illustrations, it is arguable that in societies like the contemporary US, the norm or norms that apply to dictator games (DGs) are weaker or less salient than those that apply to UGs – indeed, some would say that unless further context is provided there are no clear norms for proposer behavior in a DG. If this is correct, then on a norm-based approach, one should expect (*ceteris paribus*, of course) more variability in proposer behavior in DGs than in UGs. On the assumption that one of the norms governing UGs mandates an equal split, one should see a number of proposer offers at this value. On the other hand, on a norm-based approach, there is no particular reason to expect a spike at offers of, e.g., 0.39 of the stake in a UG, since presumably there is no norm that suggests this exact division – if such a division is observed, this will be because the proposer just happens to trade off payoff to self, the utility of norm conformity, and whatever other preferences she possesses in such a way as to generate this result. In fact, in conformity with these expectations, one does observe an increased frequency of offers around 0.5 in UGs in developed societies (although there are intermediate offers as well) and more variance in DGs than in UGs. Similarly, on the assumption that a UG corresponds to something like a "take it or leave it" offer between two strangers in real life, one might expect that there will be clearer norms governing such offers in societies with substantial experience of bargaining, trade, and market exchange with relative strangers than in societies lacking such experience. If so, and such norms influence behavior in UGs, one would expect less variance in proposer offers in the former societies than in the latter. A number of the papers in Henrich *et al.* report just this pattern.[28]

On the other hand, within a model that explains cooperative behavior in terms of norms, it will be entirely possible that even rather similar looking interactions are governed by different norms. To the extent this is so, we should not expect norm-adhering subjects to play in a similar way across such games (i.e., to exhibit any very strong stability of type across similar looking games). For example, if the norms governing play in a UG are different from those governing play in a DG, with the former cuing norms involving bargaining and the latter norms involving "giving" or unconditional charitable donation, then unless there is some reason to think that proposers who are sensitive to one norm are likely to be sensitive to the other (and that both norms require generous behavior on the part of proposers) there will be no general reason to expect proposers who make generous offers in a DG to do so in a UG. The absence of such a correlation between proposer behavior in UGs and DGs will be likely if, for example, a substantial number of proposers understand the norm governing UGs as not making considerations of generosity relevant at all – that because the context involves bargaining, any offer that the responder will accept is appropriate – while by contrast, DGs carry with them implicit expectations to exhibit some level of generosity. It is of course an empirical question to what extent such a hypothesis about the norms thought to govern UGs and DGs is correct, but to the extent

---

[28] Also relevant in this connection are results reported in Güth *et al.* (2001). The authors find that in an MUG, proposers choose highly unequal splits much more often when an even split is replaced with a slightly uneven one. For example, given a choice between (17, 3) and (10, 10), proposers choose (10, 10) half the time. When the (10, 10) option is replaced with (9, 11), responders choose (17, 3) two-thirds of the time. A natural interpretation of this result is that proposers are influenced by an equal split norm when conformity to it is possible, but that there is no obvious norm that tells them they should prefer (9, 11) to (17, 3). On a social preference approach, there must be an (unexplained) sharp discontinuity between proposer's attitudes toward (10, 10) and (9, 11).

It is also worth noting in this connection that a norm-based account might be used to provide a natural treatment of another experimental result due to Güth: that in three-person variants of an ultimatum game in which the proposer makes an offer that distributes the stake between a responder with the usual option to reject and a passive third party with no power to reject, the passive party receives very little – much less than the responder and less than the second party in a DG. If the three person UG cues norms involving bargaining (with any offer that will not be rejected being acceptable) rather than norms involving charity or fairness, this outcome is unsurprising.

that it is, we should expect little correlation in generous behavior at the individual level across the two games.[29] For similar reasons, we should expect that behavior will differ across groups which are subject to different norms, even when they play the "same" game – which of course is what is found in the Henrich *et al.* study.

More generally, to the extent that subjects are influenced by norms we should expect some degree of similarity of play in the same game across different subjects, provided these subjects are drawn from a common group or culture in which the same norms are operative, but not necessarily consistency of type of play among individuals across games governed by different norms. (The similarity of play across different subjects in the same game reflects the fact that to at least some extent motivationally heterogeneous subjects will be able to get themselves to conform to the same norm.)

By contrast, to the extent to which self-interest and norm-independent social preferences alone are sufficient to explain behavior and subjects are *not* influenced by norms, we should expect correlations in play at the individual level and (absent some special reasons for expecting otherwise) variance in individual behavior within a society. Both theoretical and empirical considerations suggest that there is no particular reason to expect everyone to have the same norm-independent social preferences. So, on the hypothesis that behavior is governed by norm-independent social preferences, then, even within a single game where there is an

---

[29] Henrich *et al.* (2004) draw attention to this sort of pattern in connection with the Orma. As noted, the Orma saw the public goods game as closely related to the harambee but, according to Henrich *et al.* saw no such connection between the ultimatum game and the harambee. The Orma also believe that wealthier households should make a larger contribution to the harambee than poorer households. Consistently with this, contributions in the public goods game but not offers in the ultimatum game are correlated with household wealth among the Orma (Henrich *et al.* 2004: 49). This is evidence that play in the public goods game is influenced by a social norm associated specifically with the Harambee and not just by a generalized social preference (e.g., favoring generosity) that is operative among the Orma and expresses itself in a stable way across different games.

Thinking of the behavior detected in the Henrich *et al.* experiments as due in part to the influence of norms also allows us to make sense of another, *prima facie* puzzling empirical result. This is that degree of integration with the market is correlated with cooperative behavior across different societies but not within societies at the level of individuals. In the nature of the case, norms must be widely shared within a given group (in part as a conceptual matter but also if they are to be effective in enhancing predictability and coordination). Provided one is a member of the relevant group, one is expected to conform to the norm, regardless of one's idiosyncratic characteristics. So norm adherence should manifest itself in substantial uniformity of behavior within a society, regardless of individual characteristics, but allows for variation across societies. In particular, societies that are more integrated into markets may adopt cooperative norms that are shared within those societies even by individuals who are not personally integrated into markets.

opportunity to express social preferences, there should be considerable variation in subject behavior. Moreover, although advocates of the social preference approach have had relatively little in general to say about what distribution of types of preferences we should expect (presumably because they regard this as an empirical matter) there is no obvious theoretical reason (at least absent additional assumptions) why this distribution should be discontinuous or clumpy in any particular way rather than relatively continuous. Thus there is no obvious reason why, in a UG, many more subjects should have norm-independent social preferences (say in the form of some degree of inequality aversion) that lead them to offer 0.5 rather than 0.45 of the stake. On the other hand, there should be some non-trivial consistency of type at the individual level across games – so that one can use behavior in ultimatum games to predict behavior in other games, as Fehr and Schmidt attempt to do.

5   Conclusion

The patterns of behavior we actually observe seem to be a mixture of the two "ideal types" just described (i.e., behavior that is purely norm-governed and behavior that is entirely due to norm-independent preferences). Although there is substantial empirical evidence for an important influence of norm-based preferences as well as theoretical arguments that people must be influenced by such preferences if they are to achieve stable cooperation, it is also obvious that norm-based preferences are not the whole story in explaining cooperative behavior. Even in cases in which there are clear norms applying to some interaction, people may vary considerably in the degree to which they adhere to them, with some being more willing than others to violate norms, particularly (but not only) in order to satisfy self-interested preferences. Both casual observation and more systematic investigation support the conclusion that people are sensitive to the costs to themselves and perhaps to others incurred in conforming to norms and are more willing to violate norms as the benefits of doing so increase.[30] Adherence to norms (and motivation by social preferences more generally) also becomes attenuated as anonymity increases and is strengthened under arrangements that make information about behavior more public. And of course, there are many situations in which there are no clear norms governing behavior at all, so that unless subjects import norms from other situations that they judge

---

[30] For example, Clark and Sefton (2001) report evidence that the rate of defection by second movers in a sequential PD after the first mover has cooperated goes up as the benefits from defection increase.

to be similar or create norms to govern the new situation, their behavior will not be norm governed.

REFERENCES

Ahn, T., E. Ostrom, and J. Walker 2003. "Heterogeneous preferences and collective action," *Public Choice* 117: 295–315.

Andreoni, J. 1988. "Why free ride? Strategies and learning in public goods experiments," *Journal of Public Economics* 37: 291–304.

Andreoni, J. and J. Miller 1993. "Rational cooperation in the finitely repeated prisoner's dilemma: Experimental evidence," *Economic Journal* 103: 570–585.

Berg, J., J. Dickhaut, and K. McCabe 1995. "Trust, reciprocity, and social history," *Games and Economic Behavior* 10: 122–142.

Bicchieri, C. 2006. *The Grammar of Society: The Nature and Dynamics of Social Norms*. Cambridge: Cambridge University Press.

Binmore, K. 1994. *Game Theory and the Social Contract, Volume 1: Playing Fair*. Cambridge, MA: MIT Press.
    2005. "Economic man – or straw man," *Behavioral and Brain Sciences* 28: 23–24.
    2007. *Does Game Theory Work? The Bargaining Challenge*. Cambridge, MA: MIT Press.

Boyd, R. and P. Richerson 1988. "The evolution of reciprocity in sizable groups," *Journal of Theoretical Biology* 132: 337–356.
    1992. "Punishment allows the evolution of cooperation (or anything else) in sizable groups," *Ethology and Sociobiology* 13: 171–195.

Burlando, R. and F. Guala 2005. "Heterogeneous agents in public goods experiments," *Experimental Economics* 8: 35–54.

Camerer, C. 2003. *Behavioral Game Theory: Experiments in Strategic Interaction*. Princeton, NJ: Princeton University Press.

Camerer, C. and E. Fehr 2004. "Measuring Social Norms and Preferences using Experimental Games. A Guide for Social Scientists." In J. Henrich *et al.* (2004), pp. 55–95.

Campbell, J. (forthcoming). "Causation in Psychiatry." In K. Kendler (ed.) *Philosophical Issues in Psychiatry: Explanation, Phenomenology, and Nosology*. Johns Hopkins University Press.

Clark, K. and M. Sefton 2001. "The sequential prisoner's dilemma: Evidence on reciprocation," *The Economic Journal* 111: 51–68.

Cox, J., D. Fredman, and S. Gjerstad 2007. "A tractable model of reciprocity and fairness," *Games and Economic Behavior* 59: 17–45.

Croson, R. 2007. "Theories of commitment, altruism, and reciprocity: Evidence from linear public goods games," *Economic Inquiry* 45: 199–216.

Ehrhart, K.-M. and C. Keser 1999. "Mobility and Cooperation: On the Run," Working Paper 99(24), CIRANO, Montreal.

Ensminger, J. 2004. "Market Integration and Fairness: Evidence from Ultimatum, Dictator, and Public Goods Experiments in East Africa." In Henrich *et al.* (2004).

Falk, A., E. Fehr, and U. Fischbacher 2003. "On the nature of fair behavior," *Economic Inquiry* **41**(1): 21–26.

Fehr, E. and U. Fischbacher 2003. "The nature of human altruism," *Nature* **425**: 785–791.

2004. "Third-party punishment and social norms," *Evolution and Human Behavior* **25**: 63–87.

Fehr, E., U. Fischbacher, and S. Gachter 2002. "Strong reciprocity, human cooperation and the enforcement of social norms," *Human Nature* **13**: 1–25.

Fehr, E. and K. Schmidt 1999. "A theory of fairness, competition, and cooperation," *Quarterly Journal of Economics* **114**: 817–868.

Gibbard, A. 1990. *Wise Choices, Apt Feelings.* Cambridge, MA: Harvard University Press.

Gintis, H. 2006. "Behavioral ethics meets natural justice," *Politics, Philosophy and Economics* **5**: 5–32.

Gintis, H., S. Bowles, R. Boyd, and E. Fehr 2005. *Moral Sentiments and Material Interests: The Foundation of Cooperation in Economic Life.* Cambridge, MA: MIT Press.

Greif, A. 2006. *Institutions and the Path to the Modern Economy: Lessons from Medieval Trade.* Cambridge: Cambridge University Press.

Guala, F. 2006. "Has game theory been refuted?," *Journal of Philosophy* **103**: 239–263.

Gunnthorsdottir, A., K. McCabe, and D. Houser 2007. "Disposition, history, and public goods experiments," *Journal of Economic Behavior and Organization* **62**: 304–315.

Güth, W., S. Huck, and W. Muller 2001. "The relevance of equal splits in ultimatum games," *Games and Economic Behavior* **37**: 161–169.

Henrich, J., R. Boyd, S. Bowles, C. Camerer, E. Fehr, and H. Gintis (eds.) 2004. *Foundations of Human Sociality: Economic Experiments and Ethnographic Evidence from Fifteen Small-Scale Societies.* Oxford: Oxford University Press.

Kitcher, P. 1993. "The evolution of human altruism," *Journal of Philosophy* **90**: 497–516.

2006. "Ethics and evolution: How to Get There from Here." In F. De Waal (ed.) *Primates and Philosophers: How Morality Evolved.* Princeton, NJ: Princeton University Press.

Mantzavinos, C. 2001. *Individuals, Institutions, and Markets.* Cambridge: Cambridge University Press.

McCabe, K., M. Rigdon, and V. Smith 2003. "Positive reciprocity and intentions in trust games," *Journal of Economic Behavior and Organizations* **52**: 267–275.

Ostrom, E. 1990. *Governing the Commons: The Evolution of Institutions for Collective Action.* Cambridge: Cambridge University Press.

2005. *Understanding Institutional Diversity.* Princeton, NJ: Princeton University Press.

Page, T., L. Putterman, and B. Unel 2005. "Voluntary Association in Public Goods Experiments: Reciprocity, Mimicry and Efficiency," *The Economic Journal* **115**: 1032–1053.

Pettit, P. 1995. "The cunning of trust," *Philosophy and Public Affairs* **24**: 202–225.

Rabin, M. 1993. "Incorporating fairness into game theory and economics," *American Economic Review* **83**: 1281–1302.

Rawls, J. 2000. *Lectures on the History of Moral Philosophy.* B. Herman (ed.) Cambridge, MA: Harvard University Press.

Samuelson, L. 2005. "Foundations of human sociality: A review essay," *Journal of Economic Literature* **43**: 488–497.

Shaked, A. 2007. "On the explanatory value of inequity aversion theory." www.wiwi.uni-bonn.de/shaked/rhetoric/. Accessed September 25 2007.

Spitzer, M., U. Fischbacher, B. Herrnberger, G. Gron, and E. Fehr 2007. "The neural signature of social norm compliance," *Neuron* **56**: 185–196.

Tullock, G. 1985. "Adam Smith and the prisoner's dilemma," *Quarterly Journal of Economics* **100**: 1073–1081.

Vanberg, V. and R. Congleton 1992. "Rationality, morality and exit," *American Political Science Review* **86**: 418–431.

Woodward, J. 2009. "Experimental Investigations of Social Preferences." In H. Kincaid and D. Ross (eds.) *The Oxford Handbook of Philosophy of Economics.* Oxford: Oxford University Press, pp. 189–222.

Woodward, J. and J. Allman 2007. "Moral intuition: Its neural substrates and normative significance," *Journal of Physiology-Paris* **101**: 179–202.

8 – Comment
Putting the Problem of Social Order Into Perspective

Werner Güth and Hartmut Kliemt

1   Cooperation and the Classical Social Order Problem

"The possibility of co-operation" (Taylor 1976, 1987) has been on the social theory agenda ever since the British Moralists (Raphael 1969 and Schneider 1967). They already perceived the "Hobbesian problem of social order"[1] as an obstacle to – what later researchers came to see as – "rational choice explanations" of cooperation and social order. The British Moralists were not only aware of the problem but from their discussion a quite convincing solution of the problem emerged.[2] The pinnacle of the ongoing discussion was reached with the solution of the order problem presented by David Hume in his *Treatise on Human Nature* (Hume 1739, 1978). Hume relied already on a, to use Woodward's term, SIRG (self-interest in repeated games) account of interaction. But he went beyond that in a direction Woodward is hinting at as well. Hume, like other of the British Moralists also insisted on the role of norm guided and rule following behavior.[3] Hume emphasized already the crucial role of conventions and their evolution in solving the social order problem (Sugden 1986).

According to Hume conventions are not simply regularities in behavior corresponding to stationary equilibrium choices in a repeated game. The relevant equilibrium notion went beyond behavior to conviction systems and the intentional pursuit of rules (as later made precise in the concept of an internal point of view (Hart 1961)). Hume, besides drawing attention to SIRG, insisted in the British Moralist tradition on what then was called "opinion." Any satisfactory account of ordered cooperation had to include a reference to the belief of the people in the legitimacy of certain acts whose performance was deemed normatively mandatory.

---

[1] So-called by Parsons (1968).
[2] See on this Mackie (1980) and Kliemt (1985).
[3] As today pointed out in particular by Hayek and his followers, see Hayek (1973–79), Vanberg (1994).

It is regrettably widely forgotten that Hobbes had already admitted that "... the power of the mighty hath no foundation but in the opinion and belief of the people" (Hobbes 1682; 1990: 16). The remark that "government was based on opinion" was the routine response to criticisms which pointed out the deficiencies of basically Hobbesian SIRG explanations of social order and cooperation.

Substituting the very broad concept of opinion we can say in more modern parlance that the British Moralists were of the opinion that without referring to behavioral guidance by accepted rules and standards the emergence of social order and inter-individual cooperation cannot be explained. In his most sophisticated version of the theory Hume discusses how the division of labor can be extended to the enforcement of rules such as to make social cooperation viable in n-person prisoner's dilemma like public goods structures.[4] To put the same thing slightly otherwise, Hume already presents a sophisticated account of how interest – or the deep structure of social interaction – is affected by and affects rule and norm orientation, or the surface structure of social interaction (for a modern reconstruction of the relationship between deep and surface structure see Güth and Kliemt 1998; 2000).

Had researchers better understood Hume, many of the later controversies could have been avoided. Still, in the wake of Darwinism and Social Darwinism (Hofstadter 1969) the emergence of order and of cooperation became problematic again. Darwinism is Thomas Hobbes rather than "Adam Smith grafted on nature." And it is therefore no surprise that, as in the older discussion among theorists of human sociality, the issue of cooperative behavior crept up again. In that discussion Peter Kropotkin pointed out that the extent to which there was mutual aid among animals and men was absolutely surprising (Kropotkin 1919). Taking what was basically a Hobbesian account, but was not necessarily seen as such, at its word, explanations within the evolutionary research program somehow had to cope with what had to appear as the "anomaly of co-operation." Against the background of the simple minded Darwinist – or for that matter economic – model of competition much less cooperation than observed, if any, should have been expected. Woodward's question why there is so much cooperation was high on the agenda then. Social Darwinists responded to challenges like Kropotkin's by suggesting that cooperation arose out of competition, along the same lines as Hobbes. In the first place people cooperate because they need to, to compete more efficiently. According to this kind of "science of society" (Sumner and Keller 1927), cooperators remain competitive toward

---

[4] See the section on government in Hume (1739/1978).

each other all the time. They restrain their non-cooperative inclinations in pursuit of their own aims, ends, or values. Selfishness is propelled to a higher level, as it were. This applies not only in the context of conflict and war but also on the most basic level of social exchange per se. For in every exchange as a form of mutually advantageous cooperation there is a prisoner's dilemma lurking in the background.[5]

For this mixed-motive cooperation, Willian Graham Sumner and Albert Galloway Keller coined the apt phrase "antagonistic co-operation" (Sumner and Keller 1927). Clearly, Sumner and Keller with their leanings toward some kind of old-fashioned Social Darwinism were a bit out of step with the spirit of their own scientific times already. It is, however, a mistake to ignore, as we all tend to do, this older literature almost entirely. Woodward's question "Why do people co-operate as much as they do?" is not a response to a puzzle that came up recently. It is misleading to put it as if the question emerged out of the efforts of (experimental) game theorists and (socio-)biologists. There has been an extended discussion of the order problem, of the problem of how human cooperation is related to "human nature" and of why human actors cooperate as much as they do ever since Hobbes. Many modern social theorists seem to be either unaware of it[6] or tend to think of it as outdated. But as the discussion by Woodward and his emphasis on the coordinative function of norms in social cooperation shows, it is sometimes useful to go back to past masters.

In stating the preceding we do not claim that previous discussions of human cooperation contained all relevant answers. Regardless of the efforts of former theorists and their often quite impressive, if largely forgotten results, it is still an open question whether all forms of cooperation can be explained as instrumental to the pursuit of some non-cooperative interest (and perhaps even motive) or whether there must be intrinsically motivated cooperative inclinations (including rule following). It is laudable that this classical controversy is center stage in Woodward's chapter. We also think that it is worthwhile to put it into the perspective of present experimental game theory research. But even though we very much agree with his choice of topic and many things he says, we have some disagreements with how Jim Woodward deals with many of the more specific problems.

---

[5] See for a nice short statement, Hardin (1982) and for a book length treatment Kliemt (1986).

[6] Though in particular Binmore (1994, 1998) dug into much of the old swamp.

2   Woodward on Cooperation

On the one hand, we find a rather thorough review of the main arguments that have been put forward as explanations of voluntary cooperation in the more recent experimental literature, on the other hand,
Woodward's presentation suffers from mixing rational choice, behavioral and evolutionary arguments in a somewhat confusing way. For
instance, when the author refers to repeated interaction (i.e., the shadow
of the future) he just assumes "indefinite length" (p. 221) without any
attempt at justifying such an empirically outrageous assumption. When
viewing the infinite horizon as an idealization of the long finite horizon
or when requiring subgame consistency in the sense that isomorphic
games must have the same solution, the Folk Theorem justification
of (subgame perfect!) equilibrium cooperation collapses (Güth et al.
1991). In finite horizon games, on the other hand, rational choice justifications of (at least initial) voluntary cooperation must either rely on
multiple equilibria of the base (social dilemma) game or on incomplete
information (Kreps et al. 1982 and Kreps and Wilson 1982). Here one
would have to argue when and why such conditions can be expected
from an empirical point of view.

Whenever parties interact repeatedly, of course, reciprocity comes
"into play." It can be an aspect of a rational choice argument. This
applies when "tit for tat" or grim strategies are embedded in the formal
proof of theorems according to what may be called the "Folk Theorem
logic." But reciprocity may also be related to widely accepted behavioral norms. Though this looks similar to rational choice arguments it is
conceptually rather different. The two animals, strategic and behavioral
reciprocity, should be kept apart more clearly than Woodward does.

On a more substantive level we object to the claim that there is a clear
cut distinction between negative and positive reciprocity. Accepting
(rejecting) a fair (meager) offer in an ultimatum experiment illustrates
that the same game can allow for both. It seems doubtful, too, whether
one can separate punishment and reciprocity. In the first place, one
should clearly distinguish between forward-looking (preventive) punishment and backward-looking or (retributive) punishment.[7] The first kind
can emerge as an aspect of subgame perfect equilibria as, for instance, in
Folk Theorems. The second form is typically non-equilibrium punishment whose presence is systematically confirmed in ultimatum experiments when actors are rejecting considerable but nevertheless unfair
offers. In both the preventive as well as the retributive case it is hard – but

---

[7] See on this instructively in a general vein, Mackie (1982).

perhaps not serving a strong purpose – to distinguish punishment from reciprocity.

It is quite unclear in many situations of interactive decision making what cooperation amounts to. This cannot be "read" off from the material payoff and move structure of the interaction situation as such. We need to combine reciprocity with social norms in the sense that only the latter provide a basis for coordinating the type and the degree of voluntary cooperation. It seems that the author is somewhat oblivious of the fact here that cooperation is often shown in the pursuit of ends that are not exactly in conformity with the common weal.[8] It seems that the author implicitly assumes that voluntary cooperation is welfare enhancing or justifiable by ethical norms across the board. Since he does not even touch on this aspect we should like to point out the obvious again, namely that most voluntary cooperation aims at exploiting other groups. Seller cartels try to exploit customers, criminal gangs with strong internal solidarity turn against their victims, and states against each other. (It is not by accident that John Mackie insists that the best teachers of morality – or rather the social workings of a cooperation enhancing code of conduct – are the criminals with their gangs (see Mackie 1977).)

More specifically, the theory part (section 1.3) starts out by criticizing some justifications of voluntary cooperation as insufficient (e.g., by arguing that tournament studies like Axelrod's are too special), that non-dyadic games in which a greater number of players interact simultaneously render cooperation and the coordination of behavior more difficult, that cooperation depends on "circumstances," and that most models of repeated interaction do not allow for opting out and, as one might add, do not provide for exclusion or ostracism as an option.[9] In a somewhat unsystematic way other theoretical accounts like social preferences (section 1.5), conditional cooperation or, more generally, reciprocity concerns (section 2.1), heterogeneity of types and endogenous grouping (section 2.3), and intentionality (section 2.7) are mentioned in an informed but somewhat loose way.

The main thread of a discussion of "Models of Cooperative Behavior" is picked up again in section 3. It presents and criticizes inequity aversion as an example of (outcome) consequentialist other regarding concerns. Rather than dancing around the corpse of inequity aversion any further, the author then rightly turns to intentionality and other aspects

---

[8] If there is such a thing at all.
[9] See for an early, quite extensive treatment of the exit option in Axelrod type studies, Schüssler (1990).

of interaction which do not fit easily into a consequentialist framework.[10] Belief dependency as inspired by so-called "psychological" games and shared social norms as a way to coordinate when and how much cooperation can be expected are introduced. The author rightly points out that there will be a lot of heterogeneity in all practical contexts and that some empirical findings might be explained by the interaction of such different types.[11]

As in the theory part the empirical, mostly experimental evidence is discussed in a story telling way. We are guilty of story telling ourselves. But we are willing to accept the obvious criticism of such a strategy: evidence is used selectively whenever it fits rather than systematically. It would be desirable to have a somewhat structured agenda of research here. Hypotheses, in the form of several stylized facts which one intends to explain by a model of voluntary cooperation, should be presented and then systematically scrutinized. Such scrutiny would clearly go beyond the experimental lab. Regarding field studies the author mentions the important contributions of Elinor Ostrom and her collaborators. Woodward also refers to the anthropological study by Henrich *et al.* pointing out cross-cultural differences in behavior. As far as this is concerned he somewhat naively uses a more or less single result to draw quite far-reaching conclusions. The results of Henrich *et al.* are indeed intriguing but might need some effort at replication. It would also be useful to put such cross-cultural heterogeneity into perspective by confronting it with intracultural heterogeneity which – as Woodward himself points out in different contexts – is considerable.[12]

In short, we think that this is an interesting chapter addressing an important topic. But we are uncertain how much closer to a convincing answer it has brought the issue of why we observe as much cooperation as we in fact do. This fact does not pose a puzzle at all if cooperation is an equilibrium outcome and it is also worthwhile to note that human behavior is very often not at all cooperative. Nevertheless, it is true that humans often cooperate even though opportunism would dictate otherwise. And as far as that is concerned we agree with Woodward's main tenet: *what cooperation means is entirely dependent on norms and cannot be explicated in terms of substantive payoffs alone.* Beliefs, intentions, and the so-called participant's attitude (Strawson 1962) to social interaction matter if we intend to understand cooperation.

---

[10] Even though that framework is much more flexible than often assumed, see Broome (1991).

[11] For further comments from our side on retributive responses, see Güth *et al.* (2001) and on the possibility of evolutionarily stable polymorphisms, see Güth *et al.* (2006).

[12] For example in large-scale newspaper experiments, see Güth *et al.* (2003, 2007).

REFERENCES

Binmore, K. 1994. *Game Theory and Social Contract. Volume I – Playing Fair.* Cambridge, MA: MIT Press.

    1998. *Game Theory and Social Contract. Volume II – Just Playing.* Cambridge, MA: MIT Press.

Broome, J. 1991. *Weighing Goods. Equality, Uncertainty and Time.* Oxford: Oxford University Press.

Güth, W. and H. Kliemt 1998. "Towards a fully indirect evolutionary approach," *Rationality and Society* **10**(3): 377–399.

    2000. "Evolutionarily stable co-operative commitments," *Theory and Decision* **49**: 197–221.

Güth, W., H. Kliemt, and S. Napel 2006. "Population-dependent Costs of Detecting Trustworthiness – An Indirect Evolutionary Analysis," *Max Planck Institute for Economics*, Working Papers, Jena.

Güth, W., H. Kliemt, and A. Ockenfels 2001. "Retributive responses," *Journal of Conflict Resolution* **45**(4): 453 – 469.

Güth, W., W. Leininger, and G. Stephan 1991. "On Supergames and Folk Theorems: A Conceptual Analysis." In R. Selten *et al.* (eds.) *Game Equilibrium Models. Morals, Methods, and Markets.* Berlin: Springer, pp. 56–70.

Güth, W., C. Schmidt, and M. Sutter 2003. "Fairness in the mail and opportunism in the internet – A newspaper experiment on ultimatum bargaining," *German Economic Review* **4**(2): 243–265.

    2007. "Bargaining outside the lab – A newspaper experiment of a three person ultimatum game," *The Economic Journal*, pp. 449–469.

Hardin, R. 1982. "Exchange theory on strategic basis," *Social Science Information* **2**: 251ff.

Hart, H. L. A. 1961. *The Concept of Law.* Oxford: Oxford University Press.

Hayek, F. A. von 1973–1979. *Law, Legislation and Liberty: A New Statement of the Liberal Principles of Justice and Political Economy.* London: Routledge & Kegan Paul.

Hobbes, T. 1682/1990. *Behemoth or The Long Parliament.* Chicago, IL: University of Chicago Press.

Hofstadter, R. 1969. *Social Darwinism in American Thought.* New York, NY: The Beacon Press.

Hume, D. 1739/1978. *A Treatise of Human Nature.* Oxford: Oxford University Press.

Kliemt, H. 1985. *Moralische Institutionan.* Freiburg and Munich: Alber.

    Kreps, D., P. Milgrom, J. Roberts, and R. Wilson 1982. "Rational cooperation in the finitely-repeated prisoners' dilemma," *Journal of Economic Theory* **27**: 245–252.

Kreps, D. M. and R. Wilson 1982. "Reputation and imperfect information," *Journal of Economic Theory* **27**: 253–279.

Kropotkin, P. 1919. *Mutual Aid: A Factor of Evolution.* New York, NY: McClure, Phillips & Co.

Mackie, J. L. 1977. *Ethics. Inventing Right and Wrong.* Harmondsworth: Penguin.

1980. *Hume's Moral Theory*. London: Routledge.

1982. "Morality and the retributive emotions," *Criminal Justice Ethics* 1982: 3–10.

Parsons, T. 1968. "Utilitarianism. Sociological Thought." *International Encyclopedia of Social Sciences*. New York and London: Macmillan, vol. 16, pp. 229–236.

Raphael, D.-D. (ed.) 1969. *British Moralists*. Oxford: Oxford University Press.

Schneider, L. (ed.) 1967. *The Scottish Moralists on Human Nature and Society*. Chicago, IL and London: The University of Chicago Press.

Schüssler, R. 1990. *Kooperation unter Egoisten: Vier Dilemmata*. Munich: Oldenberg Verlag.

Strawson, P. F. 1962. "Freedom and resentment," *Proceedings of the British Academy*, pp. 187–211.

Sugden, R. 1986. *The Economics of Rights, Co-operation and Welfare*. Oxford, New York: Basil Blackwell.

Sumner, W. G. and A. G. Keller 1927. *The Science of Society*. New Haven, CT: Yale University Press.

Taylor, M. 1976. *Anarchy and Cooperation*. London: John Wiley & Sons Ltd.

1987. *The Possibility of Cooperation*. Cambridge: Cambridge University Press.

Vanberg, V. 1994. *Rules and Choice in Economics*. London and New York: Routledge.

Situations Against Virtues:
        The Situationist Attack on Virtue Theory

Ernest Sosa

Why did Tom give up his seat on the bus to someone frail and elderly? Perhaps only to impress his girlfriend, perhaps rather to be considerate, out of concern for the elder's welfare. Or maybe it was just a random act, entirely out of character, and due only to his being in a good mood. Alternatively, it might be quite in character for him to act that way. Kindness may be one of his character traits, manifest in that act.[1]

 Or so one might think without a second thought. Such reasoning seems typical of how we constantly try to understand people's conduct. Based on a body of troubling results in social psychology, however, an intriguing critique has been pressed against such character trait attributions. Here we shall review the most striking, best known results, and the arguments based on them.[2]

1   The Attack Presented

1.1   _The Milgram Experiments_

In the early sixties, experiments conducted by the psychologist Stanley Milgram at Yale University had disturbing results (Milgram 1963, 1974). In multiple replications, moreover, the results have held up with impressive consistency. Milgram's experimental subjects believe themselves to be participating in a study of the effects of punishment on

---

[1] Virtue ethics goes back to Aristotle and to Ancient Greek philosophy more generally. Long neglected in the shadow of deontological and utilitarian approaches, despite its earlier importance, it has in recent decades regained much of its former luster and influence. The approach has both an empirical side and a normative side. It appeals to virtues, or to virtuous traits of character, in both the explanation and the assessment of human action. That an action manifests a virtue bears positively in its overall assessment, that it manifests a vice bears negatively.

[2] I come to this topic through an interest in virtue theory generally, a general approach that I prefer not only in ethics but also in epistemology, and even in parts of aesthetics. But here the focus will be more specifically on virtue ethics and on its correlated virtue psychology.

learning, and to be playing the role of "teachers" in that study. Here is the scenario. The experimental subjects are expected to administer electric shocks to "learners" (who are in fact Milgram confederates). In one version, the teacher/subject sees the learner/confederate strapped down to an electric chair in a separate room. The teacher/subject listens in as the learner/confederate is told that shocks can be quite painful. When the experiment begins, the teacher/subject is given a sample 45-volt shock from the machine, just to add realism. Then he is taken to a position in another room from which the learner/confederate is no longer visible. As the experiment proceeds, the learner/confederate repeatedly fails to answer the questions correctly, which prompts the teacher/subject to intensify the shocks in 15-volt increments. What were the results?

At 300 volts, the learner/confederate would pound on the wall, scream, and then, at 330 volts, would stop responding to the shocks. Yet most teacher/subjects continued to intensify the voltage in 15-volt increments all the way up to 450 volts. That means *ten* further voltage boosts *after* the pounding and screaming! This was done by twenty-six of the forty teachers. The remaining fourteen all went up to at least 300 volts, and stopped somewhere between 300 volts and 450 volts. If a teacher protested to the experimenter, he was prompted to continue. Under such prompting, *all* teachers administered shocks up to the severe 300 volts, and 65% went beyond that to shocks of the maximum 450-volt severity.

1.2   The Good Samaritan Experiment

A second, much-cited experiment was conducted at the Princeton Theological Seminary (Darley and Batson 1973). Seminarians were all directed to a nearby building, where they would give a corresponding talk, some about the Good Samaritan, some about vocational choices. Some were asked to hurry, while advised that they were already late, others were just told to proceed without delay, and the rest were told that they had a few minutes to spare. On the way they all came across a figure *slumped over* in a doorway, *groaning* and *coughing*. Whether they stopped to help was determined mainly by their degree of hurry. That is what was positively correlated with their willingness to help.[3] In conclusion, the experimenters and authors themselves suggest, "only hurry was a

---

[3] Sixteen out of the forty subjects offered to help. Of the eight with some time to spare, five stopped to help. Of the twenty-two who had been told to go right over, ten stopped to help. Finally, of the ten who had been told they were already late, only one stopped to help.

significant predictor of whether one will help or not." For most subjects, punctuality trumped the evident need of someone in distress.

Based on these and several other similar results, an attack has been launched on virtue psychology and virtue ethics.[4] The critics have raised questions of two sorts. First, they have challenged the notion that humans vary significantly in their traits of character *important for the explanation and prediction of human action*. The claim is that no such variation in virtues matters for explanation and prediction. Moreover, the critics have also challenged the normative ideal of human virtue held up by virtue ethics, for the reason that humans are unlikely ever to guide their conduct by any such ideal.

Leading the situationist charge within philosophy are Gilbert Harman and John Doris, whose views we consider next. [5]

1.3   *Harman's Case*

According to Harman, empirical testing has found no relevantly different character traits to account for behavioral differences. Yes, people are normally said to differ in traits and virtues.

We ordinarily suppose that a person's character traits help to explain at least some things that the person does. The honest person tries to return the wallet because he or she is honest. The person who pockets the contents of the wallet and throws the rest of the wallet away does so because he or she is dishonest. (Harman 1998–99, sec. 2)

However, people might behave differently, regularly so, without differing in character traits. The difference in behavior might derive rather from situational differences. In order to differ in character traits, people must be disposed to act differently though similarly enough situated (in their situations as viewed by them). As ordinarily conceived, moreover, traits are dispositions to issue the trait-relevant conduct across a *broad* range of relevant situations. True honesty, for example, requires honest conduct across a broad enough range of relevant situations. It will not be enough that one be honest in forbearing to shoplift, if one cheats on tests, on one's income tax returns, and in returning change.

---

[4] To appreciate the extensive relevant literature, see the masterful John Doris (2002). The three studies I cite (two here and one in my concluding footnote) are the best known and the ones I have found most striking; the philosophical issues emerge fully, as I see it, on the basis of these three studies, which are also the most cited in the relevant philosophical literature.

[5] Nisbett and Ross (1980); Ross and Nisbett (1991); Doris (1998); and Harman (1998–9). An earlier, softer challenge is due to Owen Flanagan (1991).

Harman joins psychologists Nisbett and Ross, moreover, in distinguishing traits from sustained goals or strategies, and also in finding us too often guilty of the "fundamental attribution error," the error of attributing a trait based on too paltry an evidential basis. According to Nisbett and Ross, "individuals may behave in consistent ways that distinguish them from their peers not because of their enduring predispositions to be friendly, dependent, aggressive, or the like, but rather because they are pursuing consistent goals using consistent strategies, in the light of consistent ways of interpreting their social world" (Nisbett and Ross 1980: 20).

Harman comments as follows on our two striking experiments:

The fundamental attribution error in [the Milgram] ... case consists in how readily he makes erroneous inferences about the actor's destructive obedience (or foolish conformity) by taking the behavior at face value and presuming that extreme personal dispositions are at fault. (Harman 1998–99, sec. 5.1)[6]

Standard interpretations of the Good Samaritan Parable [experiment] commit the fundamental attribution error of overlooking the situational factors, in this case overlooking how much of a hurry the various agents might be in. (1998–99, sec. 5.2)

And he sums up as follows:

Summary. We very confidently attribute character traits to other people in order to explain their behavior. But our attributions tend to be wildly incorrect and, in fact, there is no evidence that people differ in character traits. They differ in their situations and in their perceptions of their situations. They differ in their goals, strategies, neuroses, optimism, etc. But character traits do not explain what differences there are. (1998–99, sec. 8)

1.4   Doris's Case

Situationism for Doris (1998: 507) involves three main commitments:

Situationism's three central theses concern behavioral variation, the nature of traits, and trait organization in personality structure:

(i) Behavioral variation across a population owes more to situational differences than dispositional differences among persons. Individual dispositional differences are not as strongly behaviorally individuating as we might have supposed: to a surprising extent we are safest predicting, for a particular situation, that a person will behave pretty much as most others would.

---

[6] Here Harman is agreeing with Nisbett and Ross.

(ii) Empirical evidence problematizes the attribution of robust traits. Whatever behavioral reliability we do observe may be readily short-circuited by situational variation: in a run of trait-relevant situations with diverse features, an individual to whom we have attributed a given trait will often behave inconsistently with regard to the behavior expected on attribution of that trait. ...

(iii) Personality structure is not typically evaluatively consistent. For a given person, the dispositions operative in one situation may have a very different evaluative status than those manifested in another situation – evaluatively inconsistent dispositions may "cohabitate" in a single personality.

Situationism is not a Skinnerian evisceration of the person. While rejecting cross-situationally robust traits, the situationist admits local, situationally specific traits that distinguish people from one another. These traits are "local" rather than global, or frail rather than "robust": they do not reliably result in the same trait-relevant conduct across a variety of different situations.

At bottom, the question is whether the behavioral regularity we observe is to be primarily explained by reference to robust dispositional structures or situational regularity. The situationist insists that the striking variability of behavior with situational variation favors the latter hypothesis. (1998: 508)

Doris sums up as follows.

To summarize: According to the first situationist thesis, behavioral variation among individuals often owes more to distinct circumstances than distinct personalities; the difference between the person who behaves honestly and the one who fails to do so, for example, may be more a function of situation than character. Moreover, behavior may vary quite radically when compared with that expected on the postulation of a given trait. We have little assurance that a person to whom we attributed a trait will consistently behave in a trait-relevant fashion across a run of trait-relevant situations with variable pressures to such behavior: the putatively "honest" person may very well not consistently display honest behavior across a diversity of situations where honesty is appropriate. This is just what we would expect on the second situationist thesis, which rejects notions of robust traits. Finally, as the third thesis suggests, expectations of evaluative consistency are likely to be disappointed. Behavioral evidence suggests that personality is comprised of evaluatively fragmented trait-associations rather than evaluatively integrated ones: e.g., for a given person, a local disposition to honesty will often be found together with local dispositions to dishonesty". (1998: 508–509)

According to Doris, in brief: *First,* behavioral variation is due more to situational variation than to trait variation. *Second,* traits are frail across situational variation, not robust. *Third,* traits do *not* integrate into coherent characters.

Attentive reading reveals that both Harman and Doris reject Skinnerian nihilism on behavioral dispositions. Indeed, both of them *believe that there are traits*, dispositional traits operative in human conduct generally. *What they deny is that these are traits as conceived of by the folk, or by the Aristotelian tradition of virtue ethics.* Harman does consider the possibility of rejecting traits altogether, even those that are local and frail, as opposed to the global and robust. On this more extreme view, human conduct is to be explained not by traits, but perhaps through goals, policies, or strategies.

About Harman's radical alternative, one might well wonder how we are to distinguish *learned* traits generally from his suggested goals, policies, or strategies? The supposed alternative does not clearly differ more than verbally.

Among human beings, traits can be rare and distinctive or, alternatively, vulgar and widely shared. They can be local (or narrow, or frail), moreover, or else global (or broad, or robust). *Our traits, insofar as we have any, are said to be vulgar and local, or at least much more so than is usually supposed.* This is what situationism seems to boil down to, apart from the claim that human personality is normally fragmented, and falls far short of the integration proper to Aristotelian practical wisdom.

Situationists do agree with virtue theorists on one important point:

Variation There is substantial evaluatively interesting variation in human behavior. People can and do behave variably in regard to honesty, kindness, courage, temperance, etc.

This much is in keeping with the experiments, and in line with situationist writings. What explains such variation in people's behavior? For the explanation of *cross-personal* variation we must invoke situational differences and distinctive traits; generally shared traits will not help. Concerning traits, let us use the word "vulgar" to cover those that are generally shared. We can then easily see that *cross-situational* variation cannot be explained by *stably* vulgar traits. If character traits are stably vulgar, then evaluatively relevant behavioral variation must be situationally explained.

Consider now this question: What explains evaluatively interesting human behavior? This is *not* the question broached a paragraph ago. What explains cross-personal or cross-temporal variation in behavior need not be the same as what explains the behaviors severally. Compare this: The *differences* in the rolls of two round balls cannot be explained by appeal to their roundness, since they are the same in that respect. Yet, either roll might still be explicable (largely, importantly) through the roundness of the rolling object. Similarly, behavioral *variation* may not be explicable by appeal to traits, while still the behavior itself *is* so explicable, *even if* the traits are vulgar.

So, stable vulgarity will spoil a trait for the explanation of behavioral variation, but not for the explanation of behavior itself, whether individual behaviors or behavior patterns.

What about narrowness? How if at all does the narrowness of a trait impair its explanatory efficacy? A narrow or local trait is one that yields its relevant outputs in a relatively narrow or local set of circumstances. Dispositions come of course in degrees. This applies not only to simple dispositions, such as fragility and flexibility, but also to those more relevant to ethics, such as honesty and kindness. Accordingly, it is possible to explain the breaking of a vase by appeal both to its impact and to its fragility, even if a fine wine glass is *more* fragile and would have broken not only with that intensity of impact but *also* with others that would *not* have affected the vase. Moreover, narrow, local traits of honesty (in returning change, say) may amount to ways in which one can have a low degree of honesty (since one is not also honest in filling out one's income tax, nor in taking tests). Folk, commonsensical virtue psychology is content to postulate varying degrees of its recognized virtues, though it may be surprised to see just how much we nearly all fall short, and also the ways in which we fall short.[7]

2   The Attack Assessed

Thus far we have focused on the situationist attack on virtue ethics and psychology, led in philosophy by Harman and Doris, and we have focused also on the proposed situationist alternative. For their part, many advocates of virtue theory have converged on a response to that attack. Several authors have now accused situationists of adopting a crude, external, behaviorist conception of virtue psychology, one that virtue theorists reject as a caricature. Situationists are said to ignore the inner deliberative complexity so important to sophisticated virtue ethics and psychology.[8]

We are thus presented with two conceptions of virtue psychology: (a) Crude virtue psychology (Crude VP) focuses directly on

---

[7] This paper was presented at the Herdecke conference on philosophy and the social sciences, where my commentator, Steven Lukes, pointed out how surprising such surprise would be given ethnic cleansing and other familiar horrors of recent history and current affairs.

[8] Relevant here are four illuminating articles: Sreenivasan (2002), Kamtekar (2004), Webber (2006), and Hursthouse (2007). See also the recent work of Julia Annas (2005).

situation/behavior dispositions. (b) Sophisticated virtue psychology (Sophisticated VP) interposes situation/attitude dispositions between situations and behavior.

According to Sophisticated VP, it is the agent's character that holistically explains his conduct. In contrast with the Crude VP attacked by Doris, Harman, and Nisbett and Ross, the character of interest to Sophisticated VP is a broader whole that includes fundamental motives, desires, and even goals. The rational agent works to integrate these into a coherent whole. Given how often our lives are troubled by conflicts of values and hard choices, it can hardly be a surprise that we fail to be cross-situationally consistent at the level of external situations and attendant behaviors. The consistency is found, rather, internally, in the complex inner structure that is one's relevant character.

In order to possess the virtue of kindness, for example, one need not behave kindly whenever one is in a kindness-relevant situation. And the same goes for honesty and other traits. For one thing, one may not possess the virtue to the degree required in *demanding* kindness-relevant situations. For another thing, when values conflict, a value other than kindness may take priority. The virtues of Sophisticated VP are rational virtues manifest primarily in right choices made through proper rational deliberation.

Here now is a way to develop this alternative picture: When deliberating on a yes-or-no choice, one faces a rational structure of pros and cons, of reasons for and reasons against. Here I mean *good, factive* reasons. These constitute the rational structure of the situation. One can think of this as a one-dimensional vector space, with positive and negative vectors as the pro and con reasons. Additional options, beyond our simple yes-or-no case, will of course import a more complex vector space.

Corresponding to such a rational structure more or less well is one's motivational structure constituted by positive and negative motivating reasons, reasons that psychologically attract one to a certain choice or repel one from that choice, to various degrees.

Take our young man comfortably seated when the elder approaches on the bus. Among the reasons that structure his situation rationally are her risk of falling and her evident physical and emotional stress, while, by comparison, he is about as well off standing as sitting. We can surely imagine such a scenario so that the balance of reasons strongly favors his ceding his seat. Practical wisdom would then require that his motivating reasons reflect the objective reasons, so that the relative weights of his motivating reasons correspond to the relative weights of the relevant objective reasons. If the reasons to cede constituted by the risk and suffering of the oldster outweigh the reason not to cede constituted by his very slightly greater comfort, then the motivating force of those weightier reasons should also psychologically outweigh the motivating force of the less weighty reason, and should do so by a corresponding margin. Thus, the motivational structure in the mind of the agent should reflect the rational structure of the situation faced.

One manifests practical wisdom in any given situation to the degree that one's motivational structure reflects the relevant rational structure in that situation.[9] Unfortunate circumstances may then block the correspondingly appropriate action without blocking the manifestation of the subject's practical wisdom in his inner motivational structure. Of course, one cannot thereby *manifest* practical wisdom unless one *possesses* such wisdom, which is not something one can do only ephemerally and locally. Rather, one's level of practical wisdom is directly proportional to the following two factors: (a) how well and stably one is disposed to reflect in one's motivational structures the pertinent rational structures of the various situations that one enters in the course of human relations and other events, and (b) how robust and global is that disposition. One is practically wise in proportion to how well one appreciates the rational force of the pros and cons by giving them motivationally the respective weights that they deserve, and how broad is the span of situations across which one is thus sensitive.

Does this correction of crude virtue ethics give to sophisticated virtue ethics and psychology what they need in order to repel the situationist attack? Not plausibly. It does so, surely, only to the extent that subjects in situationist experiments reflect in their motivational structures the rational structures of the situations faced. But it is quite implausible that they do so. Either they give too much weight to factors that should not have so much weight, such as punctuality, or they give too little weight to factors that should have more weight, such as whether they are inflicting

---

[9] Here I am assuming that the subject has access to the relevant plain facts. But it is a nice question how extensively factual perceptiveness constitutes practical competence, and is detachable from value perceptiveness. Practical competence is a function not so much of mere factual perceptual acuity (of the ability to perceive sharply) as of the foregrounding of facts that are relevant good reasons for what is objectively required of one by the situation faced. And this is a *normatively constituted* competence: it requires systematic foregrounding of the normatively relevant as such. Here the phenomenon of inattentional blindness is highly relevant. Take the seminarians in a hurry. It is not implausible that their disregard of the fellow human in need bespeaks not so much callousness as inattention, and indeed inattentional blindness. However, what seems still an open question is the extent to which such blindness is to be be classed with culpable neglect as a moral failure. And this is, again, of a piece with the question whether failure to foreground the morally relevant is itself thus morally relevant. Compare the following: www.scholarpedia.org/article/Inattentional_Blindness

severe pain, or at least they prioritize obedience to the experimenter excessively over inflicting the pain.

The switch from crude virtue theory to sophisticated virtue theory is nevertheless important, if only for the sake of understanding properly the subject matter of our controversy. However, it is implausible to suppose that this proper understanding suffices for a satisfactory response to the situationist attack. It seems incumbent on virtue theory to grant that the experiments do raise legitimate doubt as to how global and robust is human practical wisdom, and how global and robust are its more specific component virtues, such as kindness, human decency, honesty, courage, and the rest.[10]

That being granted, it would hardly follow that humans have no practical wisdom, none of the structure of virtues that, when properly integrated, constitutes such wisdom. This sort of invalid inference is the crucial weakness in situationism to be probed here. Indeed, probing this sort of fallacy, once spotted, deflates the situationist attack even when aimed against the crude version of virtue theory. If the attack fails even against the crude version, it will be an even worse failure against the more sophisticated variant.

My defense of virtue ethics will be based on an analogy between moral competence and driving competence. Let's define driving competence as a disposition to produce driving that is *safe*, when one is at the wheel, and *efficient* in routing to one's destination upon getting directions. Recall the contrasts applied earlier to traits, between the robust and the frail (or between the broad and the narrow, or the global and the local), and between the distinctive and the vulgar. These can be seen to apply with similar plausibility to driving competence. Someone's driving

---

[10] And we must also acknowledge the distance between individual virtues and behavior, and the intervention of deliberation often required by Sophisticated VP, which renders opaque how we can appeal to, say, kindness specifically in explaining an action such as the ceding of a seat on a bus. It is always rather practical wisdom that is more directly responsible for such virtuous behavior. What accounts for our proper citing of kindness specifically in that case? This is closely related to the issues involved in the unity of the virtues, but is an additional issue that deserves its own separate attention.

A further problematic engaged by Sophisticated VP is that of the freedom of the will. A choice based on deliberation is an exercise of rational agency, and hence involves a kind of freedom simply for that reason. Can we attribute that free choice or free judgment to a virtue or a competence, while viewing such virtues as dispositional competences, with their associated conditionals, compatibly with the acknowledged "freedom" of that choice or judgment? Can that "freedom" be libertarian freedom or must we conceive of it on a compatibilist model? These are deep and subtle issues that I mention here only to put them aside, while highlighting their importance for a fully satisfactory philosophical treatment of our problematic. They are in any case not issues addressed by the situationist critics, nor even by the respondents on behalf of the more sophisticated variant of Virtue Psychology and Ethics.

competence may be limited to quiet neighborhoods, for example, and not extend to busy highways, nor to city driving. Their competence is then not as robust as it might be. The minimal driving competence required for safe and efficient driving in a sleepy village is very widely shared, moreover, and not as rare or distinctive as the physical abilities demanded by Formula One car races, or the navigational adroitness required by a reticulated old city.

Evaluatively relevant behavioral differences in instances of driving (one bad, one good) will not be explained by the shared vulgar competence of the two drivers. Any such behavioral difference explained by appeal to competence must of course appeal to some *difference* in competence. Absent any such difference, one must appeal to some difference in situation.

Consider now the factors that we have found to affect the safety or efficiency of driving:

(i)   brightness of light, even when the road is visible;
(ii) whether you are on a bridge when it is cold and wet (roadways on bridges being colder and potentially more slippery);
(iii) whether you're using a cell phone;
(iv) your blood level of alcohol;
(v)   whether you received directions orally or through a map.

And so on. For some or all such factors, it must at some point have been surprising for us to learn how good driving does depend on them. How should one's folk theory of driving respond to such discoveries? One possible response would be *driving situationism*, as follows:

(a) Situations are dominant in the explanation of evaluatively relevant differences in driving behavior.
(b) The robustness of our driving competence is rendered problematic: it is found to vary surprisingly with respect to previously unsuspected factors.
(c) Personal integration in driving competence is not as widespread as one might have thought: thus, operational competence at the wheel does not necessarily go together with navigational competence.

These three theses are respectively analogous to the three listed by Doris in presenting the essentials of situationism (and quoted above). However, there has been no tendency to adopt driving situationism. Nor is it plausible to conclude that driving competence is just an illusion, or that we make a fundamental attribution error in taking people to be competent drivers when we see them display some good long stretch of such driving.

Of course, situations *will* affect how we explain the performances of drivers. But then any competence, indeed any disposition, will issue in a certain behavior only given certain triggering conditions. Behavior will in general have a two-ply explanation, one strand being the disposition, the competence, and the other strand being the relevant triggering conditions that elicit the manifestation of the competence. This is obviously true of dispositions in general. A cube dissolves not just due to its solubility but also due to its insertion.

Of course, any factor that *surprisingly* matters to our driving competence reveals some inadequacy in our prior view of such competence. But this argues not so much for the abandonment of competence psychology as for its correction. With every such discovery we need to change our view of the conditions, inner basis, or conditional structure of a competence.

Sometimes we have a choice between reasonable alternatives. Take the different ways one could respond to four factors that bear on safe driving: the brightness of the light, the amount of traffic, one's blood alcohol level, and one's visual acuity. Plausibly, the alcohol level belongs with the acuity as an inner basis for safe driving. After all, one's competence can change as can even one's acuity. By contrast, the amount of traffic figures more plausibly in the antecedents of the conditionals associated with the competence. And the same goes for the quality of the light. How competent a driver one is at a time depends on both sorts of factors. The degree of robustness of one's competence, for example, will be directly proportional to the breadth of the span of situations wherein one would produce good driving.

It might be replied that the analogy between driving and moral competence is limited and potentially misleading. You do not improve your moral competence by avoiding situations where it will be severely tested, in strict analogy to how you improve your driving competence by avoiding bridges when it's wet and wintry. Even if you concede some force to this point, the analogy remains effective. For one thing, you need not avoid the bridges so long as you are sufficiently aware of the risk and adjust your behavior accordingly. Through such adjustments, you *thereby* become more competent as a driver. Similarly, one way to improve your moral competence, surely, is to become more sensitive to moral danger, and to proceed with corresponding care.[11]

---

[11] Three studies in the April 2007 issue of the *Journal of Experimental Psychology* report on experiments conducted in Germany that examined 198 men and ninety-two women ages sixteen to forty-five while they played various games on a Sony Playstation. Based on the results of these experiments, the psychologists think you should be extra cautious next time you get behind a real-life wheel. They say that people who play

Here are some lessons from our exercise.

The discovery of factors bearing surprisingly on our moral competence might more reasonably lead us to improve it than to reject its existence. So much for the normative lesson important for virtue *ethics*. (When we discover the bearing of sleep deprivation on good driving we tend to avoid driving when sleep deprived, thus improving our driving competence.)

As for the bearing of the experimental results on virtue *psychology*, what they call for is, again, correction, not rejection. The sort of practical wisdom that can explain someone's behavior varies in degree from agent to agent, and also somewhat in structure, given (a) how variably humans can fall short, and given (b) how implausible it is to postulate a single acceptable motivational structure with sharp outlines.

We have learned of factors with a previously unsuspected bearing on our morally relevant conduct, factors that seem to dull our discernment of the moral or other practical demands in certain situations, leaving nearly all of us with less practical wisdom than we had commonsensically claimed. Similarly, certain factors affect our driving competence in ways that once proved surprising. A driver on his cell phone while crossing a bridge in wintry twilight will still likely reach his destination without incident. We have long known that accidents under such circumstances are significantly more probable, however, so that smooth driving is then *less* explicable through driving competence than we earlier thought, and more a matter of situational luck.

There are two sorts of relevant discoveries. Some factors might be thought to reduce a driver's competence temporarily: thus sleep-deprivation and blood alcohol. Others might be conditions not covered by a driver's competence. That is to say, the competence might be more local, less robust, than we earlier thought, since the conditionals associated with driving competence are more subject to restrictive provisos than we had imagined. Thus, conditionals associated with driving competence are now known to be qualified with previously overlooked provisos. These pertain to ambient light, to cell-phone usage, and to whether you are on a wet bridge in winter, to cite just three.

On this view, the quality of driving competence is determined by a certain inner state of the agent's, and by a set of associated triggering conditions and corresponding outcomes. If the inner state changes so as to induce more restrictive provisos in the associated conditionals, the degree or robustness of the competence is reduced.

car-racing games drive more aggressively and have a greater risk of car accidents than people who play racing games less often, or who play "neutral" games.

On an alternative view, a driver's fundamental competence does not necessarily change with the noted inner changes. That the driver is in certain relevant inner states is instead built into the associated conditionals. With an increase in blood alcohol, the driver enters a different triggering condition relative to the relevant outcome. On this alternative, our driving competence is viewed as stable throughout, although the difference in expectable outcomes shows it to be less robust than previously thought, since the good outcomes are restricted relative to alcohol level and other inner states.

On one view, alcohol takes away your competence. On the other view, alcohol does not take away your competence, but it does constitute a condition wherein your behavior can fall short despite your underlying competence. This is because your underlying competence requires you to behave successfully in certain normal conditions that exclude too high an alcohol level.

Whichever of the two views we take, we have discovered through the years that we possess neither the robustness of practical wisdom nor the robustness of driving competence that we had once optimistically self-attributed. We should not jump from that fact, however, to the conclusion that driving competence and practical wisdom are just illusions.

Finally, we must ask: Is the supposed clash between situationism and virtue theory perhaps itself an illusion?

1. Virtue theory makes no claim that normatively relevant behavioral variation is to be explained exclusively or even mainly through difference in virtues. Virtue theory is quite compatible with the view that humans are pretty much alike in the degree of virtue that they normally attain. *Compatibly with this, human conduct might still be universally explicable through the attribution of virtue to the agent.* Variability of marble rolls is not explicable through the universally shared disposition of marbles to roll, but each marble roll might still be explained in essential part through that disposition and its underlying basis, the rigid sphericity of a marble.

2. Virtue theory should accept that the experiments have shown humans to be less practically wise than folk virtue theory had imagined. Our practical wisdom now seems less robust or global than we had believed. While concluding that there are "no virtues of the sort that virtue theory had imagined," situationists do not leap all the way to a Skinnerian nihilism of behavioral dispositions. But virtue theorists for their part should *accept* that there are no virtues *of exactly the sort virtue theory had imagined*, since, after all, it is now granted that we are *less* robustly, globally virtuous than we had believed commonsensically.

3. Situationists conclude: forget virtues, explain by situations! But wait: remember, behavioral explanation is two-ply, requiring when laid out fully *both* the relevant particulars of the situation, *and* the relevant non-Skinnerian dispositions. These latter look for all the world like traits, competences, virtues, however robust they may or may not turn out to be. Neither extreme position seems acceptable: neither that such traits explain with no situational help at all, nor that such situational particulars explain with no dispositional help at all. Behavioral explanation is two-ply when laid out fully.

4. In light of the above, it would seem that any remaining substantive disagreement will be largely over degrees: Just how robust are the relevant human virtues?[12]

5. In addition to such disagreement over degrees, however, situationists may also disagree with virtue theorists on whether separable competences, such as the navigational, parallel parking, and steering competences of a driver can properly be unified into an overall "driving competence" with its own coherent explanatory and predictive efficacy. We do ordinarily cluster such sub-competences into unified competences. Just think of nearly any sport and the respective concept of a good player of that sport, whether it be tennis, or football, or basketball. Undeniably there will be separable sub-competences. Yet, we also commonly unify these and appeal to overall ability or competence in making choices, in predicting, and in explaining. Does commonsense *require* correction here? This is an interesting issue for virtue theory suggested by some of the situationist critique, especially by Doris's attack on the proper "integration" of the virtues of virtue psychology. The issue presents a real problem to be faced by virtue theory. So far as I can see, the problem has not been solved with full satisfaction, but nor has it been shown to be insoluble by the situationist critique.

6. As for the situationist recommendation that we should assess the situations we enter for the relevant risks, this surely will be endorsed fully by virtue theory, which will take the discernment of and proper weighing of such risk to be among the most important virtues pertinent to any given domain of human performance.

7. And we have been given no more reason to doubt that instilling virtue is worth the effort (on the part of parents and teachers) than to

---

[12] And there is indeed a surprising array of factors that can influence our behavior subliminally and unexpectedly. For example, noise levels seem to affect helping behavior (Mathews and Cannon 1975); as can fragrances in shopping malls (Baron 1997). But the like is, again, true of our driving performance.

doubt that instilling driving competence, and requiring its certification, is worth the effort (on the part of relevant government authorities).[13] [14]

REFERENCES

Annas J. 2005. "Being Virtuous and Doing the Right Thing," *Proceedings and Addresses of the American Philosophical Association*.

Baron, R. 1997. "The sweet smell of ... helping: Effects of pleasant ambient fragrance on prosocial behavior in shopping malls," *Personality and Social Psychology Bulletin* 23: 498–503.

Darley, J. M. and C. D. Batson 1973. " 'From Jerusalem to Jericho': A study of situational and dispositional variables in helping behavior," *Journal of Personality and Social Psychology* 27: 100–108.

Doris, J. 1998. "Persons, situations, and virtue ethics," *Nous* 32.

2002. *Lack of Character: Personality and Moral Behaviour*. Cambridge: Cambridge University Press.

Flanagan, O. 1991. *Varieties of Moral Personality*. Cambridge, MA: Harvard University Press.

Harman, G. 1998–99. "Moral philosophy meets social psychology: Virtue ethics and the fundamental attribution error," *Proceedings of the Aristotelian Society* 99: 315–331.

Hursthouse, R. 2007. "Virtue Ethics." In *Stanford Encyclopedia of Philosophy*.

Kamtekar, R. 2004. "Situationism and virtue ethics on the content of our character," *Ethics* 114: 458–491.

---

[13] The third most important experiment bearing on our issues is the Stanford Prison experiment due to Philip Zimbardo, who concludes as follows in a recent retrospective: "The critical message then is to be sensitive about our vulnerability to subtle but powerful situational forces and, by such awareness, be more able to overcome those forces. Group pressures, authority symbols, dehumanization of others, imposed anonymity, dominant ideologies that enable spurious ends to justify immoral means, lack of surveillance, and other situational forces can work to transform even some of the best of us into Mr. Hyde monsters, without the benefit of Dr. Jekyll's chemical elixir. We must be more aware of how situational variables can influence our behavior." (Zimbardo 2007 a, b) This critical message is one that virtue theorists can applaud. Yes, just as alcohol can deprive us of our driving competence, so group pressure can apparently deprive us of our moral competence. This would not show such competence to be an illusion, however, nor would it tend to show that belief in it is based on a fundamental attribution error.

[14] This paper was presented at a Herdecke conference on philosophy and the social sciences, and at an Oxford conference on powers and their manifestations. I am grateful for excellent discussion at each conference. My special thanks to Steven Lukes, for illuminating formal comments at the Herdecke conference, to Jason Kawall for helpful philosophical and bibliographical comments, to Gil Harman and John Doris for correspondence that helped me to understand their views better, and to Christian Miller, for instructive written commentary. Others who have given me helpful comments related to this chapter, in conversation or by e-mail, include Laura Ashwell, Alexander Bird, Jonathan Lowe, Penelope Mackie, Chrys Mantzavinos, Jennifer McKitrick, Stephen Mumford, Diego Rios, and Christoph Schmidt-Petri. My warm thanks to all.

Mathews, K. and L. Cannon 1975. "Environmental noise level as a determinant of helping behavior," *Journal of Personality and Social Psychology* **32**: 571–577.

Milgram, S. 1963. "Behavioral study of obedience," *Journal of Abnormal and Social Psychology* 67: 371–378.

  1974. *Obedience to Authority: An Experimental View.* New York, NY: Harper and Row.

Nisbett, R. E. and L. Ross 1980. *Human Inference: Strategies and Shortcomings of Social Judgment.* Englewood-Cliffs, NJ: Prentice-Hall.

Ross, L. and R. E. Nisbett 1991. *The Person and the Situation: Perspectives of Social Psychology.* New York, NY: McGraw-Hill.

Sreenivasan, G. 2002. "Errors about errors: Virtue theory and trait attribution," *Mind* **111**: 47–68.

Webber, J. 2006. "Virtue, character, and situation," *Journal of Moral Philosophy* **3**(2): 193–213.

Zimbardo, P. G. 2007a. *The Lucifer Effect: Understanding How Good People Turn Evil.* New York, NY: Random House.

  2007b. "Revisiting the Stanford Prison experiment: A lesson in the power of situation," *The Chronicle of Higher Education* **53**(30): B6.

9 – Comment
Do People Have Character Traits?

Steven Lukes

The dispute between situationism and virtue theory has the appearance of a duck–rabbit problem – the problem being that you cannot see both the duck and the rabbit at one and the same time. Either the situationists are right and there are no character traits, only situations in which people pursue goals, policies, and strategies in convergent ways, or else they are wrong and there are character traits. If Sosa is right, and I think he is, this appearance is an illusion, and the illusion derives from the situationists' way of posing the issue. Thus Harman writes that "[e]mpirical studies designed to test whether people behave differently in ways that might reflect their having different character traits have failed to find relevant differences" and so "ordinary attributions of character traits to people may be deeply misguided, and it may even be the case that there is no such thing as character." If that is so, then "there is no such thing as character building." (Harman 1998–99: 328) Indeed the thought that children may need moral education may be as misplaced as the thought they need to be taught their first language. And Doris writes: "To put things crudely, people typically lack character" (1998: 506). In short, although people routinely explain the actions of others by appeal to robust character traits, there is no scientific evidence for the existence of the sorts of traits that people standardly attribute to others. What a person with a seemingly ideal moral character will do in a particular situation is pretty much what anyone else will do in exactly that situation, allowing for random variation.

There are, I suggest, two clues that suggest that the appearance of this dispute as duck–rabbit-like is an illusion. The first is that both sides view people as fragmented, as driven by diverse and sometimes conflicting dispositions. The situationists see these as "narrow" rather than "robust" and thus not qualifying as character traits (which are, by definition, robust). But to virtue theorists, whether they adhere to folk psychology or Aristotelian ethics, such fragmentation is no surprise, for virtue, or practical wisdom, precisely consists in overcoming such fragmentation. To be fully virtuous, after all, is an ethical achievement only attained, if at all, by some.

The second clue is that the most radical of the situationists, (Harman, cited in Kamtekar 2004: 472), allows that "People have different innate temperaments, different knowledge, different goals, different abilities, and tend to be in or think they are in different situations. All such differences can affect what people will do." But how and why are these indicated factors not character traits? How does "character" differ from "temperament"? In being non-innate? and, if so, why? And why are our "goals" not part of our character traits? Situationists want to include such subjective factors as aspects of people's situations because they think of character traits in a way that contrasts with the way that virtue theorists think of them. So, to some indefinite extent, this dispute turns on a further, unresolved issue of how to distinguish between situation and character. The point is that both parties seem to allow for a certain range of dispositions or competences, of varying robustness, that they, however, put in different boxes.

As I see it, there are five serious problems with the way that the situationists have set up their critique of what Sosa calls virtue theory (which is, in fact, an amalgam of folk psychology and Aristotelian or more generally virtue ethics).

The first is that they begin from the observation, shared with the social psychologists from whose work they draw, that most of us wrongly attribute cross-situational consistent character traits on the basis of insufficient and inadequately distributed observations. So, Harman (1998–99) writes, our ordinary convictions about differences in character traits can be explained away as due to a "fundamental attribution error" together with "confirmation bias." But even if it is so (and it probably often is), this is completely irrelevant to the question of whether people have or don't have character traits, or indeed, if they do, to the question of the extent to which they do. The fact (if it is a fact) that our attributions are often, even systematically, unwarranted doesn't have any bearing on the question of whether such attributions are, or can be, warranted. It is interesting, incidentally, to note that the social psychologists Nisbett and Ross, who draw attention to this error and seek to correct it, don't draw the conclusion that therefore character traits can be assumed not to exist, suggesting that we look for "enduring motivational concerns and cognitive schemes that guide attention, interpretations, and the formulation of goals and plans" (Nisbett and Ross 1980: 20).

Secondly, the situationists assume that character traits are individualizing, that is, as Harman (1998–99: 317) puts it, "relatively long-term stable dispositions to act in distinctive ways": they distinguish one individual from another. But why should this be a defining feature of character traits – whether within folk psychological views or in virtue ethics? As Kamtekar (2004: 408) rightly suggests, it "seems reasonable to suppose, that how distinctive people's character traits are is an empirical matter, depending on how much their education and the requirements of their society make them alike (e.g., with respect to obedience or promptness)." And Sosa makes the related point that virtue theory "is quite compatible with the view that humans are pretty much alike in the degree of virtue they normally attain" (this volume, p. 287). Explicability via the attribution of character traits or virtues does not depend on variation across individuals.

The third, fourth, and fifth problems with the situationists' critique concern their picture of what character traits are and, in particular, how to identify them. Doris (1998: 509) defines them (he uses the term "virtue") as follows: "if a person possesses a virtue, she will exhibit virtue-relevant behavior in a given virtue-relevant eliciting condition with some markedly above-chance probability p." But this is a very impoverished, external, and mechanical behavioral view of character traits. It treats them as isolable from others, as externally identifiable independently of the agent's viewpoint and as responding in an unproblematic or stereotypical way to "virtue-relevant eliciting conditions" or situations. Let me elucidate each of these briefly.

So, third, it treats character traits one by one as isolable and independently functioning, rather than as involving reflection and deliberation across a range of situations and in interaction with other considerations. So, to quote Kamtekar (2004: 474) again:

In their exclusive attention to behavior, the psychological studies implicitly conceive of a character trait as something that will, if present, manifest itself in characteristic behavior, and will do so no matter what else there is in the situation for the person to respond to – in other words, the character trait will determine behavior in isolation from other character traits, thoughts, concerns, and so on a person might have in a given situation.

Fourth, it systematically ignores how agents construe the situations they are in – or rather it assumes that the subject and observer (or experimenter) agree in their understanding of both the "virtue-relevant behavior" and the "virtue-relevant eliciting condition." So, for example, they may disagree about what constitutes honest or helpful or caring behavior in a given situation (e.g., in one well-known experiment, the subject, who pockets some stray change, may adhere to the principle "finders keepers" and so regard doing so as entirely honest). What looks inconsistent from the outside may appear consistent from within and vice versa, and so assessing consistency across situations cannot sensibly ignore how agents construe those situations. Most of the experiments cited in these debates do exactly that, making contestable assumptions about both behavior and situation and using supposedly objective measures for each.

And fifth, more deeply, it makes the mistake of looking for consistency between "relevant behavior" and a relevant situation or "eliciting condition," treating the former straightforwardly as a response to the latter. But people don't respond as if mechanically to brute situations or "eliciting conditions": they respond to a situation as they interpret it and, as Sreenivasan (2002: 59) puts it, to "some *reason* for action present within the situation." They judge, at some level of consciousness, how compelling or weighty the reason is relatively to others. There may be a reason to enact the character trait or virtue in question, but it might be defeated by another reason or other reasons also present within the situation. Moreover, to possess such a trait or virtue certainly does not require enacting it on every occasion that might seem to "elicit" it; otherwise we would count as uncharitable or uncompassionate every time we pass a beggar in the street. The Milgram and Good Samaritan experiments cited by Sosa gain their force by suggesting that such countervailing reasons can only be too weak to defeat the reason to be compassionate or charitable. That is not obvious, especially when combined with the previous point concerning the situationists' ignoring of the role of subjective construal. But Sosa is, of course, right that they do seem to show the subjects to "go wrong" in their weighing of the reasons. They do, in these particular experimental situations, behave callously and uncharitably.

I applaud Sosa's strategy of deflating the situationist attack. I think that the analogy with driving, though, like all analogies, imperfect, is revealing and supportive of his argument and I agree with his conclusion that "behavioral explanation is two-ply, requiring when laid out fully *both* the relevant particulars of the situation, and the relevant non-Skinnerian dispositions," (this volume, p. 288) which, in turn, may be more or less robust. That is why the duck–rabbit analogy fails here: in explaining behavior, we not only can but must see both.

At the end of his chapter Sosa makes a statement and raises a question, and I want to conclude by questioning his statement and pursuing his question. The statement is that "the experiments have shown humans to be less practically wise than folk virtue theory had imagined" and indeed that "it is now granted that we are *less* robustly, globally virtuous than we had believed commonsensically" (this volume, p. 287). Now, I don't know what empirical basis he would cite for these claims about folk theory and our commonsense, but, to be frank, I think anyone paying attention to the state of the world over the last century, and in particular to the litany of atrocities that stretches from the Nazi Holocaust to the Rwandan genocide and current events in Darfur, would find his account of what "we had believed" until the experiments of social psychologists came along rather surprising and would see the experiments in question as confirmatory rather than revealing.

This leads me to comment on his last footnote, which, in reference to the Stanford Prison experiment, raises the issue of the deprivation of moral competence through group pressure. This relates directly to Sosa's question: "Just how robust are the relevant human virtues?" What the Stanford experiment seems to show, like the Milgram experiment, is that under certain conditions they just may not be very robust. These social-psychological experiments can help to specify these conditions in a scientific way by trying to control for different variables, but there is a much richer body of material that cannot do this but is no less relevant, indeed, I suggest, far more so. I refer to the whole literature of memoirs and historical research concerning the massive *social* experiments across the last century, experiments whose very purpose was to undermine resistance in the shape of robust human virtues. This is the theme, just to name a few examples, of the writings of Primo Levi and Nadezhda Mandelstam, of Jan Gross's books *Neighbors* (2001), about the massacre of Jews in a small Polish town, and *Fear: Antisemitism in Poland after Auschwitz* (2006), of Tzvetan Todorov's *Facing the Extreme: Moral life in the Concentration Camps* (1995) and of Christopher Browning's extraordinarily powerful *Ordinary Men* (1992).[1]

Ordinary Men is interesting in this connection because it is in search of an explanation of how it could happen that a battalion of a few hundred ordinary reserve policemen from Hamburg, who were not even Nazi party members could, after an initial, appalling atrocity (from which they were free to withdraw, though few did) end up being responsible for 89,000 deaths. Browning cites the Milgram experiments along the way to illustrate obedience to authority and considers also group pressure among other factors, such as the effects of propaganda, racism, and so on. He cannot prioritize these various factors in terms of causal weight, of course, but he can show which were present in the situation in which the *Einsatzgruppe* Battalion 101 found itself. But it is worth drawing attention, I think, to the assumption behind his explanatory

---

[1] Another text that raises these issues is Philip Gourevitch, *We Wish to Inform You That Tomorrow We Will be Killed With Our Families: Stories from Rwanda* (1999) and, most recently Gourevitch and Morris's (2008) remarkable account of the soldier photographers at Abu Ghraib, which is based on interviews conducted by Morris for his film of the same title.

search, namely that ordinary men are vulnerable to the collapse of virtue in extraordinary situations.

By contrast, we might consider the accumulated literature about, not the perpetrators but rather the rescuers of victims, notably Jews, during the Nazi period. Here there is a consensus that there are no generalizations that satisfactorily correlate acts of rescue with sociological or political or economic variables. Rescuers belong to no specific class or status group or profession; they come from no typical family background and have no typical political or religious orientation. No identifying factor marks them out from non-rescuers. In short, individual character seems to have been what made the difference between action and inaction. Here the situations in which rescuers were placed were, in relevant respects, common in a context of massive and widespread compliance and conformity, but a range of character traits, *pace* the situationists, would seem to have played the decisive role. Here the focus is on extraordinary men and women exhibiting virtuous character traits in what came to be all-too-ordinary situations.

It is striking, when reflecting on this contrast, to see a sort of parallel with the situationist/virtue theory debate. Studies of perpetrators, such as Browning's, suggest that, under the appropriate situations, ordinary people will exhibit what Judith Shklar called "ordinary vices." Studies of rescuers, conversely, suggest that some people, however oppressive and dangerous the situation, will manifest ordinary virtues.

I want to raise a final question. Sosa concludes that explanation of behavior needs to include both the particulars of the situation and relevant dispositions. But it is short-sighted to view these dispositions simply as properties of individuals. The range of relevant dispositions that count as virtues will vary across *contexts*. The discussion in this debate all centers around character traits possessed by individuals qua moral agents. But dispositions and competences are also *social*, in more than one sense. They are socially generated, specific to and definitive of certain social categories and not others; they are shared by their members and mark out their differences from others; and they serve to perpetuate distinctions within society. These dispositions are called *habitus* by Pierre Bourdieu (1984), author of *La Distinction*. He sees *habitus* as consisting in schemes of perception, appreciation and action, generating a permanent disposition, a durable way of standing, speaking, walking, and thereby of feeling and thinking, existing below the level of consciousness and control of the will. Social agents are endowed with *habitus*, inscribed in their bodies by past experience: social norms and conventions are incorporated, or inscribed into their bodies, thereby generating dispositions that are spontaneously attuned to the social order, perceived as self-evident and natural (Bourdieu 1984). They are specific to social groups, distinguishing them one from another in their body language, their way of speaking, in their tastes in food and art, and so on: families, professions, social classes, nations, each is character-ized by a distinctive *habitus*. (This is close to what John Searle calls "the background" – he has indeed suggested this himself – the "capacities and abilities" that render one "at home in his society," and "*chez lui* in the social institutions of the society," that render one "sensitive to the rules," the "motivational dispositions" that "give sense" to people's beliefs and desires (Searle 1995: 127–147).)

Are these "social" dispositions what Sosa means to refer to when he writes of those that are "vulgar and widely shared" and of "stably vulgar traits"? Sosa's point is that "stable vulgarity" will not explain individual variation. Viewing the question narrowly, he is right that Bourdieu–Searle type dispositions will not explain why you are, for instance, more *upright* or *loyal* or *honorable* than I am, within any given context. But, taking a wider view, they will surely play an essential role in such explanations, since their instantiation will indicate in what these virtues consist in any given context: what it means *here* to be upright or loyal or honorable. What counts as virtue in some contexts does not do so in others. Thus my character may emerge as more virtuous than yours, just because the con-text favors my dispositions more than it favors yours. Moving from one milieu to another, uprightness can appear as self-righteousness and pom-posity, loyalty as clientalism or nepotism and defending one's honor can look ridiculous, outmoded, and even dangerous. This is where the driv-ing analogy breaks down. For what counts as good driving is, broadly, uncontested across contexts, although it is tested in varying situations, whereas what counts as virtuous can be highly context-dependent.

REFERENCES

Bourdieu, P. 1984. *Distinction: A Social Critique of the Judgment of Taste.* Cambridge, MA: Harvard University Press.

Browning, C. 1992. *Ordinary Men: Reserve Police Battalion 101 and the Final Solution in Poland.* New York, NY: HarperCollins.

Doris, J. 1998. "Persons, situations and virtue ethics," *Nous* **32**: 504–530.

Gourevitch, P. 1999. *We Wish to Inform You That Tomorrow We Will be Killed With Our Families: Stories from Rwanda.* New York, NY: Picador.

Gourevitch, P. and E. Morris 2008. *Standard Operating Procedure.* New York, NY: The Penguin Press.

Gross, J. 2001. *Neighbors: The Destruction of the Jewish Community in Jewabdne, Poland.* Princeton, NJ: Princeton University Press.

2006. *Fear: Antisemitism in Poland after Auschwitz*. New York, NY: Random House.

Harman, G. 1998–99. "Moral philosophy meets psychology: Virtue ethics and the fundamental attribution error," *Proceedings of the Aristotelian Society* **99** (3): 315–331.

Kamtekar, R. 2004. "Situationism and virtue ethics on the content of our character," *Ethics* **114**: 458–491.

Nisbett, R. E. and L. Ross 1980. *Human Inference: Strategies and Shortcomings of Social Judgment*. Englewood Cliffs, NJ: Prentice-Hall.

Searle, J. 1995. *The Construction of Social Reality*. New York, NY: Free Press.

Sreenivasan, G. 2002. "Errors about errors: Virtue theory and trait attribution," *Mind* **111**(441): 47–68.

Todorov, T. 1995. *Facing the Extreme: Moral Life in the Concentration Camps*. New York, NY: Henry Holt.

10   What Kind of Problem is the Hermeneutic Circle?*

C. Mantzavinos

The hermeneutic circle serves as a standard argument for all those who raise a claim to the autonomy of the human sciences.[1] The proponents of an alternative methodology for the human sciences present the hermeneutic circle either as an ontological problem or as a specific methodological problem in the social sciences and the humanities. One of the most influential defenders of interpretivism in the English-speaking world, Charles Taylor, contends for example (1985: 18):

> This is one way of trying to express what has been called the "hermeneutical circle." What we are trying to establish is a certain reading of text or expressions, and what we appeal to as our grounds for this reading can only be other readings. The circle can also be put in terms of part-whole relations: we are trying to establish a reading for the whole text, and for this we appeal to readings of its partial expressions; and yet because we are dealing with meaning, with making sense, where expressions only make sense or not in relation to others, the readings of partial expressions depend on those of others, and ultimately of the whole.

Our understanding of a society is supposed to be circular in an analogous way: we can only understand, for example, some part of a political process only if we have some understanding of the whole, but we can only understand the whole, if we have already understood the part.[2] In this chapter I would like to check the soundness of this argument. I will

[1] In this chapter I draw from material in my *Naturalistic Hermeneutics* (2005). I would like to thank especially Pablo Abitbol, Dagfinn Føllesdal, Catherine Herfeld, and Diego Rios for their comments and criticisms. I am particularly thankful also to the participants of the Witten/Herdecke conference of June 2007 for their comments and suggestions. I would also like to thank the participants of the Joined Session of the Aristotelian Society and the Mind Association in July 2005 in Manchester for their comments on a preliminary version of this chapter.
[2] In Wolfgang Stegmüller's words (1988: 103): "[T]he circle of understanding seems to be the rational core which remains after we eliminate all irrational factors from the thesis of the distinction or special position of the humanities *vis-à-vis* the natural sciences."

* I dedicate this chapter to Petros Gemtos, Professor Emeritus of the Philosophy of the Social Sciences at the University of Athens, on the occasion of his seventieth birthday.

start with listing and shortly sketching out three variations of the problem (sec. 1). I will then critically discuss these and appeal to alternative solutions (sec. 2) and I will close with a short conclusion (sec. 3).

1   The Problem of the Hermeneutic Circle

1.1   *Is the Hermeneutic Circle an Ontological Problem?*

The philologist Friedrich Ast was probably the first to draw attention to the circularity of interpretation. He pointed to "[t]he foundational law of all understanding and knowledge," which is "to find the spirit of the whole through the individuals, and through the whole to grasp the individual" (Ast 1808: 178).[3] There is a series of philosophers that present the hermeneutic circle as an ontological problem. The locus classicus that they refer to is Heidegger (1927/1962: 195): "This circle of understanding is not an orbit in which any random kind of knowledge may move; it is the expression of the existential *fore-structure* of Dasein itself. It is not to be reduced to the level of a vicious circle, or even of a circle which is merely tolerated."[4] The question arises about what is meant by that and whether in fact the hermeneutic circle is this kind of a problem (Albert 1994). According to the traditional view ontology concerns itself with what exists and ontological arguments are usually presented that the world must contain things of one kind or another as for example necessary beings, unextended things, simple things etc. Alternatively, Quine's principle of ontological commitments, according to which to be is to be the value of a bound variable, does not tell us what things exist, but how to determine what things a theory claims to exist.[5] In any case ontology deals with the issue of the existence of entities and the question at hand is whether the hermeneutic circle is an issue of ontology.

1.2   *Is the Hermeneutic Circle a Logical Problem?*

The circle of understanding can alternatively be thematized as a logical problem.[6] It could be the case that the phenomenon of the hermeneutic

---

[3] Schleiermacher characterizes as a hermeneutic principle the fact "that the same way that the whole is, of course, understood in reference to the individuals, so too, the individual can only be understood in reference to the whole." (In a talk of 1829 now reprinted in Schleiermacher (1999: 329ff.).)

[4] See also the remark of Gadamer (1959/1988: 71): "Heidegger's hermeneutic reflection has its point not so much in proving the existence of a circle as in showing its ontologically positive meaning."

[5] See e.g., Quine (1980).

[6] The locus classicus to which the literature refers is Gadamer (1959/1988: 68): "The hermeneutical rule that we must understand the whole from the individual and the

circle has something to do with a logical circle. The relationship of the meaningful whole to its elements and vice versa could be of a logical nature. Two kinds of problems of a logical character could be relevant here. The hermeneutic circle could be concerned with *circular argumentation* in a deduction, which arises because in the process of proving something one falls back on a statement that one was supposed to prove. Or it could be related to a *circular definition*, which arises because the concept, which is still to be defined, has already unreflectively been used in the text beforehand. Is the nature of the problem a logical one?

1.3   Is the Hermeneutic Circle an Empirical Problem?

The hermeneutic circle is typically either viewed as an ontological or as a logical problem and is analyzed correspondingly. However, the question arises whether the phenomenon that the hermeneuticists are thinking of and characterize as the "circle of understanding" does present an empirical problem after all. With that, I mean that the movement of understanding from the whole to the part and back to the whole is a mental operation that could be analyzed with the tools of empirical science. In this case, the circle of understanding has nothing to do with ontology or with logic, but with the representation of knowledge in the mind of the interpreter which would present the following sort of empirical problem: how does the cognitive system of the interpreter perceive, classify, and understand written signs? Is this mental operation automatized, and what sort of cognitive mechanism is activated so that the meaning of part of a written expression is only available to the interpreter in dependence of the whole and vice versa?

2   The Solution to the Problem

If the hermeneutic circle were either an ontological or a logical problem, then this might indeed have very serious consequences. If the hermeneutic circle were an issue of ontology, this could force us to think differently with respect to ontology, the hermeneutic circle being practically ubiquitous when using language and dealing with texts. On the other hand, if the hermeneutic circle were a logical problem, then this would mean that the foundations of the human sciences were insecure and their scientific character was endangered. A lot seems to be at stake in both

individual from the whole stems from ancient rhetoric and was carried over by modern hermeneutics from the art of speaking to the art of understanding. There is in both cases a circular relationship."

cases. In what follows I would like to show that the hermeneutic circle is neither a genuine ontological problem nor a logical problem and that consequently neither ontology nor the methodology of the human sciences face the danger that many philosophers and scholars in the social sciences and the humanities suggest they do.[7] Rather, it will be shown that it is an empirical problem, which has long been studied using the tools of the empirical sciences.

2.1   *Why the Hermeneutic Circle is Not an Ontological Problem*

The philosophers that stress the ontological character of the hermeneutic circle are not concerned with a regional or special ontology, say of the social world. Their investigation is not about how social facts exist and what their properties are.[8] Nor is their investigation about how social reality fits into our overall ontology i.e., how the existence of social facts relates to other things that exist. They instead claim that the hermeneutic circle is an expression of the fundamental structure of human beings. Besides, they claim that the inquiry of the fundamental structure of human beings has to take place within the framework of a special discipline, fundamental ontology, consisting of propositions of a special status, i.e. neither logical nor empirical. Heidegger stresses in his classic text, for example (1927/1962: 195): "The 'circle' in understanding belongs to the structure of meaning, and the latter phenomenon is rooted in the existential constitution of Dasein – that is, in the understanding which interprets. An entity for which, as Being-in-the-world, its Being is itself an issue, has, ontologically, a circular structure." Claims such as this, being usually left unqualified, can function as poetic descriptions of human nature, but do not constitute *problems* and not even *arguments* that could be somehow reasonably dealt with.

2.2   *Why the Hermeneutic Circle is Not a Logical Problem*

Since there are hardly any genuine *arguments* suggesting that the hermenentic circle is a problem of ontology, the question arises whether the hermenentic circle has anything to do with logic instead. As Stegmüller noted in his classic article (1979/1988: 104ff.), logically the dispute about the hermeneutic circle runs up against a series of difficulties, which burden all hermeneutic literature: the

---

[7] For textbook discussions of interpretivism see Little (1991, ch. 4), Kincaid (1996, ch. 6), and Manicas (2006, ch. 3).
[8] For such an investigation see for example Searle (1995 and 2005).

pictorial-metaphorical language, the blurring of object- and meta-levels, the lack of clarity about the status of the key hermeneutical terms (above all the ambiguity of the word "understanding"), the merely apparent distance from psychologism, and finally, the complete lack of the analysis of examples.

However, what in any case applies is that the phenomenon of the hermeneutic circle has nothing to do with a logical circle, despite frequent insinuations of hermeneuticists to the contrary. The relationship of the meaningful whole to its elements and vice versa is not of a logical nature. It is thus not concerned with *circular argumentation* in a deduction, which arises because in the process of proving something one falls back on a statement that one was supposed to prove. Nor is it related to a *circular definition*, which arises because the concept, which is still to be defined, has already unreflectively been used in the text beforehand.

It is nevertheless possible that the hermeneutic circle, while not being a case of circular logic, still presents another type of logical problem. In a detailed explication of the concept, Stegmüller maintains that it constitutes a dilemma, or more concretely, one of six specific forms of dilemmas, depending on what is meant by the "hermeneutic circle" in a particular case.[9] However, this transformation of the phenomenon into different forms of dilemmas i.e., into the types of difficulties that force the researcher to choose between two alternatives that are equally undesirable, does not seem to be correct. In principle, Stegmüller's analysis attempts to show that the hermeneutic circle is not in fact a *logical* problem, but that it still can be viewed as a *methodological* problem, which in some of its variations is by no means a narrow epistemological problem of the human sciences, but instead something that epitomises all disciplines. This applies, for example, to what is known as the dilemma of confirmation. It also applies to the dilemma in distinguishing between background knowledge and facts. In a careful analysis based on examples both from literature and astronomy, Stegmüller shows that, in testing the relative hypotheses, difficulties arise in precisely differentiating between background knowledge and facts. The testing of hypotheses requires a clear separation between hypothetical components in the observational data, on the one hand, and the theoretical background knowledge, on the other. As Stegmüller (1979/1988: 145ff.) convincingly shows, by no means does this problem just arise in the humanities. It can only be solved through critical discussions and the agreement of those in the discipline in question about what are to be considered facts and what is to be considered background knowledge in connection with the

[9] For an even more detailed explication of the concept see Goettner (1973: 132 ff.).

specific hypothesis to be tested. Føllesdal, Walløe and Elster also defend the position that the hermeneutic circle is a methodological problem. They discuss a series of methodological problems that arise during the processes of understanding and claim that they all appear in the context of the justification of an interpretation.[10]

Now, I have no objections to this treatment per se, except that it certainly is not concerned with a logical problem in any narrow sense, but rather with a methodological problem. I would, however, deny that the problem of the relationship between the meaningful whole and its elements can be plausibly transformed in this way. One central view that I share with Stegmüller and with Føllesdal *et al.* is that, in the development of the meaning of texts, interpretative hypotheses are to be tested. In testing such interpretative hypotheses, the methodological problems or the dilemmas that these authors discuss will often, if not always, arise, especially the problem of distinguishing between facts and background knowledge. However, the problem of the relationship between the meaningful whole and its elements does not arise when *testing the interpretative hypotheses* but when *formulating them*. It is concerned with a special phenomenon that arises when one does not manage to understand linguistic expressions (or other signs) immediately i.e., more or less automatically. It is then necessary to set up interpretative hypotheses, and it is in doing this that one runs up against the problem of the meaningful whole and its elements. I will subsequently deal with what this activity more concretely looks like and how it is to be explained.

In summary, it can be asserted that the way that the hermeneutic circle is presented by representatives of philosophical hermeneutics does not suggest a methodological dilemma that can be solved by means of a decision or in any other way. Rather, the inevitability of the hermeneutic situation is pointed out and a "circle" is spoken of in order to somehow dramatize the issue. Stegmüller and Føllesdal *et al.* deny the hopelessness of escaping this problem, and with the help of methodological considerations, show that there are rational ways to come to grips with this issue after all. I would like to admit this hopelessness, but to play it down by showing that the hermeneutic situation is an empirical phenomenon.

---

[10] See Føllesdal *et al.* (1996: 116ff.). They work out four variations of it: the whole and part circle, the subject–object circle, the Hypothetico Deductive Method circle and the question–answer circle. Martin (1994: 265ff.) also tries to "show that there is a problem analogous to the hermeneutic circle in the natural sciences but that has not prevented natural scientists from objectively testing their theories."

2.3   Why the Hermeneutic Circle Is an Empirical Phenomenon

"A person who is trying to understand a text is always projecting. He projects a meaning for the text as a whole as soon as some initial meaning emerges in the text. Again, the initial meaning emerges only because he is reading the text with particular expectations in regard to a certain meaning. Working out this fore-projection, which is constantly revised in terms of what emerges as he penetrates into the meaning, is understanding what is there" (Gadamer 1960/2003: 267). This is how Gadamer, the most influential representative of philosophical hermeneutics, sketches out the process of understanding a text as a series of "hermeneutic circles." The reader or the interpreter reads a text with preconceived expectations (preconceived opinions or prejudices), and in his work, he makes revisions. The understanding of the text, however, remains "permanently determined by the anticipatory movement of fore-understanding" (Gadamer 1960/2003: 293). When this activity has occurred, when understanding has already taken place, the circle of whole and parts is "not dissolved in perfect understanding," if you will, "but, on the contrary, is most fully realized" (ibid.). In this classic exposition[11] of the hermeneutic circle, it seems clear to me – in contrast to the view of most hermeneutic philosophers – that the phenomenon being described is empirical.[12]

What is, more specifically, the case? What kind of cognitive activity is linguistic understanding? Given that this cognitive activity is improvable with practice, one can become faster at it and can become more precise, it is clear that it is a skill. In general, acquiring skills is much different than learning facts.[13] For example, a violinist learns to play pieces faster and

---

[11] Classic insofar as the present discussion continually refers to this passage. See e.g., Reale (2000: 96f.).

[12] It is characteristic of the prevailing confusion that, in diverse loci, Gadamer himself says different or contradictory things about the hermeneutic circle. So, he says in *Truth and Method* (1960/2003) on p. 293: "Thus the circle of understanding is not a "methodological" circle, but describes an element of the *ontological structure of understanding*" [my emphasis]. But then, in a footnote, Gadamer reacts to the above mentioned criticism of Stegmüller (p. 266): "The objection raised from a logical point of view against the 'hermeneutic circle' fails to recognize that this concept makes no claim to scientific proof, but presents a *logical metaphor*, known to rhetoric ever since Schleiermacher" [my emphasis]. Thus, it appears, it is supposed to be both an "element of the ontological structure of understanding" and a "logical metaphor," whereby it is completely unclear what is meant by a "logical metaphor."

[13] Neurological studies with patients suffering from amnesia show that the difference between acquiring skills and learning facts is honored by the nervous system. In a classic study, for example, Cohen und Squire report on patients who were capable of acquiring a "mirror-reading skill," although they had a memory neither of the words that they read nor even of being confronted with the task. Their amnesia in relation to the specific words and the fact that they dealt with them in a laboratory experiment did not hinder the learning or exercising of a skill i.e., the reading of words that were presented in mirror images. See Cohen and Squire (1980).

to hold tones by practicing. A small child can only learn to brush his or her teeth by practice, etc. The investigation of learning processes that lead to the acquisition of these types of skills has long been an established branch of psychological research.

In our context it is significant that in acquiring skills one will not only become faster and more precise, but that it will also continue to be easier to exercise them, and in fact the skill will become automatic.[14] In everyday life an enormous number of skills are carried out in this automatized fashion. That means that they become routines, and no cognitive resources in the form of attention are required in carrying them out. The automatization of the skills implies that they are carried out without conscious effort. In the case of understanding language, which is of interest here, the stroop effect is characteristic, named after its discoverer, Ridley Stroop (1935): If people are confronted with the names of colors that are printed in other colors – "blue" printed in red, "green" printed in black, etc., and they are to name the colors in which the words are printed, then they tend to read the words, because reading is an automatized skill. We tend to pronounce the words unconsciously because we have practiced doing so for years.[15]

This automatization of learned skills is a general phenomenon, which has already been empirically investigated and explained (although there is still no consensus about the neurophysiological processes that underlie it). It is known, for example, that in the middle phase of a game, a chess master needs five to ten seconds in order to propose a good move, which is often objectively the best move (Simon 1979: 386ff.). As Simon notes when referring to this explanation (1983: 26):

[I]t does not go deeper than the explanation of your ability, in a matter of seconds, to recognize one of your friends whom you meet on the path tomorrow as you are going to class. Unless you are very deep in thought as you walk, the recognition will be immediate and reliable. Now in any field in which we have gained considerable experience, we have acquired a large number of "friends" – a large number of stimuli that we can recognize immediately. [...] We can do this not only with faces, but with words in our native language. Almost every college-educated person can discriminate among, and recall the meanings of, fifty to a hundred thousand different words. Somehow, over the years, we have all spent many hundreds of hours looking at words, and we have made friends with fifty or a hundred thousand of them. Every professional entomologist has a comparable ability to discriminate among the insects he sees,

---

[14] See on this Baron (1994).
[15] It is possible to experience the same difficulty in a similar way. Try to give the number of symbols in each group of symbols in the following list. For example, when you see YYY, answer with three, when you see 5555 answer with four, etc.:

YYY YY 5555 33 444 22 222 3333 44444 3 11 222.

and every botanist among the plants. In any field of expertise, possession of an elaborate discrimination net that permits recognition of any one of tens of thousands of different objects or situations is one of the basic tools of the expert and the principal source of his intuitions.

It thus appears that texts are not only read against the background of the reader's presumptions and prejudices, but also – and more generally – against the background of their own experience with the material. Because the corresponding skill has become routinized, the text is normally understood automatically, and not consciously. Thereby it is of course to be emphasized that, because it is a complex skill, all levels play a role in understanding language: the phonologic, the semantic, the syntactic, and the pragmatic levels. One gains experience in all of these levels over the course of time, so that sounds, words, sentences, and entire texts are automatically classified and therefore language processing under standard conditions takes place effortlessly.

If a difficulty arises in the language comprehension process and if one does not manage to understand linguistic expressions immediately, then cognitive resources for solving the problem are activated. We focus our attention in order to consciously interpret an expression: an interpretative hypothesis is consciously generated. In psycholinguistics this conscious comprehension of language is often modeled as an interactive process. The relevant levels of information processing, the phonologic, the semantic, the syntactic, and the pragmatic, are not sequentially activated i.e., one after another. Rather, the information is processed in all of these levels in parallel and simultaneously. Our language comprehension system keeps all the information available so that it is possible to have recourse to all of the information categories at any time.[16]

The discourse on the hermeneutic circle does nothing more than imprecisely depict the search process that is activated if the interpreter of a linguistic expression does not understand something immediately. Nowadays psycholinguistics does not only offer more precise descriptions of the phenomenon, it also provides explanations of the underlying search processes and mechanisms of language comprehension. We know, for example, that language recognition results from the classification of patterns and that a considerable amount of data is necessary for this classification. The explanations that are offered from psycholinguistics are formulated in a testable form and have been tested in laboratory experiments; but nobody talks about hermeneutic circles in that case.[17]

---

[16] This interactive approach of the language processing system has been experimentally studied, especially by Danks, Bohn, and Fears (1983).

[17] For an informative overview of linguistic understanding, with a further bibliography see Anderson (2005, ch. 12).

Furthermore, with respect to the completion of understanding in accord with the completion of the hermeneutic circle, I would like to point to the cognitive mechanism that lies at the basis of every "aha" experience. The "aha" experiences of diverse intensity, which an interpreter has when the process of comprehension is completed, are neither irrational nor *a priori*. The main argument why a cognitive mechanism is at work on the phenomenon at hand is the fact that only people with the appropriate knowledge have "aha" experiences (Simon 1986: 244f.). Without recognition based on previous experience, the process of comprehending new linguistic expressions cannot take place, and while performing this activity our intuition exploits the knowledge that has been gained through past searches.

Finally, it is important in this context to emphasize that in the perceptual process that underlies the overall mental process of understanding texts, first, the written expression is encoded, before, at a second stage, the syntactic and semantic analysis known as *parsing* can follow. Parsing is the process by which the words in the expression are transformed into a mental representation with the combined meaning of the words. During this procedure, the meaning of a sentence is processed phrase by phrase, and the exact formulation of the phrases is only accessed while processing its meaning (Anderson 2005: 391). People integrate both semantic and syntactic cues in order to achieve an understanding of a statement or a text. As Steven Pinker (1994: 227) has noted:

Understanding, then, requires integrating the fragments gleaned from a sentence into a vast mental database. For that to work, speakers cannot just toss one fact after another into a listener's head. Knowledge is not like a list of facts in a trivial column but is organized into a complex network. When a series of facts comes in succession, as in a dialogue or text, the language must be structured so that the listener can place each fact into an existing framework.

It thus appears that in understanding, the phenomenon called "hermeneutic circle" is at work. As soon as a word occurs, people attempt to extract as much meaning as possible out of it: they do not to wait until a sentence is completed to decide on how to interpret a word – a finding brought to light by the experiments of Just and Carpenter, among others.[18] If a sentence contains unfamiliar words, which cannot

---

[18] Just and Carpenter studied the movement of the eyes during the reading of a sentence, and since in reading a sentence subjects typically fixate on almost every word, they found out that the time that the subjects spend fixating on a word is proportional to the amount of information the respective word contains. If a sentence contains a relatively unfamiliar word, the eye movement pauses longer at this word. There are also longer pauses at the end of the phrase in which the unfamiliar word is found. See Just and Carpenter (1980).

be understood immediately, then one spends additional time at the end of the phrase or the sentence to integrate the meaning. Thus the problem of the relationship between the meaningful whole and its constitutive elements, and vice versa, does not arise when testing interpretative hypotheses, but when generating them. It refers to a phenomenon that arises when it is not possible to understand linguistic expressions immediately i.e., more or less automatically. This problem thus appears to arise *both for words and sentences, and for entire texts.* To resolve it, cognitive resources are activated. We focus our attention to consciously interpret an expression, and interpretive hypotheses are consciously generated.

It should be sufficiently obvious by now, but I would like to state it also explicitly: For my own general argument to hold, it is not necessary to accept that, for example, the mechanism of parsing constitutes the correct explanation of the phenomenon or that the relevant levels of information processing are activated simultaneously and not sequentially. What is important is only that those claims are empirical claims – *even if they are wrong, they are still empirical.*

3   Conclusion

Concluding, it is possible to assert that until now it has not been possible to show that the hermeneutic circle constitutes an ontological or a logical problem. Rather, everything indicates that it describes an empirical phenomenon, which can be studied within the framework of psycholinguistics and other empirical disciplines. It is thus not capable of serving as a legitimating argument for the separation between the natural and the human sciences and therefore cannot lend any support to the claim for autonomy of the social sciences and the humanities.

REFERENCES

Anderson, J. R. 2005. *Cognitive Psychology and Its Implications.* 6th edn. New York, NY: W.H. Freeman and Company.
Albert, H. 1994. *Kritik der reinen Hermeneutik.* Tübingen: J.C.B. Mohr (Paul Siebeck).
Ast, G. A. F. 1808. *Grundlinien der Grammatik, Hermeneutik und Kritik.* Landshut: Jos. Thomann, Buchdrucker und Buchhändler.
Baron, J. 1994. *Thinking and Deciding.* 2nd edn. Cambridge: Cambridge University Press.
Cohen, N. and L. R. Squire 1980. "Preserved learning and retention of pattern-analyzing skill in amnesia: dissociation of knowing how and knowing that," *Science* 210: 207–210.

Danks, J. H., L. Bohn, and R. Fears 1983. "Comprehension Processes in Oral Reading." In G. B. Flores d'Arcais and R. J. Jarvella (eds.) *The Process of Language Understanding*. Chichester, New York: John Wiley & Sons, pp. 193–223.

Føllesdal, D., L. Walløe, and J. Elster 1996. *Argumentasjonsteori, språk og vitenskapsfilosofi*. Oslo: Universitetsforlaget.

Gadamer, H.-G. 1959/1988. "On the Circle of Understanding." In J. M. Connolly and T. Keutner (eds.) *Hermeneutics versus Science? Three German Views*. Notre Dame, IN: University of Notre Dame Press, pp. 68–78. Originally published as "Vom Zirkel des Verstehens." In G. Neske Pfullingen (ed.) *Festschrift für Martin Heidegger zum 70. Geburtstag*, pp. 24–35.

   1960/2003. *Truth and Method*. 2nd rev. edn, trans. J. Weinsheimer and D. G. Marshall. New York: Continuum. Originally published as *Wahrheit und Methode*. Tübingen: J.C.B. Mohr (Paul Siebeck).

Goettner, H. 1973. *Logik der Interpretation*. Munich: Wilhelm Fink.

Heidegger, M. 1927/1962. *Being and Time* trans. J. Macquarrie and E. Robinson. New York, NY: Harper & Row. Originally published as *Sein und Zeit* Tübingen: Max Niemeyer Verlag.

Just, M. A. and P. A. Carpenter 1980. "A theory of reading: From eye fixations to comprehension," *Psychological Review* 87: 329–354.

Kincaid, H. 1996. *Philosophical Foundations of the Social Sciences*. Cambridge: Cambridge University Press.

Little, D. 1991. *Varieties of Social Explanation*. Boulder, CO: Westview Press.

Manicas, P. 2006. *A Realist Philosophy of Social Science. Explanation and Understanding*. Cambridge: Cambridge University Press.

Mantzavinos, C. 2005. *Naturalistic Hermeneutics*. Cambridge: Cambridge University Press.

Martin, M. 1994. "Taylor on Interpretation and the Sciences of Man." In M. Martin and L. C. McIntyre (eds.) *Readings in the Philosophy of Social Science*. Cambridge, MA: The MIT Press, pp. 259–279.

Pinker, S. 1994. *The Language Instinct*. New York, NY: Perennial Classics.

Quine, W. van Orman 1980. *From A Logical Point of View*. 2nd rev. edn. Cambridge, MA: Harvard University Press.

Reale, G. 2000. "Gadamer, ein großer Platoniker des 20. Jahrhunderts." In G. Figal (ed.) *Begegnungen mit Hans-Georg Gadamer*. Stuttgart: Reclam, pp. 92–104.

Schleiermacher, F. D. E. 1999. *Hermeneutik und Kritik*. M. Frank (ed.). Frankfurt am Main: Suhrkamp, 7th edn.

Searle, J. 1995. *The Construction of Social Reality*. New York, NY: Free Press.
   2005. "What is an institution?," *Journal of Institutional Economics* 1: 1–22.

Simon, H. 1979. *Models of Thought*. New Haven, CT and London: Yale University Press.
   1983. *Reason in Human Affairs*. Stanford, CA: Stanford University Press.
   1986. "The information processing explanation of gestalt phenomena," *Computers in Human Behavior* 2: 241–255.

Stegmüller, W. 1979/1988. "Walther von der Vogelweide's Lyric of Dream-Love and Quasar 3C 273. Reflections on the So-called 'Circle of Understanding' and on the So-called 'Theory-Ladenness' of Observations." In J.M. Connolly and T. Keutner (eds.) *Hermeneutics versus Science? Three German Views*. Notre Dame, IN: University of Notre Dame Press, pp. 102–152. Originally published as "Walther von der Vogelweides Lied von der Traumliebe und Quasar 3C 273. Betrachtungen zum sogenannten Zirkel des Verstehens und zur sogenannten Theorienbeladenheit der Beobachtungen." In *Rationale Rekonstruktion von Wissenschaft und ihrem Wandel*. Stuttgart: Reclam 1979, pp. 27–86.

Stroop, H. R. 1935. "Studies of interference in serial verbal reactions," *Journal of Experimental Psychology* **18**: 643–662.

Taylor, C. 1985. "Interpretation and the Sciences of Man." In *Philosophical Papers*, vol. 2: *Philosophy and the Human Sciences*, Cambridge University Press, pp. 15–57.

10 – Comment
Going in Circles

David-Hillel Ruben

If there is such a thing as the hermeneutic circle, it is surely at least a circle. In this usage, however, "circle" is a mere metaphor. What is a circle in the sense required by the idea of a hermeneutic circle? The purpose of this comment is to develop a typology for circles in the relevant sense, and thereby show how the idea of a hermeneutic might be shown to be intelligible and what the requirements would be for there to be such a thing. I also address the issue of the need for a hermeneutic circle at all, of any kind, albeit briefly, at the end of the comment.

The hermeneutic circle must have something to do with a characteristic of the relationship between the items that are joined by the circle. So there must be (1) the items so related, and (2) the relations between them. By all accounts, the hermeneutic circle says something about explanation or understanding. So the relation of (2) must be the relation of explaining.

Is the relation in which we are interested really the explaining relation? There is an awful lot of talk in Chrys Mantzavinos's chapter, and in the literature he cites, about understanding. I know of no plausible distinction between understanding and explanation, in advance of a thesis about the irreducible differences between knowledge in, or the methodology of, the natural and social sciences. We cannot start by assuming that understanding and explanation are different ideas. I will therefore use them as interchangeable, until or unless we find a convincing reason to introduce the distinction between them.

To an earlier draft of this comment, Mantzavinos replied that unlike me, he starts by taking the idea of the difference between understanding and explanation "seriously." If we focused on understanding rather than explanation, and agreed that they were distinct, I am not sure in any case that it would make any difference to the results of this comment. We could say that something was understandable by virtue of another thing's being understandable. We might take the relation of something's making a second thing understandable as a primitive relation, or we might try and give it some account different from the account given to the idea of explanation. But in either case, it seems to me that all the problems I raise here regarding explanation could be re-raised using the alternative terminology of understanding and the relationship between one item's being understandable and another's being understandable.

Claims about circles, or their near relations, wholes, are not uncommon in philosophy, but different claims about circles might not all be using the idea of a circle in the same way. One might start by trying to distinguish between circles of particulars and circles of concepts, and finally between them and "hybrid" circles which have both.

A circle purely of particulars is a circle embracing a definite number of particulars. Without some such restriction, there would be an endless series of particulars rather than a circle of them. There are of course an infinite number of points on a circle. But the circles we are considering are circles that join various "discrete" items (whether propositions, texts, facts, people, concepts, or whatever). It is to these items that the limit restriction applies.

Perhaps there are particulars a, e, i, ... u, and y, and some relation R such that a has R to e, e has R to ... u has R to y, and y has R to a. In this case, the particulars form a circle purely of particulars, and the same particular "reappears" as one "travels" around the circle repeatedly. (The ideas of travelling and repetition are to be understood metaphorically, since the relata might not even be ones with spatio-temporal location.)

A simple example of a circle purely of particulars might be the example of a so-called love triangle. Arnold might love Betty and Betty might love Charles and Charles might love Arnold, who loves Betty, who loves... etc. Circles purely of particulars such that the particulars on them were temporally dated items like events can be problematic, depending on the nature of the relation R. Arnold and all his mates above have temporal location but, in combination with the loving relation, this presents no problem. But where "R" stands for some other relation, this might not be so. For example, if the relation in question were the causal relation and if all causes must occur before their effects, such a circle of token events would require a particular to happen at two distinct times, which is an absurdity. So if causes must occur before their effects and if no particular can occur wholly at two distinct times, then there can be no such circle purely of particulars as one in which event e causes i and i causes o and o causes e. Event e cannot occur both before and after i.

The connection between holism and circularity is complicated.[1] Does Davidson's "holism of the mental" involve the idea of a circle? "...

---

[1] Charles Taylor says: "... The circle can also be put in terms of part–whole relations: we are trying to establish a reading for the whole text, and for this we appeal to the readings

we cannot intelligibly attribute any propositional attitude to an agent except within the framework of a viable theory of his beliefs, desires, intentions, and decisions" (Davidson 2001: 221f.). A minimalist interpretation of what this means seems to be that for the agent to have some propositional attitude, he must have other propositional attitudes; the existence of any one propositional attitude entails the existence of many others. But if that is true, each of the further entailed propositional attitudes will require still others. If there are a finite number of propositional attitudes that an agent holds at a time, and these are token states of an individual, this might suggest that some sort of circle is lurking in Davidson's holism. The claim might be understood as asserting that there is a circle purely of particulars, where the relata are the individual propositional states of the agent. The claim is that if the agent has one

of its partial expressions; and yet because we are dealing with meaning, with making sense, where expressions only make sense or not in relations to others, the readings of partial expressions depend on those others, and ultimately of the whole." The circle that Taylor speaks about here is a circle of a whole and its parts.

"Understanding a part depends on understanding the whole" has a certain ambiguity about it. Let "S" stand for the totality of social facts. It could be that what we have already said exhausts the content of this claim. S may have gained its rightful place in the explanatory circle (if there were one) just in case every part of S figures in the circle at some point or other. On this view, the whole, S itself, would not figure in any one point on the circle. Only every one of its parts would be on the circle somewhere, if there were such a circular chain.

There is another view, one which might be attributed to Hegel and which seems to be what Taylor has in mind above, that to understand any one thing (say, a part of a text), one must first understand not just each part *seriatim*, but the totality of them, S, say the whole text: "Das Ganze is...not formed by composition, but by development out of its Concept. The whole is prior to its parts, and the parts can only be understood in terms of the whole. Each part serves the purpose of the whole" (Inwood 1992: 309). Some of Mantzavinos's quotes seem to say much the same: quoting Friedrich Ast, "... to find the spirit of the whole through the individuals, and through the whole to grasp the individuals." Indeed, the problem that Mantzavinos sets himself as the major issue in the chapter, and to which he thinks turns out to be an empirical question, concerns "... the movement of understanding from the whole to the part and back to the whole" (p. 301).

Before I could decide whether the problem was empirical, conceptual, or logical (the three alternatives Mantzavinos offers us), I should like to understand just what the problem is a bit better. Perhaps the best that can be made of this Hegel–Taylor–Ast view is in terms of a distinction between merely adequate and full explanation or understanding. If we switch now from texts to society, the view might be this: one might be able to merely adequately explain one social fact in terms of another, but such explanation as that provides lacks something, falls short in some way. A truly full explanation of any social fact can only be achieved after one has a merely adequate explanation of them all, as a whole, and this may bring one to revise in some way some of the earlier understandings one had offered previously. In light of an understanding of the whole (text or society), the explainer may readjust his understanding of some of the constituent parts. This view also seems to bear some affinities to Rawls's method of reflexive equilibrium (Rawls 1999:18–19, 42–45). I am not sure that this way of representing the dialectic between whole and part – whether the meaning of texts or the explanation of society – is best expressed as a circle at all.

such propositional attitude, then there are many others he has, although which other attitudes there are may differ from agent to agent. Each agent will have his own circle of propositional attitudes that may not be quite like any other agent's circle.

In Davidson's holism of the mental, what is the relation that holds between the relata understood as an agent's token mental states? It certainly cannot be the causal relation, for the reason given above. The relation in the quote above seemed to be the relation of requiring the existence of, or some such. Davidson also says: "… we make sense of particular beliefs only as they cohere with other beliefs, with preferences, with intentions, hopes, fears, expectations, and the rest … the content of a propositional attitude derives from its place in the pattern." The relation between the propositional attitudes in this last quote is a relation that fixes their content, whatever that might be. Finally, he says that "… the attribution of mental phenomena must be responsible to the background of reasons, beliefs, and intentions of the individual." Just what the idea of responsibility is that he employs here, and what relation Davidson thinks that involves, he never makes clear. But it certainly seems that the three quotes are not making exactly the same claim.

Clarification of the connection between wholes and circles requires careful analysis of the meaning of both a circle and a whole, in the relevant senses. If there are an infinite or an indefinitely large number of well-formed sentences in a language (considered as abstract objects, whether they had ever been uttered by anyone or not), one might believe that each sentence gains its sense from its connections with all the rest. Such a belief could be described as a belief about the holistic nature of meaning, although the idea of a circle of meaning would seem inappropriate. A "picture" of such a non circular whole might be an infinitely long straight line drawn through an infinite number of nodes, each node representing a sentence, with additional curved lines connecting each sentence, each node, directly with all others to which it is not already directly connected on the straight line. The picture would be akin to a straight line with lots of humps, indeed an infinite number of such humps, but certainly not a picture of a circle. Without the restriction, the idea of a circle is lost. In a circle, something must return to its point of origin in some way or other. That return will not ever happen in the case of a line that has nodes on it which represent an infinite or indefinitely large number of things.

There can also be circles purely of concepts. Quine's claim, for example, that there is a circle of concepts, embracing synonymy, meaning, possibility, definition, semantic rules, and so on, is a claim about such a circle (Quine 1961). The Quinean thesis might be expressed in this way: there exists a set of concepts, such that each can be explicated using some of the others, and no one of which can be explicated using concepts not in that set. The picture for such a circle might be of a circle with lots of straight lines inside the circle, joining each point on the circle's circumference directly with other points, including points distant from one another on the circumference itself. The curved and straight lines converging from many concepts to each concept on the circle (including the concepts next to it on the circumference) represent the concepts needed in the explication of the former.

The thought behind circles purely of concepts need not make use of the idea of a particular at all. Quine's claim can be set out using the explication relation and concepts. In a circle purely of concepts, there is a set containing a definite number of concepts and there is some relation R that relates one concept to another or to others in the set. His claim about these concepts is that there is, as it were, no way out of the circle.

But a third type of circle, the "hybrid" as I called it earlier, needs the idea of a particular as well as the idea of a concept. First, for any hybrid circle, there must of course be some relation, R, such that the items or nodes have that relationship to one another. In all hybrid circles that we will consider, there will be a definite number of concepts, F, G, H, J. I will distinguish two kinds of such hybrid circles. In the first kind, there will be a definite number of particulars as well as a definite number of concepts. In the second kind, there will be a definite number of concepts, but an unrestricted, inexhaustibly or indefinitely large, number of particulars, a, e, i, o, u ... Read "Fa," "Ge," etc., as the fact that a is or becomes F. It is true that one might even think of a fact as a particular, since the relata of relations are generally taken to be particulars, but I prefer to distinguish these hybrid circles from cases of purely particular circles, because of the inclusion of concepts in them but not in purely particular circles. Both concepts and particulars matter to the identity of facts, and that is why circles of singular facts can be thought of as circles with structured items or nodes, with both a particular and a concept playing a part in that structure.[2]

In the first sort of hybrid circle, for example, Fa has R to Ge, and Ge has R to Hi, and Hi has R to Mo, and Mo has R to Fa. In this first kind of hybrid circle, since there will be a definite number of particulars and of concepts, there will only be a definite number of combinations of particulars and concepts, in short, a definite number of facts. In the case in which the definite number of facts are about temporally locatable items, at some point, somewhere, some first fact will at one time have whatever

---

[2] For identity conditions for facts, see my *Explaining Explanation* (1990, ch. V).

the specified relation is to a second fact, and yet also some third fact will have that same relation to the first fact, but at a different time. This will place the particular, or the particular's having some concept true of it, which is what that first fact is about, at two different times. The above assumes that the particular or the concept's being true of the particular occurs at a time. The story will have to be made more complicated to cover cases of temporal duration or extension rather than occurrence at a time, but the lessons of the story will not thereby substantially change.

Whatever temporal problems there might be with some purely particular circles regarding double temporal location would carry over to some hybrid of this kind, if the facts are facts about temporally dated items, and depending of course on what relation R is. For example, if the causal relation related facts, rather than events, the remarks above about difficulties for a purely particular circle of events joined by the causal relation would apply, *mutatis mutandis*, to a hybrid circle of the first kind that was allegedly a circle of facts joined by that same relation.

However, in a more interesting, second kind of hybrid circle, the concepts will repeat, since there is only a definite number of them, but the particulars will not, there being an inexhaustible supply of the latter. (I do not say that there must be an infinitely large number; let "inexhaust- ible" or "indefinitely large" serve us here.) For example, the hybrid cir- cle of the second kind might be as above but Mo will have R to Fu (NOT to Fa), where u is not identical to a, and then Fu have R to Gw (NOT to Ge), where w is not identical to e, and so on. It is the latter, second sort of hybrid circle on which I will focus.

(Might there be hybrid circles with a finite number of particulars but an indefinitely large supply of concepts? It is difficult for me to see how there could be a science with an infinitely large conceptual repertoire, but I shall not address this possibility further.)

Think of each "journey" around the circle as a revolution of the cir- cle. On each revolution of a hybrid circle of what I have called the more interesting, second kind, the same concepts constantly reappear but the particulars change on each circular revolution (or perhaps only change in relation to which concept is true of them). There are no restrictions on the number of particulars which a hybrid circle includes; the only restriction is that the number of concepts must be finite. The number of particulars is indefinitely large, or inexhaustible, so that no specific singular fact needs to appear more than once in the repeated circular revolutions.

If one wants to think about hybrid circles pictorially, perhaps the fol- lowing would help. Hybrid circles of the first kind might be represented by two interlocking circles: a circle of a definite number of particulars and a circle of a definite number of concepts. The circles interlock because the concepts are true of the particulars. Hybrid circles of the second kind might be represented by a straight line which is indefinitely long in at least one direction, and a single continuously-revolving circle above the line at all points. The indefinitely long straight line represents the particulars; the single continuously-revolving circle represents the definite number of concepts, at least one of which is true of each of the indefinitely many particulars. The picture is not quite right as described: some particular o might recur, once as an F, later as a G, and so on. In truth it is no singular fact that can appear more than once. If both concepts and particulars were finite, then some fact would eventually reappear; the indefinite largeness of the number of the particulars prevents any singular fact from ever recurring.

For cases in which the relation R is either not reflexive or not symmetrical, one might have thought that circles in general would saddle us with unwanted reflexivity and symmetry. Suppose that R is a transitive relation. (But is explanation or understanding transitive? Views about this differ.) If there is transitivity, then if there is a circle such that aRb, bRc and cRa, it follows both that aRc and cRa (and similarly for every pair of particulars). So R must be symmetrical. But using transitivity again, if aRc and cRa, it also follows that aRa. So R must be reflexive. Surely this cannot be right for explanation or understanding. It is contentious whether or not explanation is transitive but no one, I take it, thinks that explanation must be reflexive (but it might be non-reflexive rather than irreflexive) or that it must be symmetrical (surely it must be asymmetrical).

On purely particular circles, if R is transitive, these results would follow and this shows something deeply unwelcome about the idea of some alleged purely particular circles that employ transitive relations. (It would not disturb the participants in our love triangle, since the loving relation is, alas, not transitive.) However, on the second kind of hybrid we are considering, each singular fact making its appearance only once due to the inexhaustible supply of particulars, these will not be genuine issues for us. There can be no reflexivity or symmetry issues on the types of circles we are considering, even assuming the transitivity of explanation. No singular fact will explain itself or be explained by what it explains, since no singular fact ever reappears anyway (even if the particular involved in it does).

Let's return to our original question: what kind of circle might a hermeneutic circle be? If there is a hermeneutic circle, there are items on it which are joined by certain relations. The relationship is that of understanding or explanation: something explains a second thing, or the second thing is understandable in the light of the first. But what are the items joined by this relation? Are they particulars or concepts or facts? What kind of circle would a hermeneutic circle be, if there were one: purely of particulars, purely of concepts, or hybrid?

First, let's consider what the items or nodes would be on a hermeneutic circle. Are texts or other linguistic items that for which we seek understanding or explanation? Charles Taylor says that it is "a certain reading of text or expressions ... ." The end of Mantzavinos's chapter makes it clear that he is thinking of a hermeneutic circle in the understanding of a text ("words, and sentences, and for entire texts," p. 309). For those who wish to draw out the implications for the social sciences, he adds that the objects are, for example, "some part of a political process" (p. 299, bottom).

Now, it does seem to me to be important to decide which we are going to discuss. I do not think that whatever lessons we might learn about the understanding of texts necessarily carries over to the understanding of items such as parts of the political process. The analytic tradition of Mill, Hempel, and Popper, has always focused on the explanation of laws and singular events. "Understanding the meaning of a text" was never thought to be the sort of thing that non-hermeneutic philosophers were intending to analyse. In a well-known passage, Hempel says:

... to put forward the covering-law models of scientific explanation is not to deny that there are other contexts in which we speak of explanation, nor is it to assert that the corresponding uses of the word "explain" conform to one or another of our models. Obviously, these models are not intended to reflect the various senses of "explain" that are involved when we speak of explaining the rules of a contest, explaining the meaning of a cuneiform inscription or of a complex legal clause or of a passage in a symbolist poem, explaining how to bake Sacher torte or how to repair a bike ... Hence to deplore, as one critic does, the "hopelessness" of the deductive-nomological model on the ground that it does not fit the case of explaining or understanding the rules of Hanoverian succession is simply to miss the intent of the model. (Hempel 1970; 412–413)

I do not think that the analytic tradition concerned with explanation has ever had much to say about the issue of understanding or explaining a text or a passage in a text. One might feel: *tant pis* for the analytic tradition if that is so. But still, it is so.[3]

---

[3] For what it is worth, I don't think there is a circle in understanding a text, in one obvious meaning of "understanding" at any rate. Take the understanding of a contemporary text rather than an ancient or historical one, for the same lessons ought to apply in either case. Suppose I read a contemporary novel. Don't I understand the whole novel by understanding each chapter and each chapter by understanding each page, and so on down to the basic units of meaning? So one understands the sentence if one understands

But happily we need not decide any of these issues here. For what we are interested in is a position in the philosophy of social science, not in the philosophy of language, or in the philosophy of literature. To revert to Mantzavinos's own example, understanding some part of the political process – say, understanding the voting system in western European countries – is a different matter from understanding a text. But "understanding the voting system" is surely an elliptical expression. Something is missing. But what is it to understand or explain the voting system? To put the point in a different way, the items that stand in the explaining relation in which we are interested are not substance-like items, particulars, such as "the voting system in western European countries." So the alleged hermeneutic circle is not just a circle purely of particulars. What exactly are the items that explain and get explained, that provide understanding and get understood?

Well, in the case at hand, it could be a lot of different things: understanding what caused the voting system to be the way it is, what function it has within a society, why it is in danger of breaking down, how it is perceived by voters, what its overall significance in the society is, and so on. I do not think that just repeating the phrase solemnly, "understanding the voting system," advances us. The question needs answering: understanding what about it? Its causes, its function, its meaning for the participants, its significance, its likely future, and so on, are some of the possibilities. In short, we might want any of a number of different and distinct facts about that voting system explained to us. Let's call facts of this kind, like the fact that the voting process in western European countries has certain causes, or that it performs certain functions in those societies, or that it has a certain point for its participants, "social facts."

The social facts I will be considering will be singular facts about particular social events, states, processes, or whatever, having certain

---

the words in the sentence and the ways in which they are combined; understands the paragraph if one understands the sentences that make it up, and so on.

However, it may be that the sense of "understanding the meaning of" that interests those who find a hermeneutic circle thesis about texts illuminating is something stronger. Perhaps by "understanding the meaning of a text" they are referring to the text's real significance, its point, its real message, its interpretation in the sense that a literary critic might use that expression. To really understand a Brecht play, one has to understand not just the words, sentences, and acts that make it up, but the political context in which it is written, the kind of intervention it was trying to make in the political life of Germany in the period in which it was written, and so on. And once we do that, it may be that, even though we have already fully understood the semantic meaning of all the sentences in the play, we can see that they have significance and a point which we did not see until we understood the point, etc., of the whole play. But all of this, plausible though it might be, takes us far beyond the confined realms of understanding semantic meaning.

features or properties, or about non-social particulars having certain social properties.[4] There are also social facts about laws, for example, which are not singular facts, but I disregard the extension of my discussion to cases of non-singular social facts.

Second, now that we know that social facts are the items that are needed to explain and be explained, and that the relation in question is the relation of understanding or explanation, it follows that if there is a hermeneutic circle of understanding in the social sciences, it would have to be a hybrid circle. Since, in a hermeneutic circle, the relation R is the relation of explanation or understanding, the circle cannot be a Quinean-like purely concept circle, because it is not concepts that explain concepts. Nor can it be a purely particular circle, since pure particulars do not get explained or understood. The objects of explanation are certainly not just tokens or particulars, but facts like the fact that a certain token or particular has a certain characteristic or feature. Moreover, the hermeneutic circle would have to be a hybrid circle of the second kind. Had the alleged hermeneutic circle been a hybrid circle of the first kind, with a definite number of both concepts and particulars, there would have been some very powerful objections to the very idea of a hermeneutic circle, because of the double temporal location issue.

Assuming that explanation requires certain temporal assumptions (e.g., in general, earlier things explain later ones) and with only a definite supply of both concepts and particulars, singular facts would eventually have had to reappear on a circle, and the reappearance of the same singular fact would have demanded that we date it, or rather what it is about, at two distinct times. But, given an indefinitely large supply of particulars, no singular fact needs appear on the endless different "revolutions" of a hybrid circle of the second kind. On these circles at least, no fact or particular the fact is about will have to occur at two distinct times, whatever R might be.

In the case of the explanation of social facts, the inexhaustibly large number of particulars seems to be an easily satisfiable requirement. The number of particulars that social facts are about might not be an infinite number. Assuming that society started at some time and will end at some time, I also assume that the particulars of which social concepts are true are finite in number as well. But as long as society exists, there will always be new particulars, of which social concepts can be true – more mayors, more presidents, more banks, etc. So for as long as society continues, there will be an inexhaustible supply of particulars for social

---

[4] I say a lot more about the distinction between social and non-social properties in my *The Metaphysics of the Social World* (1985, ch. 3).

facts to be about. The number of such particulars is indefinitely large, its limits set only by the limits on the duration of society itself.

Explanations (of facts by facts) come in chains. (Or at least so I will assume here.) One fact explains another, which explains a third, and so on. (On some views of course, non-singular facts would enter the picture at this juncture.) What do such chains look like? Let's assume that every social fact is capable of explanation. (This, by the way, is not a trivial assumption.) Then:

1. Suppose a social fact can be explained only by another social fact.
2. If (1) is true, then either the explanatory chain of social facts is indefinitely long or looping or a hybrid with both characteristics.
3. If (1) is not true, then either some social facts are explained by something other than another social fact or they explain themselves.

(1) is ambiguous: is the supposition that a social fact can be explained only by other social facts, and by nothing else, or is it that in the full explanation of a social fact, some more social facts always play a part, but so might other facts as well? This is an ambiguity that needs more attention, but on either option, the chains will be either indefinitely long or looping. There is no need to disambiguate this for my purposes here. If the chains loop, we obtain a circle. Explanatory circles are looping explanatory chains that are hybrid circles.

Again, consider the simplest example of the second sort of hybrid circle: e's being F explains i's being G, and i's being G explains o's being F. The root idea in such a circle is that something's being an F can explain something else's being a G, and that thing's being a G can explain a third thing's being F. How could that be? Explanatory punch is carried by concepts, not by particulars, and it might not seem possible that F can have that punch in respect of G and G also have the same punch in respect of F, even though the particulars may shift from punch to punch. One might call this the "alleged impossibility of explanatory punch reciprocity of concepts."

However, we *know* from experience that there are legitimate examples of such circles. Consider the following sort of case of reciprocal causal interaction. Wage increases explain higher inflation and higher inflation explains wage increases. These apparently symmetrical cases are not really symmetrical at all once the different tokens and hence the different social facts are introduced: a certain wage increase w at t_1 explains a certain rise in inflation i at t_2, and that rise in inflation i at t_2 in turn explains another wage increase $w^\star$ at t_3. w is not identical to $w^\star$. Nothing is double temporally located. Such reciprocal causal generalisations are not uncommon, especially in social science.

Call a hybrid circle of the second type "tight" if it has only two concepts (and of course any number of particulars or tokens). Call a hybrid circle of the second type "loose" if it has more than two concepts. It seems to me that tight hybrid circles of the second type are mysterious without further information. How could an F explain a G and some G also explain some F? I postulate that the explanatory force that might attach to reciprocally related concepts of a hybrid circle of the second type, a hermeneutic circle, can only arise if we can show that there are different intermediary steps that interpose themselves between the F and the G and the G and the F. There must be some links in the chain that connect the F with the G and some different links that connect the G with the F. Perhaps the F explains the G only because there is some F★ such that the F explains the F★ and the F★ explains the G. And the G explains the F only because there is some G★ such that the G explains the G★ and the G★ explains the F. So at the more mediate or direct explanatory level, the F★ and not the F explains the G and the G★ and not the G explains the F. This is certainly the case with the example to hand: the path by which wage rises cause inflation is a different pathway than that by which inflation leads to more wage rises. The mystery of how Fs explain Gs and Gs also explain Fs is then dispelled. If this is so, then hybrid circles of the second type must all be loose circles.

But are there really any hybrid circles of the second type in general or anyway in social science? Let's assume that there are no inexplicable social facts. Moreover, although there may be facts of some kind that are self-explanatory, social facts do not strike me as a terribly plausible candidate for this status. (Explanation in general may not be irreflexive, but this does not seem relevant for the case of the explanation of singular social fact.) So we are left with four options: (a) there are long chains of social facts that stretch indefinitely, or (b) there are chains of social facts that loop or circle, or (c) there are hybrid chains of social facts that stretch and circle, or (d) some social facts are explicable by something other than social facts.

Of course it would take some very strong reasons to rule out (d), which is not necessarily a reductive position at all: it does not say that some, let alone all, social facts can be reduced to non-social ones. Partly, this will hinge on the connection a theorist makes between reduction and explanation. Even if Taylor were right in the case of texts, that one reading can only be supported by another reading, it is not obvious that the same is true in the case of the explanation of social facts. If at least some social facts can be given full explanations by non-social facts, then both the alleged regress and the circle would be broken. So I think that in order to know that there is a hermeneutic circle (of the only sort I consider plausible), we also need to establish that no social facts can be given full explanations in terms of non-social facts and nothing in Mantzavinos's chapter or in the literature he cites convinces me that this is so.

REFERENCES

Davidson, D. 2001. *Essays on Actions and Events*. Oxford: Clarendon Press.
Hempel, C. 1970. *Aspects of Scientific Explanation*. New York, NY: Free Press.
Inwood, B. 1992. *A Hegel Dictionary*. Oxford: Blackwell.
Quine, W. 1961. "Two Dogmas of Empiricism." In *From A Logical Point of View*. Cambridge, MA: Harvard University Press, pp. 20–46.
Rawls, J. 1999. *A Theory of Justice*. rev. edn. Cambridge, MA: Harvard University Press.
Ruben, D.-H. 1985. *The Metaphysics of the Social World*. London: Routledge & Kegan Paul.
  1990. *Explaining Explanation*. London: Routledge.

Epilogue

C. Mantzavinos

Every discipline is defined by the problems that it deals with; and since the nature, complexity, and breadth of problems are not settled once and for all but change over time, so do the disciplinary boundaries. Philosophy of the social sciences as a discipline is also changing over time, according to the problems it deals with and the solutions it works out. Despite the constant change of the discipline amidst constant debates on a variety of topics, there are some problem areas that possess a certain prominence and somehow define the field: I mean the problem areas that any interested person approaching the field for the first time legitimately expects to see addressed. These are roughly three: the class of problems dealing with sociality (or to put it very broadly, with social ontology); the problem area of the methodology of the social sciences; and lastly the issues dealing with the interaction between philosophy and the social sciences.

This volume was designed to address these three problem areas in its three respective parts. The basic tenet of the book is that the discipline of the philosophy of the social sciences can only produce fruitful results if it takes the findings and the practices of the social scientists into account. In a sense, the volume takes a *middle ground position* between two extremes. Dictating *a priori* truths to the social scientists without any knowledge of scientific results and any contact with their practices, a stance that many philosophers adopt, is both arrogant and unfruitful, since social scientists – rightly – do not take them seriously. Systematically ignoring philosophical developments and methodological discussions and focusing on what "evidence" is telling us, a stance that many social scientists adopt, is equally arrogant and naïve, because it already presupposes a philosophical position, namely crude empiricism. The middle ground position adopted here is that philosophy of the social sciences as a discipline is only valuable if it evolves in constant interaction with the theoretical developments in the social sciences, taking into consideration the specific scientific practices according to which scientific work is conducted.

The role of the philosophy of the social sciences as a discipline is not, however, to serve an *apologetic function* by providing ex post legitimization of *any* theoretical developments and *all* specific practices of the social sciences. On the contrary, the discipline should also productively criticize the theoretical developments and scientific practices of the social sciences; but in order to accomplish this aim, it must constantly keep track of them. As simple as this might seem, it is of fundamental importance if philosophy of the social sciences is successfully to serve its *normative function*.

Name Index

Subject Index